Climate Change and Environmental Sustainability-Volume 2

Editors

Bao-Jie He
Ayyoob Sharifi
Chi Feng
Jun Yang

MDPI • Basel • Beijing • Wuhan • Barcelona • Belgrade • Manchester • Tokyo • Cluj • Tianjin

Editors
Bao-Jie He
Chongqing University
China

Ayyoob Sharifi
Hiroshima University
Japan

Chi Feng
Chongqing University
China

Jun Yang
Northeastern University
China

Editorial Office
MDPI
St. Alban-Anlage 66
4052 Basel, Switzerland

This is a reprint of articles from the Topics published online in the open access journal *Atmosphere* (ISSN 2073-4433), *Buildings* (ISSN 2075-5309), *Land* (ISSN 2073-445X), *Remote Sensing* (ISSN 2072-4292), *Sustainability* (ISSN 2071-1050) (available at: https://www.mdpi.com/topics/Climate_Environmental).

For citation purposes, cite each article independently as indicated on the article page online and as indicated below:

LastName, A.A.; LastName, B.B.; LastName, C.C. Article Title. *Journal Name* **Year**, *Volume Number*, Page Range.

ISBN 978-3-0365-2670-6 (Hbk)
ISBN 978-3-0365-2671-3 (PDF)

188 k words

Contents

About the Editors . . . . **vii**

Preface to "Climate Change and Environmental Sustainability-Volume 2" . . . . **ix**

Tatiana Andrikopoulou, Ralph M. J. Schielen, Chris J. Spray, Cor A. Schipper and Astrid Blom
A Framework to Evaluate the SDG Contribution of Fluvial Nature-Based Solutions
Reprinted from: *Sustainability* **2021**, *13*, 11320, doi:10.3390/su132011320 . . . . **1**

Mariana Madruga de Brito, Danny Otto and Christian Kuhlicke
Tracking Topics and Frames Regarding Sustainability Transformations during the Onset of the
COVID-19 Crisis
Reprinted from: *Sustainability* **2021**, *13*, 11095, doi:10.3390/su131911095 . . . . **27**

Mohammad Ali Yamin
Investigating the Drivers of Supply Chain Resilience in the Wake of the COVID-19 Pandemic:
Empirical Evidence from an Emerging Economy
Reprinted from: *Sustainability* **2021**, *13*, 11939, doi:10.3390/su132111939 . . . . **47**

Alessio Cimini
Evolution of the Global Scientific Research on the Environmental Impact of Food Production
from 1970 to 2020
Reprinted from: *Sustainability* **2021**, *13*, 11633, doi:10.3390/su132111633 . . . . **63**

Tingchen Wu, Xiao Xie, Bing Xue and Tao Liu
A Quantitative Modeling and Prediction Method for Sustained Rainfall-$PM_{2.5}$ Removal Modes
on a Micro-Temporal Scale
Reprinted from: *Sustainability* **2021**, *13*, 11022, doi:10.3390/su131911022 . . . . **85**

Hongbin Dai, Guangqiu Huang, Huibin Zeng and Fan Yang
$PM_{2.5}$ Concentration Prediction Based on Spatiotemporal Feature Selection Using
XGBoost-MSCNN-GA-LSTM
Reprinted from: *Sustainability* **2021**, *13*, 12071, doi:10.3390/su132112071 . . . . **103**

Akvile Cibinskiene, Daiva Dumciuviene, Viktorija Bobinaite and Egidijus Dragašius
Competitiveness of Industrial Companies Forming the Value Chain of Wind Energy
Components: The Case of Lithuania
Reprinted from: *Sustainability* **2021**, *13*, 9255, doi:10.3390/su13169255 . . . . **127**

Jiajie Tang, Jie Zhao, Hongliang Zou, Gaoyuan Ma, Jun Wu, Xu Jiang and Huaixun Zhang
Bus Load Forecasting Method of Power System Based on VMD and Bi-LSTM
Reprinted from: *Sustainability* **2021**, *13*, 10526, doi:10.3390/su131910526 . . . . **147**

Savitri Jetoo and Varvara Lahtinen
The Good, the Bad and the Future: A SWOT Analysis of the Ecosystem Approach to
Governance in the Baltic Sea Region
Reprinted from: *Sustainability* **2021**, *13*, 10539, doi:10.3390/su131910539 . . . . **167**

Lakshmy Naidu and Ravichandran Moorthy
A Review of Key Sustainability Issues in Malaysian Palm Oil Industry
Reprinted from: *Sustainability* **2021**, *13*, 10839, doi:10.3390/su131910839 . . . . **183**

Hannah Barker, Peter J. Shaw, Beth Richards, Zoe Clegg and Dianna Smith
What Nudge Techniques Work for Food Waste Behaviour Change at the Consumer Level? A
Systematic Review
Reprinted from: *Sustainability* **2021**, *13*, 11099, doi:10.3390/su131911099 . . . . **197**

Sugey de Jesús López Pérez and Xavier Vence
When Harmful Tax Expenditure Prevails over Environmental Tax: An Assessment on the 2014
Mexican Fiscal Reform
Reprinted from: *Sustainability* **2021**, *13*, 11269, doi:10.3390/su132011269 . . . . **215**

**Somayeh Mohammadi Hamidi, Christine Fürst, Hossein Nazmfar, Ahad Rezayan and
Mohammad Hassan Yazdani**
A Future Study of an Environment Driving Force (EDR): The Impacts of Urmia Lake
Water-Level Fluctuations on Human Settlements
Reprinted from: *Sustainability* **2021**, *13*, 11495, doi:10.3390/su132011495 . . . . **233**

Ziye Huang, Anmin Huang, Terence P. Dawson and Li Cong
The Effects of the Spatial Extent on Modelling Giant Panda Distributions Using Ecological Niche
Models
Reprinted from: *Sustainability* **2021**, *13*, 11707, doi:10.3390/su132111707 . . . . **253**

Wei Ji, Chengpeng Lu, Jinhuang Mao, Yiping Liu, Muchen Hou and Xiaoli Pan
Public's Intention and Influencing Factors of Dockless Bike-Sharing in Central Urban Areas: A
Case Study of Lanzhou City, China
Reprinted from: *Sustainability* **2021**, *13*, 9265, doi:10.3390/su13169265 . . . . **269**

About the Editors

Bao-Jie He is a research professor of urban climate and built environment at the School of Architecture and Urban Planning, Chongqing University, China. Prior to Chongqing University, Bao-Jie He was a PhD researcher at the Faculty of Built Environment, University of New South Wales, Australia. Bao-Jie works for Cool Cities and Communities and Net Zero Carbon Built Environment. Bao-Jie has strong academic capability, with about 80 peer-reviewed papers published in high-ranking journals and oral presentations given at reputable conferences. Bao-Jie acts as Topic Editor-in-Chief, Leading Guest Editor, Associate Editor, Editorial Board Member, Conference Chair, Sessional Chair and on the Scientific Committee of a variety of international journals and conferences. Dr. He received the Green Talents Award (Germany) in 2021 and the National Scholarship for Outstanding Self-Funded Foreign Students (China) in 2019. Dr. He was ranked as one of the top 100,000 global scientists (both single-year and career top 2%) by Mendeley, 2021.

Ayyoob Sharifi is with the Graduate School of Humanities and Social Sciences, Hiroshima University. He also has a cross-appointment at the Graduate School of Advanced Science and Engineering. Ayyoob's research is mainly at the interface of urbanism and climate change mitigation and adaptation. He actively contributes to global change research programs, such as Future Earth, and is currently serving as a lead author for the Sixth Assessment Report (AR6) of the Intergovernmental Panel on Climate Change (IPCC). Before joining Hiroshima University, he was the Executive Director of the Global Carbon Project (GCP)—a Future Earth core project—leading the urban flagship activity of the project, which is focused on conducting cutting-edge research to support climate change mitigation and adaptation in cities.

Chi Feng received his joint PhD training in South China University of Technology (China) and KU Leuven (Belgium). He is now a research professor in the School of Architecture and Urban Planning, Chongqing University (China), and is leading a research group of more than 10 members. His research topics cover the coupled heat and moisture transfer in porous building materials as well as the hygrothermal performance of building envelopes and built environments. He has led eight international and national research projects, including a China–Europe round robin campaign on material property determination (nine countries participated), the National Natural Science Foundation of China and the National Key R&D Program of China. He has published more than 60 peer-reviewed journal/conference papers at home and abroad. He has been drafting two Chinese standards and participating in another nine international/national ones.

Jun Yang is working at the Urban Climate and Human Settlements Lab, Northeastern University (Shenyang China). His research expertise involves urban climate zones, urban ecology, urban human settlements and sustainability. As PI or Co-PI, he has been involved in 50 research projects, receiving a total of 15 million in RMB from EGOV.CN (e.g., NFC, MOST and MOE) since 2002. He has authored and co-authored more than 160 papers and book chapters and published more than 50 English papers as well as more than 110 Chinese papers in academic journals. He is now an Associate Editor of *SN Social Sciences* and the *International Journal of Environmental Science and Technology*, on the Editorial Board of *PLOS One*, *PLOS Climate* and *Frontiers in Built Environment*, the Lead Guest Editor of *Complexity* and a Guest Editor of the *IEEE Journal of Selected Topics in Applied Earth Observations and Remote Sensing*.

Preface to "Climate Change and Environmental Sustainability-Volume 2"

The Earth's climate is changing; the global average temperature is estimated to already be about 1.1 °C above pre-industrial levels. Indeed, we are now living in conditions of a climate emergency. Climate change leads to many adverse events, such as extreme heat, flooding, bushfire, drought, and many other associated economic and social consequences. Further warming is projected to occur in the coming decades, and climate-induced impacts may exceed the capacity of society to cope and adaptive in a 1.5 °C or 2 °C world. Therefore, urgent actions should be taken to address climate change and avoid irreversible environmental damages.

Climate change is interrelated with many other challenges such as urbanisation, population increase and economic growth. For instance, cities are now the main settlements of human being and are major sources of greenhouse gas emissions that are key contributors to climate change. Moreover, rapid and unregulated urbanisation in some contexts further causes urban problems such as environmental pollution, traffic congestion, urban flooding and heat island intensification. In the absence of well-designed measures, increasing urbanisation trends in the next two–three decades are likely to further aggravate such problems. Overall, climate change and many other challenges have deteriorated the sustainable development of the world.

The United Nations proposed the Sustainable Development Goals in 2015. Goal 13, Climate Action, emphasises the need for urgent action to combat climate change and its impacts in order to enhance sustainability. To achieve this, there is a need to develop a holistic framework that considers mitigation—the decarbonisation of society—to address the challenge of climate change from the root, and adaptation—an immediate action—to increase the resilience of and protect society from climate-induced hazards. The framework prioritises the transformation of the traditional methods of environmental modifications in various fields, including transportation, industry, building, energy generation, agriculture, land use and forestry, towards sustainable ones to limit greenhouse gas emissions. The framework also highlights the significance of sustainable environmental planning and design for adaptation in order to reduce climate-induced threats and risks. Moreover, it encourages the involvement and participation of all stakeholders to accelerate climate change mitigation and adaptation progress by developing sound climate-related governance systems.

The framework also calls for the support and engagement of all societal stakeholders. To support the achievement and implementation of the framework, this book focuses on climate change and environmental sustainability by covering four key aspects, including climate change mitigation and adaptation, sustainable urban–rural planning and design, decarbonisation of the built environment in addition to climate-related governance and challenges. Climate change mitigation and adaptation covers topics of greenhouse gas emissions and measurement, climate-related disasters and reduction, risk and vulnerability assessment and visualisation, impacts of climate change on health and well-being, ecosystem services and carbon sequestration, sustainable transport and climate change mitigation and adaptation, sustainable building and construction, industry decarbonisation and economic growth, renewable and clean energy potential and implementation in addition to environmental, economic and social benefits of climate change mitigation.

Sustainable urban–rural planning and design deals with questions of climate change and regional economic development, territorial spatial planning and carbon neutrality, urban overheating mitigation and adaptation, water-sensitive urban design, smart development for urban habitats, sustainable land use and planning, low-carbon cities and communities, wind-sensitive urban planning and design, nature-based solutions, urban morphology and environmental performance in addition to innovative technologies, models, methods and tools for spatial planning. Decarbonisation of the built environment addresses issues of climate-related impacts on the built environment, the health and well-being of occupants, demands on energy, materials and water, assessment methods, systems and tools, sustainable energy, materials and water systems, energy-efficient design technologies and appliances, smart technology and sustainable operation, the uptake and integration of clean energy, innovative materials for carbon reduction and environmental regulation, building demolition and material recycling and reusing in addition to sustainable building retrofitting and assessment. Climate-related governance and challenges concerns problems of targets, pathways and roadmaps towards carbon neutrality, pathways for climate resilience and future sustainability, challenges, opportunities and solutions for climate resilience, the development and challenges climate change governance coalitions (networks), co-benefits and synergies between adaptation and mitigation measures, conflicts and trade-offs between adaptation and mitigation measures, mapping, accounting and trading carbon emissions, governance models, policies, regulations and programs, financing urban climate change mitigation, education, policy and advocacy of climate change mitigation and adaptation in addition to the impacts and lessons of COVID-19 and similar crises.

Overall, this book aims to introduce innovative systems, ideas, pathways, solutions, strategies, technologies, pilot cases and exemplars that are relevant to measuring and assessing the impact of climate change, mitigation and adaptation strategies and techniques in addition to public participation and governance. The outcomes of this book are expected to support decision makers and stakeholders to address climate change and promote environmental sustainability. Lastly, this book aims to provide support for the implementation of the United Nations Sustainable Development Goals and carbon neutrality in efforts aimed at achieving a more resilient, liveable and sustainable future.

Our world is facing many challenges, such as poverty, hunger, resource shortage, environmental degradation, climate change, and increased inequalities and conflicts. To address such challenges, the United Nations proposed the Sustainable Development Goals (SDG), consisting of 17 interlinked global goals, as the strategic blueprint of world sustainable development. Nevertheless, the implementation of the SDG framework has been very challenging and the COVID-19 pandemic has further impeded the SDG implementation progress. Accelerated efforts are needed to enable all stakeholders, ranging from national and local governments, civil society, private sector, academia and youth, to contribute to addressing this dilemma. This volume of the *Climate Change and Environmental Sustainability* book series aims to offer inspiration and creativity on approaches to sustainable development. Among other things, it covers topics of COVID-19 and sustainability, environmental pollution, food production, clean energy, low-carbon transport promotion, and strategic governance for sustainable initiatives. This book can reveal facts about the challenges we are facing on the one hand and provide a better understanding of drivers, barriers, and motivations to achieve a better and more sustainable future for all on the other. Research presented in this volume can provide different stakeholders, including planners and policy makers, with better solutions for the implementation of SDGs.

Prof. Bao-Jie He acknowledges the Project NO. 2021CDJQY-004 supported by the Fundamental Research Funds for the Central Universities. We appreciate the assistance from Mr. Lifeng Xiong, Mr. Wei Wang, Ms. Xueke Chen and Ms. Anxian Chen at School of Architecture and Urban Planning, Chongqing University, China.

Bao-Jie He, Ayyoob Sharifi , Chi Feng, Jun Yang
Editors

 *sustainability*

Article

A Framework to Evaluate the SDG Contribution of Fluvial Nature-Based Solutions

Tatiana Andrikopoulou [1,*], Ralph M. J. Schielen [1,2], Chris J. Spray [3], Cor A. Schipper [2] and Astrid Blom [1]

[1] Faculty of Civil Engineering and Geosciences, Delft University of Technology,
    2628 CN Delft, The Netherlands; r.m.j.schielen@tudelft.nl (R.M.J.S.); astrid.blom@tudelft.nl (A.B.)
[2] Ministry of Infrastructure and Water Management (Rijkswaterstaat), P.O. Box 2232,
    3500 GE Utrecht, The Netherlands; ralph.schielen@rws.nl (R.M.J.S.); Cor.schipper@rws.nl (C.A.S.)
[3] UNESCO Centre for Water Law, Policy and Science, University of Dundee, Dundee DD1 4HN, UK;
    C.J.Spray@dundee.ac.uk
* Correspondence: tatianaandrikopoulou@gmail.com; Tel.: +30-69-7180-8702

Abstract: Nature-based solutions (NBSs) are measures reflecting the 'cooperation with nature' approach: mitigating fluvial flood risk while being cost-effective, resource-efficient, and providing numerous environmental, social, and economic benefits. Since 2015, the United Nations (UN) 2030 Agenda has provided UN member states with goals, targets, and indicators to facilitate an integrated approach focusing on economic, environmental, and social improvements simultaneously. The aim of this study is to evaluate the contribution of fluvial NBSs to the UN 2030 Agenda, using all its components: Sustainable Development Goals (SDGs), targets, and indicators. We propose a four-step framework with inputs from the UN 2030 Agenda, scientific literature, and case studies. The framework provides a set of fluvial flooding indicators that are linked to SDG indicators of the UN 2030 Agenda. Finally, the fluvial flooding indicators are tested by applying them to a case study, the Eddleston Water Project, aiming to examine its contribution to the UN 2030 Agenda. This reveals that the Eddleston Water Project contributes to 9 SDGs and 33 SDG targets from environmental, economic, societal, policy, and technical perspectives. Our framework aims to enhance the systematic considerations of the SDG indicators, adjust their notion to the system of interest, and thereby enhance the link between the sustainability performance of NBSs and the UN 2030 Agenda.

Keywords: nature-based solutions; river; flooding; sustainability; sustainable development goals; indicators

Citation: Andrikopoulou, T.; Schielen, R.M.J.; Spray, C.J.; Schipper, C.A.; Blom, A. A Framework to Evaluate the SDG Contribution of Fluvial Nature-Based Solutions. *Sustainability* **2021**, *13*, 11320. https://doi.org/10.3390/su132011320

Academic Editor: Antonio Miguel Martínez-Graña

Received: 1 August 2021
Accepted: 5 October 2021
Published: 13 October 2021

Publisher's Note: MDPI stays neutral with regard to jurisdictional claims in published maps and institutional affiliations.

1. Introduction

According to the United Nations (UN) 2015 report, 'The Human Cost of Weather-Related Disasters' [1], flooding has negatively affected 2.3 billion people over the last 20 years. This accounts for 56% of all those negatively affected by weather-related disasters such as droughts, storms, landslides, and extreme temperatures (64.4 million/year). Especially for fluvial floods, the number of affected people under the most extreme river flooding scenario and without further adaptation may rise from 39 million people per year to 134 million people per year by 2050. Approximately two-thirds of this increase can be attributed to increases in the severity and frequency of flooding due to climate change and the remainder due to population growth in flooding-prone areas [2].

Rapid development combined with the expansion of infrastructure, agricultural intensification, transport, and other linked socioeconomic systems has increased society's vulnerability to environmental disasters, especially in floodplain areas [3]. At the same time, climate change is an important driver for implementing sustainable practices in protecting and managing river ecosystems. In this context, the UN 2030 Agenda [4] has provided international and national governments with goals, targets, and indicators to facilitate an integrated approach focusing on economic, environmental, and social improvements simultaneously. Since 2015, all UN member states are expected to pursue these Sustainable Development Goals (SDGs), tailoring a path towards a peaceful and prosperous planet.

Nature-based solutions (NBSs) can help in addressing many of the SDGs as established in the UN 2030 Agenda. The inclusion of natural elements could create manifold benefits for all the three pillars – 'People', 'Planet', and 'Prosperity' which reflect the three sustainability principles (society, environment, economy) and are adopted by the UN 2030 Agenda. From a societal perspective, they could provide access to nature and recreation while adding cultural and heritage value to the landscape. From ecological and environmental perspectives, they could enhance biodiversity and contribute to water and air purification. From an economic viewpoint, they could promote sustainable and responsible resource management, resulting in cost-effective practices. In Europe, nature-based protection measures (green-blue-hybrid) have already gained increasing prominence in application [5–8]. Green/Blue infrastructure indicates a strategically planned network of natural and semi-natural areas with other environmental features designed and managed to deliver a wide range of ecosystem services. It incorporates green spaces (or blue if aquatic ecosystems are concerned) and other physical features in ter-restrial (including coastal) and ma-rine areas. Hybrid solutions mix hard infrastructure with ecosystem-based infrastructure (https://portals.iucn.org/library/sites/library/files/documents/2016-036.pdf, accessed on 10 October 2021). In global relevance, the effort to learn, implement and promote NBSs is worldwide and supported by many programs and data pools [9].

To date, several NBSs frameworks have been developed to comprehensively describe, analyze, and assess the planning, implementation, and operationalization of NBSs projects. Typically, they include indicators for benchmarking, assessing, or measuring the performance or (co-)benefits of NBSs under several hydro-meteorological hazards (HMHs). Kumar et al. [10], Ruangpan et al. [11], Shah et al. [12], and Albert et al. [13], for instance, developed a single NBS framework covering at least four HMHs. The type of environment under consideration typically differs; Kumar et al. [10], Ruangpan et al. [11], Calliari et al. [14], and Nesshöver et al. [15] do not focus on a single environment, in contrast to many other studies wherein a specific type of environment is the focus. The environments most studied are urban, large rivers (250–300 km), and coasts [11,13,16–21].

Sustainability is addressed in recent NBSs-related frameworks either with the inclusion of the three pillars in the assessment of the NBSs' performance or their (co-)benefits or by measuring the sustainability performance according to the components of the UN 2030 Agenda. Initially, studies such as those by ones of Artmann et al. [22], Pakzad et al. [23] Raymond et al. [20] showed that NBSs interact across and within society, economy, and environment. Building on that, subsequent studies (e.g., [18,21,24]) examined the potential contributions of NBSs to the UN 2030 Agenda by examining the SDGs and/or their targets Schipper et al. [21] developed the Sustainability Impact Score (SIS) Assessment Framework which uses a selection of SDGs and SDG targets to score the sustainability performance of coastal management projects. Whilst it is apparent that some of the recent frameworks address the SDGs and/or SDG targets from the UN 2030 Agenda, they omit consideration of the SDG indicators.

However, in scrutinizing the UN 2030 Agenda, it is noticeable that (i) often, SDG targets refer to multiple elements which are broken down into SDG indicators, and (ii) 12 SDG indicators are repeated (some with slight amendments) under different SDG targets. The SDG indicators (rather than the SDG targets) seem to have the right abstraction level to serve as a connection between the NBSs and the UN 2030 Agenda. Although reaching the SDGs in itself is a promising achievement in preserving our planet, as stated by the United Nations, using the SDG indicators could bring a new perspective to the effort of linking the NBSs to SDGs and hence to assessing the contribution of NBSs to the achievement of the SDGs.

To bridge the gap identified in the research to date, the aim of this study is to evaluate the sustainability performance of NBSs projects with respect to the UN 2030 Agenda involving all its three components: SDGs, SDG targets, and SDG indicators. In other words, we seek the SDGs, SDG targets, and SDG indicators to which NBSs projects could contribute. We focus on NBSs projects for fluvial flood risk mitigation (FFRM) implemented in riverine ecosystems up to 100 km^2, which is a smaller scale than that examined in riverine environments to date. Specifically, we aim to evaluate the sustainability performance of NBSs projects for FFRM by:

(a)　Creating a set of fluvial flooding indicators that reflect the interactions of NBSs for FFRM projects with societal, environmental, economic, policy, and technical perspectives;

(b)　Establishing a link between the set of fluvial flooding indicators and the SDG indicators;

(c)　Testing the fluvial flooding indicators by selecting a specific case study with the necessary project metadata.

We consider case studies from countries with high-income economies only, as NBSs projects in countries with upper-middle, lower-middle, and low-income economies typically aim to cover more fundamental needs, such as water quality and scarcity, and flood mitigation is seldom the main driver for NBSs implementation. Interaction of the river ecosystem with the coastal environment is out of the scope of this research.

2. Methodology

The methodology of this study uses the SIS Assessment Framework in a way that looks to build on the systematic methodology introduced by Schipper et al. [21], but with focus on river ecosystems, recognizing and introducing new elements reflecting the scope of the research. With the UN 2030 Agenda as a starting point, four steps are considered that eventually lead to the formation of the framework. Subsequently, the framework is presented along with the four steps through which our aim is accomplished and, ultimately, a case study to test its applicability.

2.1. The Sustainability Performance Evaluation Framework

The Sustainability Performance Evaluation Framework presented here is derived from the SIS Assessment Framework with the necessary alterations. It encompasses a systematic methodology for creating a set of fluvial flooding indicators, linking them to the SDG indicators, and evaluating the sustainability performance of an FFRM NBSs project through four steps. Starting with the components of the UN 2030 Agenda as input (Figure 1), Step I defines the fluvial flooding indicators which relate to NBSs for FFRM. Subsequently (Step II), the SDGs relevant to NBSs for FFRM, along with the respective SDG targets and SDG indicators, are selected from the UN 2030 Agenda. In Step III, Step I and Step II are brought together, creating a set of NBSs fluvial flooding indicators that demonstrate the FFRM NBSs' contribution to the UN 2030 Agenda. The definition of fluvial flooding indicators and their subsequent connection to the SDG indicators (instead of directly using the SDG indicators) makes it possible to overcome an apparent lack of conceptual clarity inherent in some of the SDG indicators due to their universal nature. Finally, Step IV consists of the assessment of the fluvial flooding indicators based on their application to a case study with available project metadata, with the sustainability performance of the project as the main output.

2.2. Step I—Definition of Dimensions and Fluvial Flooding Indicators

The scope of the sustainability evaluation is defined through the identification of the NBSs dimensions. The word 'dimension' is chosen above terminology such as 'property' or 'aspect' to emphasize the broadness of the NBSs (co-)benefits. The NBSs dimensions express sectors that are affected by FFRM NBSs projects in river ecosystems. For instance, floodplain ponds will, in addition to temporarily storing water during floods, provide habitats for wildlife and support biodiversity. Therefore, wildlife and biodiversity are two sectors that are affected by floodplain ponds and are expressed by the Environmental dimension in our study.

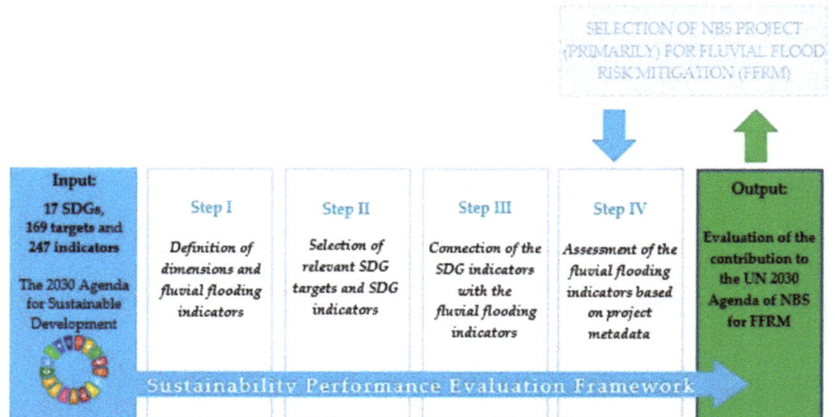

Figure 1. The Sustainability Performance Evaluation Framework, using a selection of SDGs, SDG targets, and SDG indicators published in the 2030 Agenda for Sustainable Development. The Sustainability Performance Evaluation Framework output shows the contribution of an NBSs project for FFRM (and the reason for it) to the UN 2030 Agenda (Figure adapted from [21]).

The identification of the dimensions is based upon an analysis of 7 existing NBSs frameworks or reviews ([14,15,20,22,25–27]), complemented by the examination of three case studies (Table 1). Case studies are used in order to validate that the framework/review findings are realistic and to add any relevant findings that might have been neglected in the literature. The selection of the case studies was based on the following five independent criteria. The criteria are not prioritized; the order is indicative.

1. The main objective of the NBSs should be fluvial flood risk mitigation;
2. Coverage of different geographical regions and scales;
3. Availability of documentation (language, type, and number);
4. Accessibility to relevant data, information, documentation;
5. Availability of grey literature relevant to the case studies, to be used as an additional source of information, including published articles and videos.

Table 1. The selected projects examined as case studies. They contribute to the definition of the dimensions and their fluvial flooding indicators (Step I).

Project	Location	Scale	References
Wave-attenuating willow forest	Noordwaard polder, The Netherlands	~44.50 km^2 polder area	[28–33]
Colorado front range: recovery from 2013 floods	United States of America (USA)	~105 km river and floodplain improvements	[28,34,35]
Belford natural flood management scheme	Belford, Northumberland, United Kingdom (UK)	~6 km^2 catchment size	[36–47]

Each dimension consists of fluvial flooding indicators, as shown in Figure 2, that act as a metric that condenses complexity and provides relevant information [48]. The aim of the fluvial flooding indicators is to list specific effects that might occur in a dimension when implementing an NBSs project for FFRM. For instance, biodiversity abundance is a fluvial flooding indicator that can be found under the Environmental dimension (Figure 2). A preliminary list of fluvial flooding indicators was created by:

(i) Collecting existing indicators from literature. The collection of indicators comes from the 7 frameworks reviewed for the dimensions. However, starting with the already identified frameworks and using snowballing techniques, three additional frameworks were identified that also revealed additional indicators [23,49,50].

(ii) Using the indicators derived from (i) in analyzing the three case studies (Table 1), chosen to reflect a reasonable geographical coverage, spread in surface and in geomorphological aspects and with enough information at hand to quantify the indicators. In this process, the case studies gave rise to several new indicators that were not included in Step (i).

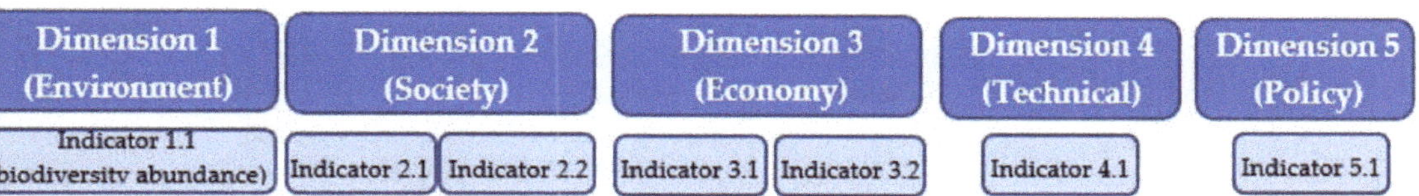

Figure 2. Within each of the five dimensions, the fluvial flooding indicators represent potential effects of NBSs projects for FFRM in the respective dimension. This figure aims to show the structure of the dimensions and their fluvial flooding indicators. The exact dimensions and their fluvial flooding indicators are fully explained in the Results section (Steps I and III).

2.3. Step II—Selection of Relevant SDG Targets and SDG Indicators

The Sustainable Development Goals, targets, and indicators constitute very broad but versatile milestones that users may need to adapt depending on the context and their precise area of interest. As the goals themselves are very broad, the starting point for examination in this study is the 169 targets, followed by the 247 indicators. In reviewing these, the aim is to establish what they address and then select those SDG targets and SDG indicators that are relevant to NBSs for FFRM. For this purpose, a screening process has been developed (Figure 3) to help select relevant SDG targets and SDG indicators according to (i) the boundary conditions (high-income economies, river ecosystem) and (ii) potential NBSs' contribution to them for FFRM. The latter, in particular, has been developed from insights gained from the literature review and case studies. The selection process is presented in Figure 3:

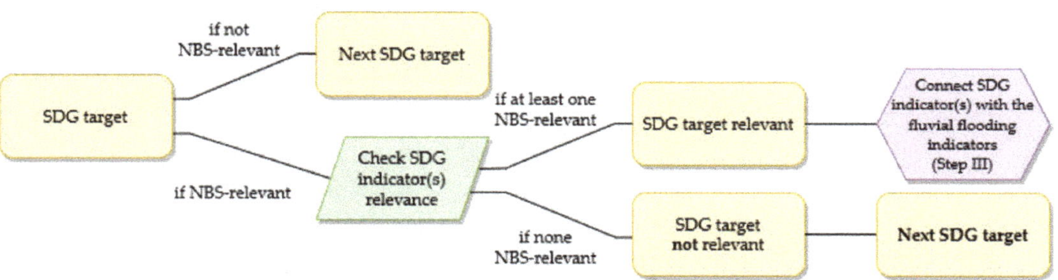

Figure 3. The screening process that selects SDG targets and SDG indicators relevant to NBSs for FFRM. Eventually, the relevant SDG indicators can be linked to the fluvial flooding indicators, as explained in Step III (hexagonal purple box).

2.4. Step III—Connection of the SDG Indicators with the Fluvial Flooding Indicators

Having selected the relevant SDG targets and SDG indicators from the UN 2030 Agenda (Step II), the SDG indicators were connected with the preliminary list of fluvial flooding indicators, as formed in Step I. The connection was made at a conceptual level: matching the description of the fluvial flooding indicator with the SDG indicators. For instance, the biodiversity abundance fluvial flooding indicator was connected to an SDG indicator that addresses the presence and diversity of species.

2.5. Step IV—Assessment of the Fluvial Flooding Indicators Based on Project Metadata

Finally, the connected fluvial flooding indicators (Step III) are applied to a selected case study using project metadata as input to them. Through examination of the fluvial flooding indicators, the sustainability performance of a selected FFRM NBSs project with respect to the UN 2030 Agenda is evaluated.

3. Results

3.1. Step I—Definition of Dimensions and Fluvial Flooding Indicators

Five dimensions are defined: Environment, Economy, Society, Policy—Procedural and Technical. The Environment, Economy, and Society dimensions represent the three pillars of sustainability. These three pillars, along with the Policy—Procedural dimension can all be found as broad divisions within the UN 2030 Agenda [51]. The Environment Economy, and Society dimensions are also used by other frameworks addressing either NBSs sustainability or the additional benefits that NBSs bring [17,19–21,52]. The Technical dimension is a new addition that is considered highly relevant because it refers to the fulfillment of the objective of the intervention (flood protection) and to characteristics that the intervention should comply with, including structural integrity, reliability, ease of implementation, adaptability, and resilience. The Technical dimension has recently been introduced in the literature. The study of Pugliese et al. [19] uses the framework introduced by PHUSICOS [53], where the Technical dimension is used as an ambit to examine the NBSs' technical and economic feasibility aspects. The Policy—Procedural dimension is usually found in the Society dimension. This is the case both in the EC Handbook for Practitioners [52], which places the 'Participatory Planning and Governance under the People pillar of Sustainable Development, and in the PHUSICOS framework [53] that includes the 'Community Involvement and Governance', in the Society ambit. In the Sustainability Performance Evaluation Framework, we distinguished the Policy—Procedural dimension from the Society one, aligning with the UN 2030 Agenda [4], which devotes entire goals to partnerships (Goal 17) and inclusive collaboration (Goal 16).

A preliminary list of 32 fluvial flooding indicators was identified spread across the five dimensions (Environment 8, Society 5, Economy 5, Technical 6, Policy—Procedural 8 to describe the effects of NBSs for FFRM. Overall, 24 of the fluvial flooding indicators were collected from the NBSs frameworks and reviews examined; three emerged from the case studies, and five technical fluvial flooding indicators were introduced by the authors adjusted after Slinger, J.H. [54]. A detailed table with all the dimensions, their fluvial flooding indicators, and their use is presented in the description of Step III to avoid repetition.

3.2. Step II—Selection of Relevant SDG Targets and SDG Indicators

Table 2 shows the selection of 10 SDGs, 42 SDG targets, and 51 SDG indicators as relevant to NBSs for FFRM, as derived from the screening process. The selection starts by examining, one by one, all the SDG targets following the screening process (Figure 3). For each SDG target that was considered relevant to the FFRM NBSs, at least one of its SDG indicators also had to be FFRM-NBSs-relevant. The explanation as to which SDG indicator is considered relevant is shown in the fourth column of Table 2 and is derived from our examination of the literature and case studies.

3.3. Step III—Connection of the SDG Indicators with the Fluvial Flooding Indicators

The preliminary list of fluvial flooding indicators was coupled with the relevant SDG indicators, resulting in 21 out of 32 fluvial flooding indicators being linked to various of the 51 SDG indicators. It was expected that not all the fluvial flooding indicators would be linked to the relevant SDG indicators since the intention was to match a targeted—to fluvial flooding—list specifically derived from the authors' examination of literature and practice to a universal solid agenda. However, since the UN 2030 Agenda is a universally recognized policy framework and the aim of this study is to examine the FFRM NBSs' contribution to it, the preliminary list was extended to cover all the relevant SDG indicators. A total of 12 fluvial flooding indicators were added in the Policy—Procedural dimension, which can be seen in Table 3, rows #29–34 and #39–44. Therefore, the final list comprises 33 fluvial flooding indicators coupled with all the 51 relevant SDG indicators.

Table 2. In total, 10 SDGs, 42 SDG targets, and 51 SDG indicators are considered relevant to NBSs projects for fluvial flood risk mitigation. These 51 relevant SDG indicators will be connected with the fluvial flooding indicators in Step III. For a full description of the SDG targets and SDG indicators, reference should be made to the UN 2030 Agenda.

SDGs Identified	Relevant SDG Targets	Relevant SDG Indicators	Explanation
GOAL 1 End poverty in all its forms everywhere	Target 1.5 Disaster Resilience	1.5.1 Casualties due to disasters	Protect from/reduce exposure of people to flooding
		1.5.2 GDP economic losses due to disasters	Prevent or minimize economic losses due to flooding
		1.5.3 Strategies in line with Sendai	Introduce to/become part of the national flood risk reduction strategies
		1.5.4 Alignment of local and national strategies	Make the alignment with national flood risk reduction strategies feasible
GOAL 3 Ensure healthy lives and promote well-being for all at all ages	Target 3.9 Pollutions and Contaminations	3.9.1 Air pollution mortality	Can contribute to air purification due to the natural elements used/enhanced
		3.9.2 Unsafe water mortality	Protect from/reduce exposure of people to poor quality water
GOAL 6 Ensure availability and sustainable management of water and sanitation for all	Target 6.3 Water Pollution	6.3.2 Water quality	Can contribute to water purification due to the natural elements used/enhanced
	Target 6.5 Management and Cooperation	6.5.1 Integrated water resources management	Require integrated water resources management
		6.5.2 Transboundary water cooperation	Can potentially achieve it
	Target 6.6 Water Quantity and Quality	6.6.1 Extent of water-related ecosystems	By enhancing the natural processes, the ecosystem expands
	Target 6.b Community Participation	6.b.1 Community engagement	They require inclusive processes and stakeholder participation in the management of water resources
GOAL 8 Promote sustained, inclusive, and sustainable economic growth, full and productive employment, and decent work for all	Target 8.1 Economic Growth	8.1.1 Economic growth per capita	Overall economic growth shared over the local population due to new jobs, increased income, or production due to intervention
	Target 8.2 Economic Productivity	8.2.1 Economic growth per employed person	Potential increase of the income per employed person due to jobs created or enhanced by the intervention
	Target 8.3 Development-Oriented Policies	8.3.1 Employment	Ameliorate existing jobs by providing opportunities and better prevailing conditions
	Target 8.4 Resource Efficiency	8.4.2 Domestic material consumption per GDP	Use of locally available materials and limited cost compared to grey materials
	Target 8.5 Employment	8.5.2 Unemployment rates	New job opportunities
	Target 8.9 Tourism-Oriented Policies	8.9.1 Economic growth due to tourists	Money and jobs due to the touristic attractiveness of the area
GOAL 9 Build resilient infrastructure, promote inclusive and sustainable industrialization, and foster innovation	Target 9.4 CO_2 Emissions Reduction	9.4.1 CO_2 emissions	C02 sequestration through use of the natural material chosen
	Target 9.5 Research and Development Expenditure	9.5.1 Research and development expenditure	Research and pilot projects needed for the implementation of the NBSs
GOAL 11 Make cities and human settlements inclusive, safe, resilient, and sustainable	Target 11.3 Participation and Management	11.3.2 Public engagement strategies	Stakeholder involvement in NBSs design and implementation
	Target 11.4 Expenditure on Preserving Heritage	11.4.1 Expenditure on culture and heritage	Protection of cultural heritage is an additional aspect to the NBSs' main function in flood risk mitigation
	Target 11.5 Economic Losses Due to Disasters	11.5.1 Casualties due to disasters	Protect from/reduce exposure of people to flooding

Table 2. *Cont.*

SDGs Identified	Relevant SDG Targets	Relevant SDG Indicators	Explanation
		11.5.2 Damages to infrastructures and services	Prevent or minimize economic losses due to flooding
	Target 11.7 Green and Public Spaces	11.7.1 Use of public areas	Accessibility, recreation, and leisure space are additional aspects to the NBSs' main function in flood-risk mitigation
	Target 11.A Economic, Social, and Environmental Links	11.a.1 Development plans accounting for future projections	Part of the NBSs design
	Target 11.B Holistic Disaster Risk Management	11.b.1 Strategies to protect development gains from the risk of disaster (Sendai framework)	NBSs are part of flood-risk-reduction strategies which align with Sendai FDRR
		11.b.2 Alignment of local and national strategies	Make the alignment with national flood risk reduction strategies feasible
GOAL 12 Ensure sustainable consumption and production patterns	Target 12.1 Consumption and Production	12.1.1 Sustainable production and consumption plans	NBSs include sustainable use/consumption of naturally available materials
	Target 12.2 Domestic Material Consumption	12.2.2 Domestic material consumption	Use of locally available materials
	Target 12.6 Sustainability in Companies	12.6.1 Sustainability reports by companies	NBSs involve the three sustainability pillars and thus could evoke sustainable activities in the companies
	Target 12.7 Procurement Practices	12.7.1 Sustainable action plans	NBSs involve inclusive strategies, actions, and the three sustainability pillars
	Target 12.8 Education and Awareness	12.8.1 Education for sustainability	Offer education through the enrichment of the area and close contact with nature
GOAL 13 Take urgent action to combat climate change and its impacts	Target 13.1 Resilience and Adaptive Capacity	13.1.1 Casualties due to disasters	Protect from/reduce exposure of people to flooding
		13.1.2 Strategies in line with Sendai	NBSs are part of flood-risk-reduction strategies that align with Sendai FDRR
		13.1.3 Alignment of local and national strategies	Make the alignment with national flood risk reduction strategies feasible
	Target 13.2 Operationalization of Climate-Related Policies	13.2.1 Climate adaptation plans and strategies	Offer multi-benefit approach that applies at NBSs design and implementation
	Target 13.3 Development Action	13.3.2 Technology knowledge, transfer, and development in countries	NBSs result from and contribute to development
	Target 13.B Capacity for Planning and Management	13.b.1 Support for climate-related actions	Strengthening the evidence and experience in NBSs would spread their application for climate resilience
GOAL 15 Protect, restore and promote sustainable use of terrestrial ecosystems, sustainably manage forests, combat desertification, and halt and reverse land degradation and halt biodiversity loss	Target 15.1 Protected Areas	15.1.2 Protected areas	Protection and conservation of designated sites (including Natura 2000) are considered during NBSs design and implementation
	Target 15.3 Land Degradation	15.3.1 Degraded areas	NBSs can contribute to halting erosion
	Target 15.5 Threatened Species	15.5.1 Red List Index	Generation of wildlife habitat and population viability are addressed by NBSs
	Target 15.6 Access to and Sharing of Benefits	15.6.1 Policies for sharing of benefits	NBSs are designed in order to provide as many benefits as possible to multiple stakeholders
	Target 15.8 Prevention of Invasive Alien Species	15.8.1 Policies for control of invasive non-native species	Contribute to awareness and prevention of spread of non-native invasive species in riverine ecosystems
	Target 15.9 Ecosystem and Biodiversity into Policies	15.9.1 Aichi biodiversity target 2	Enhance biodiversity as part of the NBSs goals

Table 2. *Cont.*

SDGs Identified	Relevant SDG Targets	Relevant SDG Indicators	Explanation
	Target 15.A Assistance and Expenditure on Biodiversity and Ecosystems	15.a.1 Use and conservation of biodiversity and ecosystems	Part of the NBSs project goals
GOAL 17 Strengthen the means of implementation and revitalize the Global Partnership for Sustainable Development	Target 17.6 Cooperation between Countries	17.6.1 Cooperation between countries	NBSs could enhance science and technology cooperation between countries
	Target 17.14 Policy Coherence	17.14.1 Mechanisms for sustainable development	The broad involvement needed in NBSs projects could lead to policy coherence for sustainable development
	Target 17.15 Use of Domestic Development Tools	17.15.1 Use country-owned resources	NBSs intervention aligned with national policies and development plans
	Target 17.16 Partnerships and Stakeholder Engagement	17.16.1 Reporting progress in SDG	NBSs can contribute to SDG progress through the multi-benefit approach, which includes society, environment, and economy
	Target 17.17 Money to Partnerships	17.17.1 Partnerships	Partnerships and coalitions formed/enhanced through NBSs
	Targets 17.18 Data and Indicators	17.18.1 Production of SD indicators per country	NBSs can create trackable indicators

Table 3. All the 44 fluvial flooding indicators. The first 32 fluvial flooding indicators that belong to the preliminary list come from (i) literature review [14,15,20,22,25–27,54] and (ii) case study examination [28–47]. From these, 21 link to SDG indicators whilst the 11 remaining do not (-). In rows #29–34 and #39–44, there are 12 new fluvial flooding indicators. On this base and with project metadata, the final list of 33 fluvial flooding indicators will show the contribution of an NBSs project for FFRM to the UN 2030 Agenda (Step IV).

#	Fluvial Flooding Indicator	(Fluvial Flooding Indicator) General Description	Ref.	Dimension	Relevant SDG Indicator	(SDG Indicator) Short Description
1	Biodiversity abundance	Animals using the site, vegetation cover, designation as a protected site (e.g., inclusion in the EU 'Natura 2000' network)	[14,15,20,22,25–27]		15.1.2	Protected areas
2	Wildlife habitat	Creation of habitat for flora and fauna	[14,15,20,22,25–27]		15.5.1	Red List Index
3	Population viability	Expresses either lifetime of a species in time or natural elements that enhance fauna abundance	[14,15,20,22,25–27]		15.5.1	Red List Index
4	Endogeneity	Presence of non-native invasive species	[14,15,20,22,25–27]	Environment	-	-
5	Continuity of water and sediment flux	Erosion, sediment traps, amount of sediment captured	[14,15,20,22,25–27]		15.3.1	Degraded areas
6	Water quality	Nitrates, phosphorus, and suspended sediments, water discharge	[14,15,20,22,25–27]		6.3.2	Water quality
7	CO_2 emissions	CO_2 captured by the vegetation/natural elements used	[14,15,20,22,25–27]		9.4.1	CO_2 emissions
8	Extent of water-related ecosystems	Spatial extent of the water-related ecosystem since the NBSs' implementation	[14,15,20,22,25–27]		6.6.1	Extent of water-related ecosystems

Table 3. *Cont.*

#	Fluvial Flooding Indicator	(Fluvial Flooding Indicator) General Description	Ref.	Dimension	Relevant SDG Indicator	(SDG Indicator) Short Description
9	Well-being	Mortality rate, numbers of people affected by water pollution, air pollution, flooding	[14,15,20,22,25–27]	Society	1.5.1/13.1.1/ 11.5.1 3.9.1 3.9.2	Casualties due to disasters Air pollution mortality Unsafe water mortality
10	Physical and mental health	People frequently using the NBSs area	[14,15,20,22,25–27]		-	-
11	Cultural heritage/educational value	Protected or (newly) created value by the intervention	[14,15,20,22,25–27]		12.8.1	Education for sustainability
12	Recreation/leisure value	(New) walking/running/biking paths, activities	[14,15,20,22,25–27]		11.7.1	Use of public areas
13	Enhance attractiveness	Improvement of 'spatial quality', accessibility of the area, number of tourists (tourist accommodation)	[14,15,20,22,25–27]		11.7.1	Use of public areas
14	Exploitation	A measure of Net Present Value from the stakeholders' perspective, e.g., income per exploitation activity (irrigation, recreation, cattle farming, agriculture, tourists)	[14,15,20,22,25–27]	Economy	8.1.1 8.2.1 8.9.1	Economic growth per capita Economic growth per employed person Economic growth due to tourists
15	Investment	A measure of Net Present Value from the intervention's perspective, e.g., less money spent compared to a traditional measure	[14,15,20,22,25–27]		8.4.2/ 12.2.2 9.5.1 11.4.1	Domestic material consumption per GDP Research and development expenditure Expenditure on culture and heritage
16	Employment	Additional jobs created (pruning of trees, mowing, renting canoes, selling local growing products)	[14,15,20,22,25–27]	Economy	8.3.1 8.5.2	Employment/ Unemployment rates
17	Value of flood damage avoided	Value of assets that would have been destroyed in case of flood avoided relocation	[14,15,20,22,25–27]		1.5.2 11.5.2	GDP economic losses due to flooding disasters Damages to infrastructure and services
18	Maintenance	Money spent for maintenance	[28–47]		-	-
19	Flood protection	Attenuation of the flood due to the natural components of the intervention, delay of the travel time of the peak flow	[14,15,20,22,25–27]		-	-
20	Structural integrity	Proof of structural stability whilst using natural materials	[54]	Technical	-	-
21	Reliability	Repairs or replacements needed since construction	[54]		-	-
22	Ease of implementation	Availability (and use) of resources and materials available on site	[54]		8.4.2/ 12.2.2 12.1.1	Domestic material consumption Sustainable production and consumption plans

Table 3. *Cont.*

#	Fluvial Flooding Indicator	(Fluvial Flooding Indicator) General Description	Ref.	Dimension	Relevant SDG Indicator	(SDG Indicator) Short Description
23	Adaptability	Future changes in function	[54]		11.a.1	Development plans accounting for future projections
					13.2.1	Climate adaptation plans and strategies
24	Resilience	Whether another major intervention will be needed in due course (long-term perspective with respect to safety)	[54]		-	-
25	Different stakeholders/disciplines involved	Different stakeholders/disciplines involved	[14,15,20,22,25–27]	Policy—Procedural	11.3.2	Public engagement strategies
					12.7.1	Sustainable action plans
26	Planning/participatory processes	Types of participatory/planning process used: top-down/bottom-up, formal/informal rule-oriented, trust-based, consultation processes, collaborative learning, learning by performing, workshops, meetings	[14,15,20,22,25–27]		6.b.1	Community engagement
					12.7.1	Sustainable action plans
27	Hierarchy relations (e.g., communication, transparency)	Gap between local stakeholders and projects managers/central bosses (committed and accessible project managers)	[28–47]		-	-
28	Environmental agendas, frameworks, directives	Different legislations that need to be considered: assessments, (water, floods, birds) directives, Natura 2000	[14,15,20,22,25–27]	Policy—Procedural	1.5.3/ 11.b.1/ 13.1.2	Strategies in line with Sendai
					1.5.4/ 11.b.2/ 13.1.3	Alignment of local and national strategies
					15.8.1	Policies for invasive alien species
29	Integrated water resources management	Newly added dimension indicator to cover the relevant SDG indicator	[4]		6.5.1	Integrated water resources management
30	Transboundary water cooperation	Newly added dimension indicator to cover the relevant SDG indicator	[4]		6.5.2	Transboundary water cooperation
31	Capacity-building for development actions	Newly added dimension indicator to cover the relevant SDG indicator	[4]		13.3.2	Technology knowledge, transfer, and development in countries
32	Sharing of benefits	Newly added dimension indicator to cover the relevant SDG indicator	[4]		15.6.1	Policies for sharing of benefits
33	Aichi biodiversity target 2	Newly added dimension indicator to cover the relevant SDG indicator	[4]		15.9.1	Aichi Biodiversity target 2

Table 3. *Cont.*

#	Fluvial Flooding Indicator	(Fluvial Flooding Indicator) General Description	Ref.	Dimension	Relevant SDG Indicator	(SDG Indicator) Short Description
34	Conservation of biodiversity and ecosystems	Newly added dimension indicator to cover the relevant SDG indicator	[4]		15.a.1	Use and conservation of biodiversity and ecosystems
35	Expectations-outcomes alignment	Alignment of project aims with the expectations of stakeholders and with the outputs and outcomes delivered	[28–47]		-	-
36	Long-term data consistency	Existence and/or maintenance of databases relevant to the project info	[14,15,20,22,25–27]		-	-
37	Raising and sharing nbss awareness	Virtual visits on respective sites/forums, publications in social media, citations/newspapers, public consultations about how the people feel after the completion of the intervention (public engagement meeting)	[14,15,20,22,25–27]		-	-
38	Promoting collaboration	Coalition and partnerships formed and sustained	[14,15,20,22,25–27]	Policy—Procedural	17.14.1 17.17.1	Mechanisms for sustainable development Partnerships
39	Sustainability reporting	Newly added dimension indicator to cover the relevant SDG indicator	[4]		12.6.1	Sustainability reports by companies
40	Climate-related support	Newly added dimension indicator to cover the relevant SDG indicator	[4]		13.b.1	Support for climate-related actions
41	(Types of) cooperation between countries	Newly added dimension indicator to cover the relevant SDG indicator	[4]		17.6.1	Cooperation between countries
42	Country-owned resources	Newly added dimension indicator to cover the relevant SDG indicator	[4]		17.15.1	Use country-owned resources
43	Progress in SDGs	Newly added dimension indicator to cover the relevant SDG indicator	[4]		17.16.1	Reporting progress in SDG
44	Production of national indicators	Newly added dimension indicator to cover the relevant SDG indicator	[4]		17.18.1	Production of SD indicators per country

Table 3 presents the five dimensions with their respective fluvial flooding indicators (first four columns) and their connection with the SDG indicators (last two columns).

3.4. Case Study: The Eddleston Water Project

The Eddleston Water Project was selected as a representative case study to test the use of the fluvial flooding indicators. Importantly for our assessment, the Eddleston Water Project adopts a multi-benefit approach to the use of NBSs aiming at (i) exploring whether flood risk can be reduced by means of NBSs, (ii) the use of NBSs for improving the ecological condition of the river, and (iii) working with landowners and communities to maintain and enhance sustainable land management practices and farm businesses. Furthermore, since the measures were implemented in 2013, preliminary outcomes from the monitoring campaigns are already available. Finally, it is also part of the EU North

Sea Region (NSR) Interreg Building with Nature (BwN) program (https://northsearegion. eu/building-with-nature/), providing good links with experts if further consultation was needed (see also Appendix A).

The details of the Eddleston Water Project are summarized in the first column of Figure 4. A full description is available on the project website (https://tweedforum.org/ our-work/projects/the-eddleston-water-project/). The Eddleston Water Project started as a learning-by-doing project, which is successfully evolving and revealing valuable insights as to how a catchment approach reduces flood risk, involving both structural measures and natural flood management (NFM), and may help improve resilience to climate change. A key element throughout the project has been close stakeholder consultation because uptake of NBSs measures is voluntary, and all the locations for NBSs measures within the project catchment are privately owned. Local land managers and the wider community had been engaged from the very beginning of the project, and these and other stakeholders are still actively involved through regular meetings and surveys, ensuring productive continuation and uptake of the project.

Eddleston Water	Outcomes of the measures implemented to date
Location: Scottish Borders	• The installation of woody debris dams, ponds and riparian tree planting in the headwaters has a significant effect on delaying flood peaks by some 3-7 hours, and the flood peak in the upper catchment has reduced by c.30% post-implementation of these measures
River Characteristics: tributary of the River Tweed, about 18km long, flowing south and joining the main river at Peebles town	• Soil infiltration rates are 6-8 times more effective underneath old native woodland compared to adjacent pasture, and also much greater than under conifer plantations
Catchment Characteristics: c. 70 km², small size that favors an integrated hillslope to floodplain natural flood risk management approach	• The morphological diversity (pool, riffles, etc.) of the river channel has increased markedly since the historically straightened and channelized river was re-meandered
Examined period of the project: 2010-2021, from the production of the Scoping Study (2010) up to the end of the Building with Nature phase	• The assemblage of aquatic invertebrates has increased in number and diversity since restoration, reflecting the increase in habitat variety and extent in the channel
Current project state: More measures implemented, monitoring and evaluation carried out with the support from the EU NSR Interreg program and now by Scottish Government.	• The numbers of fish have increased in line with the increase in channel length (average 30% increase)
Measures: • 207 ha of native tree planting in the headwaters • 116 features of large woody in-stream structures • 29 off-line ponds (wetlands) • 2.9 km reach re-meandering	• A cost benefits analysis showed that the Net Present Value of flood damages avoided in downstream communities was just under £1 million, whilst the complementary wider benefits of these NFM measures from improvements to carbon management, biodiversity, water quality and recreation amounted to another £4.2 million • Working with local farmers and the community was essential to achieving sustainable benefits for all

Figure 4. Eddleston Water Project key information. Photos of the Lake Wood site, Eddleston Water Project: the previously straightened reach (top photo), the site immediately after the completion of the re-meandering works (middle photo), and the site one year after the completion of the re-meandering works with small consecutive floodplain ponds and the new re-meandered reach (bottom photo). All were retrieved from the Tweed Forum website.

The present year (2021) was the end of the 5 years of matched funding for the Eddleston Water Project from the EU NSR Interreg BwN program. The Interreg program focused on assessing the costs and benefits of implementing NFM through improved monitoring and modeling. With ongoing support for the current phase (2021–2024) from the Scottish Government, and the participation of local farmers and landowners, the study continues with the implementation of different types of NFM measures across the catchment, alongside detailed hydrological and ecological monitoring. Some of the headline outcomes from the Eddleston Water Project so far are summarized in the second column of Figure 4. For a more extended description of the project, reference is made to the Eddleston Water Project Report [55]; to the paper on flood risk reduction [56]; and to the Tweed Forum website (see above) where all the reports, including those from the Building with Nature program, are made publicly available.

3.5. Step IV—Assessment of the Fluvial Flooding Indicators Based on Project Metadata

To assess the utility and effectiveness of the fluvial flooding indicators, they were applied to the Eddleston Water Project to examine its sustainability performance in terms of

- Whether the Eddleston Water Project contributes to the attainment of the SDGs, and if so;
- To which SDGs;
- How and why.

The application was performed with input metadata from the Eddleston Water Project to the fluvial flooding indicators. The metadata for the Eddleston Water Project in Table 4 were collected from the Project Reports and [56–59]. The output of the evaluation is presented in Table 4, with three columns and the following format: each fluvial flooding indicator (Column I) contributes to none/one/or more SDGs and SDG targets (Columns II and III) as justified by the Eddleston Water Project metadata (Column IV). The contribution of the Eddleston Water Project to the UN 2030 Agenda is presented in Table 4 in terms of SDGs because this is more practical and easier to remember as a take-home message. However, by referring back to Table 3, it is possible to see the derivation and connection between the relevant SDG targets and respective SDG indicators. For instance, in Table 4, it can be seen that the Eddleston Water Project contributes to SDG 15 and SDG target 15.1, as assessed by the biodiversity abundance (fluvial flooding) indicator according to available project metadata. By referring back to Table 3 and the biodiversity abundance indicator (#1), it can be seen that the Eddleston Water Project contributes to SDG indicator 15.1.2. Therefore, the Eddleston Water Project contributes to the SDG indicator 15.1.2, SDG target 15.1, and SDG 15 of the UN 2030 Agenda in terms of biodiversity abundance.

Table 4. Contribution of the Eddleston Water Project to the relevant SDGs and SDG targets (second, third column), examined per fluvial flooding indicator (first column) according to Eddleston Water Project's metadata (fourth column). Divisions are according to the dimensions, as in Table 3.

#	Fluvial Flooding Indicator	Contribution to SDGs	SDG Targets	Eddleston Project Metadata
1	Biodiversity abundance	15	15.1	EU Special Area of Conservation (SAC) for its salmon, lampreys, otters, and aquatic plants Macroinvertebrate: a rapid recolonization of re-meandered channels by aquatic macroinvertebrates. Species richness and diversity increased post-restoration Salmonids: Eddleston is important for breeding salmon and as a nursery habitat. Improved salmonid habitat due to restorations in terms of the provision of suitable micro habitat and overall physical diversity. Total available habitat area increased due to the increased channel length and width
2	Wildlife habitat	15	15.5	An increase in overall physical diversity of habitats within re-meandered sections and an increase in habitat area, both greater where there has been a greater degree of re-meandering

Table 4. *Cont.*

#	Fluvial Flooding Indicator	Contribution to SDGs	SDG Targets	Eddleston Project Metadata
3	Population viability	15	15.5	Potential increase in the number and extent of spawning habitats for salmon, as indicated by changes in the spatial distribution of favored micro-habitats for salmonids
4	Continuity of water and sediment	15	15.3	Morphological units: generally, there is much greater morphological diversity through the reach because of restoration, with the most significant change happening at the Lakewood reach with the biggest increase in length (re-meandering). Generally, restoration has resulted in much more diverse channel morphology, with all morphological unit types present in 2015/2016 compared to only three in 2009 Grain size per geomorphic unit: following restoration, the overall grain size and variation was seen to decrease, with units post-restoration being better sorted and grain sizes more distinctive and specific per geomorphic unit
5	Water quality	-	-	Water quality is generally good in Eddleston, apart from some isolated incidents of diffuse organic pollution and increased nitrate levels in recent decades. Generally, it was not an objective, aim, or constraint of the project (Spray et al., 2017)
6	CO_2 emissions	9	9.4	Tree planting reduces carbon; however, no specific measurements were taken because it is not a key project issue. However, more research is currently being completed in this direction
7	Extent of water-related ecosystems	6	6.6	Re-meandering (approximately 3 km): the new courses increase the existing individual lengths of channel by between 8% and 56%, reducing the gradient and adding some 300 m (approximately 3000 m^2) of new in-channel habitat
8	Well-being	1, 11, 13	1.5, 11.5, 13.1	Modeling from SEPA (SEPA's flood risk assessments) shows 521 properties in Peebles, 61 in Eddleston, and 7 rural dwellings are at risk from a 1:200 year flood event. To date, catchment communities escaped the 2015/2016 and late 2016 winter floods
9	Cultural heritage/educational value	12	12.8	The project works as a living laboratory, open to public and to schools for raising awareness of flooding in the area and encouraging pupils and teachers to take an active part in the project and learn about their catchment. Additionally, interpretation boards enhance the commercial use of the area. Finally, as a publicly funded Research Platform, the river is the location for many research projects from universities and academic institutions
10	Recreation/leisure value	11	11.7	Soon, a multi-use track (biking, walking) will be constructed on the old railway line, which will attract even more people for recreation
11	Enhance attractiveness	11	11.7	Interpretation boards to be produced along this new path will improve the recreational side of the Eddleston
12	Exploitation	8	8.1, 8.2, 8.9	Full details of the economic costs and benefits of the implementation of NBSs measures have been analyzed as they impact farm income and profitability in the case study area based on land use data, agricultural and environmental support subsidies, and foregone farm income The salmon fishery of the Tweed is worth a total of over 24 million GBP a year to the local economy and supports over 500 jobs, so any improvement to fish habitat is important. Although salmon fishing is predominantly on the main Tweed River, Eddleston Water and similar tributaries are vital as breeding and nursery locations for salmon

Table 4. *Cont.*

#	Fluvial Flooding Indicator	Contribution to SDGs	SDG Targets	Eddleston Project Metadata
13	Investment	8, 9,11,12	8.4, 9.5, 11.4, 12.2	Modeling for a range of climate change scenarios shows a positive net present value from NFM tree planting, indicating that the riparian woodland is worth implementing. Annual benefits of c. 80 k GBP per year were estimated, with a high average benefit–cost ratio for the riparian woodland, though full benefits will not be realized for some 15 years after implementation Direct measurements on the ground of the value of a range of ecosystem services/multiple benefits already delivered as part of the NFM measures is an additional GBP 4.2 million Net Present Value (NPV) over and above the NPV from flood damages avoided (GBP 950 k) from the implementation of the same measures. The total cost of physical works amounts to GBP 1.3 million across 20 different landholdings, with the majority of that attributed to river and pond excavations, fencing, and planting. Monitoring and evaluation have cost some GBP 925 k on top of that
14	Employment	-	-	No additional jobs created yet. Maybe some slight vegetation management, but nothing bigger. If the track is realized, then it is possible that there will be more additional jobs (such as renting bicycles)
15	Value of reduced flood damage	1, 11	1.5, 11.5	The value of flood damages avoided by the current NFM features is GBP 950 k Net Present Value (100 years) The value of other ecosystem services/multiple benefits delivered as part of the NFM measures is an additional GBP 4.2 million NPV over and above the NPV from flood damages avoided (GBP 950 k) from the implementation of the same measures
16	Implementability	8, 12	8.4, 12.1, 12.2	Where possible, interventions are made of local timber from recently felled trees in the forest. An exception was for the rocks protecting the meander where it approaches the road, which were imported Large woody structures: on the Middle Burn, nearby conifers were felled and pinned across the channel Woodland and riparian woodland planting with native trees: species included oak, ash, willow, birch, aspen, and hazel
17	Adaptability	11, 13	11.a, 13.2	Although measures put in are seen as permanent, they are all subject to natural ecological and hydrological processes, and thus they will eventually need replacement. Potential change in the land of the area must be feasible, and project managers must be willing to facilitate and work with the landowners for land-use changes
18	Different stakeholders/disciplines involved	11, 12	11.3, 12.7	Landowners are key, and to date, 25 farmers and landowners have been involved, and 19 have hosted measures on their land. The Tweed Forum acts as project managers with Scottish Government, SEPA, Scottish Borders Council, Dundee University, and British Geological Survey. Others include Peebles Community Council, Forest Commission Scotland, Environment Agency, Scottish Natural Heritage, and National Farmers Union (Scotland)
19	Planning/participatory processes	6, 12	6.b, 12.7	Shared policy development and implementation: as a 'Trusted Intermediary', Tweed Forum spent significant time and effort informing and engaging with the local community and landowners in framing the project prior to implementation; regular meetings and presentations with the Peebles Community Council; interviews with landowners; leaflet to locals outlining and explaining aims of the project before the start of it; hands-on participatory engagement at local shows; questionnaire survey for the implemented measures

Table 4. *Cont.*

#	Fluvial Flooding Indicator	Contribution to SDGs	SDG Targets	Eddleston Project Metadata
20	Environmental agendas, frameworks, directives	1, 11, 13	1.5, 11.b, 13.1	Tweed EU Special Area of Conservation (SAC); Water Environment and Water Services (Scotland) Act 2003; Flood Risk Management (Scotland) Act 2009; Eddleston Water forms part of the River Tweed, which has been designated as a HELP basin following the UNESCO program; Scottish Rural Development Programme (SRDP) scheme
21	Integrated water resources management	6	6.5	The Eddleston Project adopted an integrated catchment approach across all aspects of water resource management since this underpins the project approach to address the 'sources—pathways and receptors' contributing to flood risks
22	Transboundary water cooperation	-	-	Not a transboundary water project
23	Capacity-building for development actions	13	13.3	Eddleston is a small catchment where a specialized focus and strengthening of locals' interest and involvement was needed for the realization of the project. This was achieved through participatory processes and engagement strategies
24	Sharing of benefits	15	15.6	Participatory processes and engagement strategies were a way of ensuring equitable share of benefits over the sectors considered in the project, including recognition of potential impacts of NBSs on farm businesses
25	Aichi biodiversity target 2	15	15.9	Monitoring campaigns are running, aiming at evaluating the effect of the measures on biodiversity and hydro-morphology, creating evidence for strengthening biodiversity strategies
26	Conservation of biodiversity and ecosystems	15	15.a	Monitoring campaigns are running, aiming at evaluating the outputs of the measures on biodiversity and ecosystems
27	Promoting collaboration	17	17.14, 17.17	Generally, a partnership approach has been followed, and Tweed Forum has brought together the landowners, the community, and the project experts
28	Sustainability reporting	-	-	No sustainability reports by companies
29	Climate-related support	-	-	No climate-related support. Currently, more research is being carried out to examine the effects of the interventions on climate change projections
30	(Types of) cooperation between countries	-	-	No cooperation between countries in the beginning. Many countries were involved when the project became part of the Interreg North Sea Region Program
31	Country-owned resources	17	17.15	The Eddleston Project Managers are Tweed Forum, and they, along with Scottish Government and SEPA and the main science provider, Dundee University, are all based in Scotland, and thus the project was generated and developed by country-owned institutions before attracting wider interest
32	Progress in SDGs	17	17.16	Very detailed and wide-ranging monitoring campaigns are running, aiming at evaluating the outcomes of the measures on multiple sectors and thus progress on SDGs
33	Production of national indicators	-	-	No production of SDG indicators

Overall, the Eddleston Water Project contributes to 9 SDGs: 1, 6, 8, 9, 11, 12, 13, 15, and 17, and to 33 SDG targets, as can be seen graphically in Figure 5. Figure 5 complements Table 4 since it shows all the relevant SDGs and SDG targets (as established from Step II). The SDG targets in bold black color are the ones that the Eddleston Water Project contributes to, while in red, the ones that it does not. We showed that the Sustainability Performance Evaluation Framework follows a systematic methodology that allows to identify the interactions of the Eddleston Water Project within the five dimensions and define fluvial flooding indicators, which showed the Eddleston Water Project's contribution to the UN 2030 Agenda. Table 5 presents the Eddleston Water Project's SDGs under their respective dimensions. As expected, most of the Eddleston Water Project's SDGs contribute to multiple dimensions at the same time due to the repetition of some of the SDG indicators under several SDG targets.

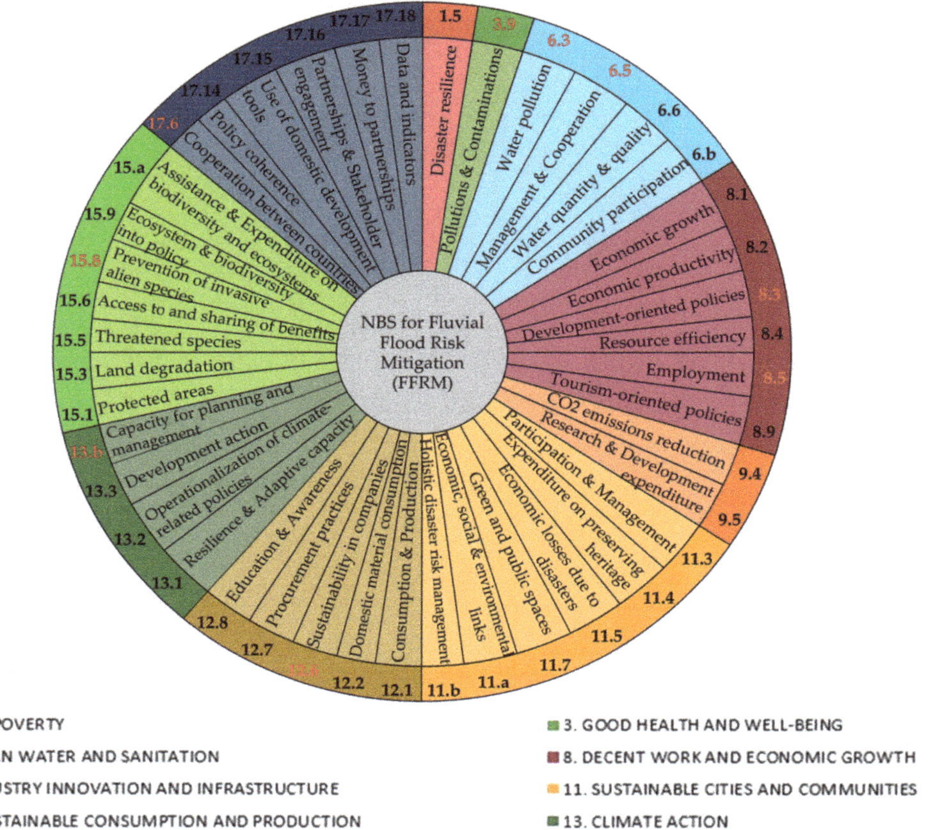

Figure 5. NBSs for FFRM could address 10 SDGs (legend) and 42 SDG targets (circle), as set by the UN 2030 Agenda. The framework application to the Eddleston Water Project revealed that the Eddleston contributes to 9 out of these 10 SDGs (not to the third) and to 33 SDG targets (not to the ones in bold red color in the outer circle).

Table 5. The Eddleston Water Project's SDGs under their respective dimension.

Dimension	Policy—Procedural	Economy	Society	Environment	Technical
	SDG 1	SDG 1	SDG 1	SDG 6	SDG 8
	SDG 11	SDG 11	SDG 11	SDG 9	SDG 12
Sustainable Development Goals	SDG 13	SDG 8	SDG 13	SDG 15	SDG 13
	SDG 6	SDG 9	SDG 12		SDG 11
	SDG 12				
	SDG 17				

4. Discussion

4.1. Regarding the Framework Itself: From Structure to Output

The Sustainability Performance Evaluation Framework was developed from literature review, insights from three case studies, and then applied to one independent case study. On the one hand, this enabled us to assess the performance of the fluvial flooding indicators in depth, as we tried to find and access metadata for each fluvial flooding indicator. On the other hand, application to more case studies would have provided more insights that could extend or alter the NBSs–SDG coupling. For instance, the research showed that SDG seven, which assesses energy resources, could potentially be linked to the fluvial flooding indicators. More specifically, the Noordwaard polder project provided some clues about energy production from biomass, but the other case studies did not (https://www.ecoshape.org/en/cases/wave-attenuating-willow-forest-noordwaard-nl/), and thus SDG seven was not included in the link between NBSs and SDGs. A broader case study examination will provide further insights regarding the potential of energy production in NBSs for FFRM interventions.

For data-dependant frameworks, such as the Sustainability Performance Evaluation Framework, data accessibility and/or method availability are key factors. Literature indicators such as 'population viability', although meaningful in the context of NBSs, are difficult to measure in practice. Similarly, within the Tier Classification for the Global SDG Indicators [60], Tier three includes the need for new or re-examination of existing measuring methods. Shah et al. [12] recognize the need for making the level of information even more local and specific alongside primary data collection for local NBSs or their indicators. From a more general point of view, Kumar et al. [10] state that the challenge of inadequate or insufficient data hinders the acceptance, assessment, and potentially successful operationalization of NBSs. Recent studies have addressed this challenge; Schröter et al. [9] provide extensive lists of online data pools on NBSs, and the EC's Handbook for Practitioners [52] devotes a chapter to types of data, data sources, and data generation techniques for NBSs monitoring and impact assessment. Therefore, current sources seem to allow for the effective use of such data-based frameworks while acknowledging the need for new or enhanced measuring methodologies.

The output of the Sustainability Performance Evaluation Framework is qualitative. We acknowledge, however, the potential to extend it to a quantitative one. Quantitative outputs, such as scored evaluation against pre-defined targets, enhance the evidence base of NBSs effectiveness [13]. To date, several studies provide quantitative results regarding effectiveness, co-benefits, and NBSs' sustainability contribution. Schipper et al. [21] provide a methodology for scoring the sustainability performance of coastal management projects using numeric data. Pugliese et al. [19] apply a multi-criteria tool to assess the effectiveness of NBSs for a specific case study compared to a grey alternative. Martín et al. [18] use qualitative analysis of Fuzzy Cognitive Maps (FCM) alongside semi-quantitative analysis of the co-benefits to examine the effectiveness of different NBSs and their co-benefits. Liquete et al. [17] perform an ex-post assessment of the environmental, social, and economic benefits of multi-purpose NBSs for water pollution control based on Multi-Criteria Analysis (MCA). The proposed framework is used to couple to the SDGs but can (with minor changes) also be used as an independent framework for the ex-post evaluation of individual NBSs projects, as a tool to compare grey–green alternatives for an NBSs project, or even as a tool to compare different NBSs projects. Andrikopoulou [61] describes the development of such a framework. We speculate that application of such a framework to a 'Room for the River' project, where pre-defined targets for the Rhine's conveyance capacity were set (https://www.tandfonline.com/doi/abs/10.1080/02508060508691839), would have resulted in a better understanding regarding the potentials of the Sustainability Performance Evaluation Framework to derive a scored evaluation. A Multi-Criteria Analysis would both bring the framework output closer to reality and better assist decision makers in prioritizing river management options. In this research, flood safety was the primary river function under consideration, while ecosystem development, water quality, and recreation issues were also examined. However, rivers typically have a larger number of functions (e.g., water supply navigation, water quality, nature development), the importance of which may have greater or lesser weighting at any one time and location, and to different stakeholders, depending on the focus of the project. Therefore, prioritization of the fluvial indicators based on the primary river functions of a specific case would greatly benefit the framework output.

4.2. How Do the Outcomes of Our Framework Relate to Other Relevant Studies?

We have set up a framework to create a set of fluvial flooding indicators to evaluate the ex-post contribution of the Eddleston Water Project to the UN 2030 Agenda. Ligtvoet [2] and Ge et al. [62] have also explored the relationship between rivers and the SDGs. Ge et al. [62] have defined SDGs for river basin scale in terms of water, ecosystems, and socioeconomic capabilities. Ligtvoet [2] has identified those SDGs related to people and the economy that are negatively affected by river flooding.

Framework application to the Eddleston Water Project shows that it contributes to six out of the nine SDGs mentioned in Ligtvoet [2]. In the study of Ligtvoet [2], Ward and Winsemius have identified the negative effects of fluvial flood risk on SDGs 1, 2, 3 6, 8, 9, 10, 13, and 16 for people and the economy (specifically on agriculture). Although Ward and Winsemius established a broader link (river flood risk and SDGs) than ours (NBSs for FFRM and SDGs), the outcomes of our framework application show that the Eddleston Water Project can ameliorate most of the negatively affected SDGs by river flooding. Exceptions constitute SDGs 2, 10, and 16. SDG two addresses food security issues and agricultural practices, which were considered out of scope for the functions of fluvial flood risk mitigation NBSs considered in this research. SDG 10 refers to the reduction of inequalities between countries in terms of providing the same means of coping with flood risk between high-income, upper-middle, lower-middle and low-income economies. NBSs could address such an aspect; however, it needs examination in a broader context combined with geopolitical considerations. Similarly, SDG 16 talks about justice and inclusivity in societies, which can be promoted as part of the general NBSs conceptual framing but are out of scope for this study. Leaving these aside then, it is apparent that the Eddleston Project positively affects the following 6 SDGs: 1, 3, 6, 8, 9, 13.

The SDGs to which the Eddleston Water Project contributes align with seven out of eight SDGs proposed by Ge et al. [62]. Ge et al. [62] link river basins (e.g., Amazon, Nile, and Heihe river basins) to SDGs in terms of water, ecosystem, socioeconomic, and ability-related issues, and they find all SDGs relevant apart from SDG seven. The water-related SDGs (6 11, 12, 14) and the ecosystem-related SDGs (14, 15) coincide with the SDGs derived from the present Eddleston Water study, apart from SDG 14 (which was out of scope for the current research context because it focuses on the coastal environment). The socioeconomic-related SDGs were omitted because they focus mostly on food security, justice, and inequalities which, whilst they might be added in other situations, were not within the scope of the impacts of the fluvial NBSs examined for the present research. However, although they were not specifically considered, it could be argued that by including and elaborating on the third aim in the Eddleston Water Project's objectives (working with landowners and communities to maintain and enhance sustainable land management practices and farm businesses), the project contributes to an element of this SDG. The ability-related SDGs (9 11, 13, 17) coincide with the ones derived from our research because they refer to structural actions and strategies for conserving and protecting the rivers. Hence, SDGs 6, 9, 11, 12 13, 15, 17 are shared between Ge et al. [62] and the Eddleston Water Project with respect to water, ecosystem, and ability-related issues.

4.3. Standardization and Scale of Sustainability Assessments

As mentioned by Pohle et al. [63], standards, in general, have a twofold role: they can serve as a source of information and enabler for the development and transfer of technology. The EU Research and Innovation program Horizon 2020 has recognized standardization as a measure that underpins innovation since it bridges the gap between research and the market but also facilitates the propagation of research outcomes to the European and international markets [64]. That said, standardization of sustainability assessments could provide harmonization in indicators, reliability and transparency in calculation methods, and comparability of results [65]. To date, standardization for sustainability assessments has already been discussed in the literature [66–68]. Although most of the studies recognize the aforementioned benefits of standardization, they also recognize the risk of compromising agendas, contexts, and needs when treating larger scales. Similarly, in the present study, although the aim was to derive indicators as widely applicable as possible, given the scope of the study, we acknowledge that most of the indicators would need further consideration when used in projects of different scales and contexts. Some of the indicators, e.g., the environmental and technical indicators, could relatively easily be standardized – some with slight amendments—for larger scales. For others, such as the Social, Policy—Procedural, and Economic indicators, this is more difficult because they are geopolitically dependent. For instance, in a transboundary water body apart from the international legislation, the in-between the country-member treaties and arrangements should also be considered. Such a view was out of scope in our research and thus requires further research.

Sustainability assessments should be able to be carried out at any scale. However, the necessary data are not always available or accessible at any scale. For instance, it is likely that the Sustainability Performance Evaluation Framework cannot be applied globally to any NBSs project due to the lack of data. Most of the data collected for such projects either suit national aims, which do not always align with the global SDG indicators, or come from private sources. Therefore, it seems that currently, the main challenge lies in finding adequate, available, and accessible data to upscale the sustainability assessments, and although there is still a lot to accomplish in this direction, the EC's Handbook for Practitioners [52] and Schröter et al. [9] have made promising steps (as discussed in paragraph 4.1).

5. Conclusions

The aim of this study is to propose a ready-to-use methodology to evaluate the sustainability performance of Nature-Based Solutions (NBSs) with respect to the United Nations (UN) 2030 Agenda, involving all its components (i.e., Sustainable Development Goals (SDGs), targets and indicators). This was achieved by building on the Schipper et al. [21] systematic framework and adjusting it to fit our needs. The focus is on NBSs for Fluvial Flood Risk Mitigation (FFRM) in river basins of sizes up to 100 km^2. The derived framework is called the Sustainability Performance Evaluation Framework. It encompasses four steps through which the end-user creates a set of fluvial flooding indicators that can then be linked to the SDG indicators, and by applying the fluvial flooding indicators to a specific FFRM NBSs project, it is possible to ascertain the project's contribution to the UN 2030 Agenda. The Eddleston Water Project was used as a case study to test the effectiveness of the fluvial flooding indicators. Application to the Eddleston Water Project has shown that it contributes to 9 SDGs and 33 SDG targets. In developing the Sustainability Performance Evaluation Framework and testing its fluvial flooding indicators, the findings are:

1. The Sustainability Performance Evaluation Framework can systematically consider SDG indicators by exploring potential interactions of NBSs for FFRM projects within five chosen dimensions: economy, environment, society, policy, and technical.
2. Through the Sustainability Performance Evaluation Framework, it is possible to adjust the SDG concept to the system of interest and qualitatively measure its alignment with and progress towards the SDGs.
3. Data availability and accessibility play a crucial role in the Sustainability Performance Evaluation Framework. Although potentially challenging in some situations, many NBSs programs and projects have been funded by the European Union (EU) or national governments and agencies, and data are typically available either publicly or upon request.

To further develop the Sustainability Performance Evaluation Framework, a key recommendation is its trial application in other areas and by different end-users. This should focus on three aspects:

- Application to projects where quantified targets pre-exist would help enable the derivation of some form of scored evaluation.
- Application to different case studies in terms of scale, location, and type of measures (e.g., projects in upland rivers and transboundary projects) is suggested.
- Application to case studies in countries with upper-middle, lower-middle, and low income economies, with different cultural contexts, legal frameworks, governance structures, challenges of environmental justice, and data scarcity would add value. This recommendation, combined with the previous one, would also shed light on the potentials of the proposed indicators to be standardized.
- Application with end-users, stakeholders, or even people unfamiliar with NBSs and SDGs, to examine whether the framework would yield the same indicators and/or the same result regarding the evaluation of the NBSs project (regarding its contribution to the SDGs).

We recognize that the more the framework is reviewed and applied, the more insights will be gained with respect to its biases, limitations, and gaps, including opportunities that could extend or alter the NBSs–SDG coupling.

Author Contributions: Conceptualization, T.A. and R.M.J.S.; methodology, T.A., R.M.J.S., C.A.S. A.B. and C.J.S.; formal analysis T.A. and R.M.J.S.; investigation T.A., R.M.J.S. and C.J.S.; data curation T.A., R.M.J.S. and C.J.S.; writing—original draft preparation, T.A.; writing—review and editing T.A. R.M.J.S., C.A.S., A.B. and C.J.S.; supervision R.M.J.S., C.A.S., A.B. and C.J.S. All authors have read and agreed to the published version of the manuscript.

Funding: This research received no external funding.

Institutional Review Board Statement: Not applicable.

Informed Consent Statement: Not applicable.

Data Availability Statement: For the framework application, data were mostly collected from publicly available sources, with a few of them from expert consultation. Details regarding data availability and accessibility for the Eddleston Water Project are available in Appendix A.

Acknowledgments: The research behind this paper was a collaboration of TU Delft and Rijkswaterstaat in The Netherlands. It was funded by Rijkswaterstaat and greatly benefited from discussions with TU Delft and RWS colleagues, examining nature-based projects and their potentials with respect to the UN 2030 Agenda. The EU Interreg Building with Nature program's projects were an inspiration and key evidence-contributor to the paper. We would also like to thank the Tweed Forum that provided all the information needed for the Eddleston Water Project, as well as the external reviewers for their time devoted to this manuscript since we feel their comments helped to improve the quality of the manuscript.

Conflicts of Interest: The authors declare no conflict of interest.

Appendix A. Data Availability and Accessibility for the Eddleston Water Project

One of the main reasons for choosing the Eddleston Water Project was the data availability and accessibility. Indeed, most of the fluvial flooding indicators (85%) were filled in with data publicly available online, while only 15% of the indicators needed a project specialist —either to verify data found online or to provide additional information. Expert consultation was needed for the following five fluvial flooding indicators: CO_2 emissions, recreation/leisure value, enhance attractiveness, employment, and adaptability. Both qualitative and quantitative data were gathered since not all indicators required a numerical value – for example, those covering the Policy—Procedural dimension. On the contrary, fluvial flooding indicators such as 'well-being' or 'extent of water-related ecosystems' could be filled in with quantitative data.

During the framework application, a few overlaps of the project metadata per fluvial flooding indicator were observed. For example, in looking at the 'recreation/leisure value' and 'enhance attractiveness', similar data were used for both fluvial flooding indicators. Although the attractiveness of the area has been enhanced and it is being used by the public, new plans for a cycleway will further increase its recreational value. To this end, there is a limited extent of data for these two indicators, leading to their current overlap. However, for a case study where all the interventions had been finalized, these two indicators might provide different information.

References

1. UN. The Human Cost of Weather-Related Disasters. Available online: https://www.unisdr.org/files/46796_cop21weatherdisastersreport2015.pdf (accessed on 26 July 2021).
2. PBL Netherlands Environmental Assessment Agency. Scientific Justification of the Information Produced for the Chapter 'Flooding' of 'The Geography of Future Water Challenges'. 2018. Available online: https://www.pbl.nl/sites/default/files/downloads/pbl-2018-the-geography-of-future-water-challenges-river-flood-risk_3147.pdf (accessed on 9 October 2021).
3. Best, J. Anthropogenic stresses on the world's big rivers. *Nat. Geosci.* **2019**, *12*, 7–21. [CrossRef]
4. Global Indicator Framework for the Sustainable Development Goals and Targets of the 2030 Agenda for Sustainable Development. Available online: https://unstats.un.org/sdgs/indicators/Global%20Indicator%20Framework%20after%202021%20refinement_Eng.pdf (accessed on 27 July 2021).
5. Nature-Based Solutions for Climate Change, People and Biodiversity. Available online: https://www.gla.ac.uk/media/Media_790171_smxx.pdf (accessed on 9 October 2021).
6. Debele, S.E.; Kumar, P.; Sahani, J.; Marti-Cardona, B.; Mickovski, S.B.; Leo, L.S.; Porcu, F.; Bertini, F.; Montesi, D.; Vojinovic, Z.; et al. Nature-based solutions for hydro-meteorological hazards: Revised concepts, classification schemes and databases. *Environ. Res.* **2019**, *179*, 108799. [CrossRef] [PubMed]
7. Nature-Based Solutions in Europe: Policy, Knowledge and Practice for Climate-Change Adaptation and Disaster Risk Reduction. Available online: https://www.eea.europa.eu/publications/nature-based-solutions-in-europe (accessed on 26 July 2021).
8. Nature-based Solutions for Climate Change in the UK: A Report by the British Ecological Society. Available online: https://www.iucn-uk-peatlandprogramme.org/sites/default/files/2021-05/NbS-Report-Final-Designed.pdf (accessed on 9 October 2021).
9. Schröter, B.; Zingraff-Hamed, A.; Ott, E.; Huang, J.; Hüesker, F.; Nicolas, C.; Schröder, N.J.S. The knowledge transfer potential of online data pools on Nature-based Solutions. *Sci. Total.Environ.* **2020**, *762*, 143074. [CrossRef]
10. Kumar, P.; Debele, S.E.; Sahani, J.; Aragão, L.; Barisani, F.; Basu, B.; Bucchignani, E.; Charizopoulus, N.; Di Sabatino, S.; Domeneghetti, A. Towards an operationalisation of nature-based solutions for natural hazards. *Sci. Total.Environ.* **2020**, *731*, 138855. [CrossRef]
11. Ruangpan, L.; Vojinovic, Z.; Plavšić, J.; Doong, D.J.; Bahlmann, T.; Alves, A.; Tseng, L.-H.; Randelovic, A.; Todorovic, A.; Kocic, Z. Incorporating stakeholders' preferences into a multi-criteria framework for planning large-scale Nature-Based Solutions. *Ambio* **2020**, *50*, 1514–1531.
12. Shah, M.A.R.; Renaud, F.G.; Anderson, C.C.; Wild, A.; Domeneghetti, A.; Polderman, A.; Votsis, A.; Pulvirenti, B.; Basu, B.; Thomson, C. A review of hydro-meteorological hazard, vulnerability, and risk assessment frameworks and indicators in the context of nature-based solutions. *Int. J. Disaster Risk Reduct.* **2020**, *50*, 101728. [CrossRef]
13. Albert, C.; Brillinger, M.; Guerrero, P.; Gottwald, S.; Henze, J.; Schmidt, S.; Ott, E.; Schröter, B. Planning nature-based solutions: Principles, steps, and insights. *Ambio* **2020**, *50*, 1446–1461. [CrossRef] [PubMed]
14. Calliari, E.; Staccione, A.; Mysiak, J. An assessment framework for climate-proof nature-based solutions. *Sci. Total. Environ.* **2019**, *656*, 691–700. [CrossRef]
15. Nesshöver, C.; Assmuth, T.; Irvine, K.N.; Rusch, G.M.; Waylen, K.A.; Delbaere, B.; Hasse, D.; Jones-Walters, L.; Keune, H.; Kovacs, E. The science, policy and practice of nature-based solutions: An interdisciplinary perspective. *Sci. Total.Environ.* **2017**, *579*, 1215–1227. [CrossRef]
16. Arkema, K.K.; Griffin, R.; Maldonado, S.; Silver, J.; Suckale, J.; Guerry, A.D. Linking social, ecological, and physical science to advance natural and nature-based protection for coastal communities. *Ann. N. Y. Acad. Sci.* **2017**, *1399*, 5–26. [CrossRef] [PubMed]
17. Liquete, C.; Udias, A.; Conte, G.; Grizzetti, B.; Masi, F. Integrated valuation of a nature-based solution for water pollution control. Highlighting hidden benefits. *Ecosyst.Serv.* **2016**, *22*, 392–401. [CrossRef]
18. Martín, E.G.; Giordano, R.; Pagano, A.; van der Keur, P.; Costa, M.M. Using a system thinking approach to assess the contribution of nature based solutions to sustainable development goals. *Sci. Total. Environ.* **2020**, *738*, 139693.
19. Pugliese, F.; Caroppi, G.; Zingraff-Hamed, A.; Lupp, G.; Giugni, M. Nature-Based Solutions (NBSs) Application for Hydro-Environment Enhancement. A Case Study of the Isar River (DE). *Environ. Sci. Proc.* **2020**, *2*, 30. [CrossRef]
20. Raymond, C.M.; Frantzeskaki, N.; Kabisch, N.; Berry, P.; Breil, M.; Nita, M.R.; Geneletti, D.; Calfapietra, C. A framework for assessing and implementing the co-benefits of nature-based solutions in urban areas. *Environ. Sci. Policy* **2017**, *77*, 15–24. [CrossRef]

21. Schipper, C.A.; Dekker, G.G.; de Visser, B.; Bolman, B.; Lodder, Q. Characterization of SDGs towards Coastal Management Sustainability Performance and Cross-Linking Consequences. *Sustainability* **2021**, *13*, 1560. [CrossRef]
22. Artmann, M.; Sartison, K. The role of urban agriculture as a nature-based solution: A review for developing a systemic assessment framework. *Sustainability* **2018**, *10*, 1937. [CrossRef]
23. Pakzad, P.; Osmond, P. Developing a sustainability indicator set for measuring green infrastructure performance. *Procedia-Soc. Behav. Sci.* **2016**, *216*, 68–79. [CrossRef]
24. Seifollahi-Aghmiuni, S.; Nockrach, M.; Kalantari, Z. The potential of wetlands in achieving the sustainable development goals of the 2030 Agenda. *Water* **2019**, *11*, 609. [CrossRef]
25. Weber, C.; Åberg, U.; Buijse, A.D.; Hughes, F.M.; McKie, B.G.; Piégay, H.; Roni, P.; Vollenweider, S.; Haertel-Borer, S. Goals and principles for programmatic river restoration monitoring and evaluation: Collaborative learning across multiple projects. *Wiley Interdiscip. Rev. Water* **2018**, *5*, e1257. [CrossRef]
26. Evaluating Nature-Based Solutions. Best Practices, Frameworks and Guidelines. Available online: https://northsearegion.eu/media/6959/report_pr3812_evaluatingnbs_final_29112018.pdf (accessed on 26 July 2021).
27. Schipper, C.A.; Vreugdenhil, H.; De Jong, M.P.C. A sustainability assessment of ports and port-city plans: Comparing ambitions with achievements. *Transp. Res.* **2017**, *57*, 84–111. [CrossRef]
28. Engineering With Nature: An Atlas. Available online: https://ewn.el.erdc.dren.mil/img/atlas/ERDC-EL_SR-18-8_Ebook_file.pdf (accessed on 26 July 2021).
29. Taming the Floods, Dutch-Style. Available online: https://www.theguardian.com/environment/2014/may/19/floods-dutch-britain-netherlands-climatechange (accessed on 26 July 2021).
30. Juarez, A. Unveiling Rules-In-Use in Eco-Engineering Projects. MSc: Wageningen, The Netherlands, 2013; Available online: https://www.google.com/url?sa=t&rct=j&q=&esrc=s&source=web&cd=&cad=rja&uact=8&ved=2ahUKEwj6pcuXwr3zAhU5wQIHHf80AiMQFnoECAMQAQ&url=https%3A%2F%2Fwww.wur.nl%2Fweb%2Ffile%3Fuuid%3D9ad1cc8e-514f-445e-a445-21023ba438f3%26owner%3Dfb4cacd6-0b7a-4091-ae5f-7fd83160b8f1&usg=AOvVaw2l3qxR7krpZOVddZDVrc43 (accessed on 9 October 2021).
31. Engineering: Building with Nature 101x MOOC—Hints and Information for Building with Nature Design—Case 1: Climate-proof Noordwaard. Available online: https://ocw.tudelft.nl/wp-content/uploads/Hints_for_Noordwaard.pdf (accessed on 26 July 2021).
32. Warner, J.F.; van Buuren, A.; Edelenbos, J. (Eds.) *Making Space for the River*; IWA Publishing: London, UK, 2012; pp. 56–62.
33. Tailor Made Collaboration: A Clever Combination of Process and Content. Available online: https://issuu.com/ruimtevoorderivier/docs/tailor_made_collaboration_a_clever_/17 (accessed on 26 July 2021).
34. Colorado Watershed Flood Recovery. Available online: https://coyotegulch.blog/2018/09/09/cwcb_dnr-colorado-watershed-flood-recovery-summary/ (accessed on 26 July 2021).
35. Post-Flood Recovery Assessment and Stream Restoration Guidelines for the Colorado Front Range. Available online: https://cpw.state.co.us/Documents/Research/Aquatic/pdf/PostFloodAssessmentandGuidelines.pdf (accessed on 26 July 2021).
36. Nicholson, A.R.; O'Donnell, G.M.; Wilkinson, M.E.; Quinn, P.F. The potential of runoff attenuation features as a Natural Flood Management approach. *J. Flood Risk Manag.* **2020**, *13*, e12565. [CrossRef]
37. Nicholson, A.R.; Wilkinson, M.E.; O'Donnell, G.M.; Quinn, P.F. Runoff attenuation features: A sustainable flood mitigation strategy in the Belford catchment, UK. *Area* **2012**, *44*, 463–469. [CrossRef]
38. Case study 16. Belford Natural Flood Management Scheme, Northumberland. Available online: https://ewn.el.erdc.dren.mil/symposiums/May2019/FridaySiteVisit/case_study_16_belford.pdf (accessed on 26 July 2021).
39. Reducing Flood in Belford [Video File]. Available online: https://www.youtube.com/watch?v=zqHLm4V_c58 (accessed on 26 July 2021).
40. Belford: A Case Study of Catchment Scale Natural Flood Management. Available online: https://www.parliament.uk/globalassets/documents/post/QuinnPOST.pdf (accessed on 26 July 2021).
41. Belford Burn Runoff Attenuation Scheme—Northumberland. Case Study 14. Available online: https://www.therrc.co.uk/sites/default/files/projects/sc120015_case_study_14.pdf (accessed on 26 July 2021).
42. Potential Use of Runoff Attenuation Features in Small Rural Catchments for Flood Mitigation. Available online: https://research.ncl.ac.uk/proactive/belford/newcastlenfmrafreport/reportpdf/June%20NFM%20RAF%20Report.pdf (accessed on 26 July 2021).
43. Natural Flood Management Handbook. Available online: https://www.sepa.org.uk/media/163560/sepa-natural-flood-management-handbook1.pdf (accessed on 26 July 2021).
44. Wilkinson, M.E.; Quinn, P.F. Belford Catchment Proactive Flood Solutions: A toolkit for managing runoff in the rural landscape. *Clim. Water Soil Sci. Policy Pract.* **2010**, *103*. Available online: https://research.ncl.ac.uk/proactive/belford/papers/SAC2010.pd (accessed on 9 October 2021).
45. Wilkinson, M.E.; Quinn, P.F.; Welton, P. Belford catchment proactive flood solutions: Storing and attenuating runoff on farms. In Proceedings of the BHS 10th National Hydrology Symposium, Exeter, UK, 15–17 September 2008.
46. Wilkinson, M.E.; Quinn, P.F.; Welton, P. Runoff management during the September 2008 floods in the Belford catchment, Northumberland. *J. Flood Risk Manag.* **2010**, *3*, 285–295. [CrossRef]
47. Wilkinson, M.E.; Quinn, P.F.; Barber, N.J.; Jonczyk, J. A framework for managing runoff and pollution in the rural landscape using a Catchment Systems Engineering approach. *Sci. Total. Environ.* **2014**, *468*, 1245–1254. [CrossRef] [PubMed]

8. Haase, D.; Larondelle, N.; Andersson, E.; Artmann, M.; Borgström, S.; Breuste, J.; Baggethun-Gomez, E.; Gren, A.; Hamstead, Z.; Kabisch, N. A quantitative review of urban ecosystem service assessments: Concepts, models, and implementation. *Ambio* **2014**, *43*, 413–433. [CrossRef]

9. Den Dekker-Arlain, J. *Sustainable Rivers*; MSc: Avans Hogeschool, The Netherlands, 2019.

0. Kabisch, N.; Frantzeskaki, N.; Pauleit, S.; Naumann, S.; Davis, M.; Artmann, M.; Hasse, D.; Knapp, S.; Korn, H.; Stadler, J. Nature-based solutions to climate change mitigation and adaptation in urban areas: Perspectives on indicators, knowledge gaps, barriers, and opportunities for action. *Ecol. Soc.* **2016**, *21*. 39. Available online: https://www.ecologyandsociety.org/vol21/iss2/art39/ (accessed on 9 October 2021). [CrossRef]

1. Development and Globalization Facts and Figures. Available online: https://stats.unctad.org/Dgff2016/DGFF2016.pdf (accessed on 27 July 2021).

2. Evaluating the Impact of Nature-Based Solutions: A Handbook for Practitioners. Available online: https://op.europa.eu/en/publication-detail/-/publication/d7d496b5-ad4e-11eb-9767-01aa75ed71a1 (accessed on 27 July 2021).

3. Comprehensive Framework for NBS Assessment. Available online: https://phusicos.eu/wp-content/uploads/2019/05/D4.1_Task4.1_UNINA_14052019_Final_withAppendicies.pdf (accessed on 27 July 2021).

4. Distilling Engineering Design Principles. Available online: https://data.4tu.nl/articles/dataset/Engineering_Building_with_Nature_101x_video_06_-_Engineering_design_principles_Dikes_/12693668 (accessed on 27 July 2021).

5. Spray, C.; Baillie, A.; Chalmers, H.; Comins, L.; Black, A.; Dewell, E.; Ncube, S.; Perez, K.; Spray, C.; Ball, T. Eddleston Water Project Report 2016. Tweed Forum: Melrose, UK. 2017.

6. Black, A.; Peskett, L.; MacDonald, A.; Young, A.; Spray, C.; Ball, T.; Thomas, H.; Werritty, A. Natural flood management, lag time and catchment scale: Results from an empirical nested catchment study. *J. Flood Risk Manag.* **2021**, *14*, e12717. [CrossRef]

7. Dittrich, R.; Ball, T.; Wreford, A.; Moran, D.; Spray, C.J. A cost-benefit analysis of afforestation as a climate change adaptation measure to reduce flood risk. *J. Flood Risk Manag.* **2019**, *12*, e12482. [CrossRef]

8. Werritty, A.; Spray, C.; Ball, T.; Bonell, M.; Rouillard, J.; MacDonald, A.; Comins, L.; Richardson, R. Integrated catchment management: From rhetoric to reality in a Scottish HELP basin. In Proceedings of the British Hydrological Society, Third International Conference. Role of Hydrology in Managing Consequences of a Changing Global Environment, Newcastle-upon-Tyne, UK, 19-23 July 2010; pp. 1–15.

9. Working with Natural Processes—Evidence Directory. Available online: https://assets.publishing.service.gov.uk/media/6036c5468fa8f5480a5386e9/Working_with_natural_processes_evidence_directory.pdf (accessed on 27 July 2021).

0. Tier Classification for Global SDG Indicators. Available online: https://unstats.un.org/sdgs/files/Tier%20Classification%20of%20SDG%20Indicators_29%20Mar%202021_web.pdf (accessed on 27 July 2021).

1. Andrikopoulou, T. Nature based solutions for fluvial flood mitigation: An integrated assessment framework. Master's Thesis, Technical University of Delft, Delft, The Netherlands, 2020.

2. Ge, Y.; Li, X.; Cai, X.; Deng, X.; Wu, F.; Li, Z.; Luan, W. Converting UN sustainable development goals (SDGs) to decision-making objectives and implementation options at the river basin scale. *Sustainability* **2018**, *10*, 1056. [CrossRef]

3. Pohle, A.; Blind, K.; Neustroev, D. The impact of international management standards on academic research. *Sustainability* **2018**, *10*, 4656. [CrossRef]

4. Veugelers, R.; Cincera, M.; Frietsch, R.; Rammer, C.; Schubert, T.; Pelle, A.; Renda, A.; Montalvo, C.; Leijten, J. The impact of horizon 2020 on innovation in Europe. *Intereconomics* **2015**, *50*, 4–30. [CrossRef]

5. Clarke, R.Y. Measuring success in the development of smart and sustainable cities. In *Managing for Social Impact*; Springer: Cham, Switzerland, 2017; pp. 239–254.

6. Huovila, A.; Bosch, P.; Airaksinen, M. Comparative analysis of standardized indicators for Smart sustainable cities: What indicators and standards to use and when? *Cities* **2019**, *89*, 141–153. [CrossRef]

7. Ávila-Gutiérrez, M.J.; Martín-Gómez, A.; Aguayo-González, F.; Córdoba-Roldán, A. Standardization framework for sustainability from circular economy 4.0. *Sustainability* **2019**, *11*, 6490. [CrossRef]

8. Caprotti, F.; Cowley, R.; Datta, A.; Broto, V.C.; Gao, E.; Georgeson, L.; Herrick, C.; Odendaal, N.; Joss, S. The New Urban Agenda: Key opportunities and challenges for policy and practice. *Urban Res.Pract.* **2017**, *10*, 367–378. [CrossRef]

 *sustainability*

Article

Tracking Topics and Frames Regarding Sustainability Transformations during the Onset of the COVID-19 Crisis

Mariana Madruga de Brito [1,*], Danny Otto [1] and Christian Kuhlicke [1,2]

[1] Department Urban and Environmental Sociology, Helmholtz Centre for Environmental Research, 04318 Leipzig, Germany; danny.otto@ufz.de (D.O.); christian.kuhlicke@ufz.de (C.K.)
[2] Institute for Environmental Sciences and Geography, University of Potsdam, 14468 Potsdam-Golm, Germany
* Correspondence: mariana.brito@ufz.de

Abstract: Many researchers and politicians believe that the COVID-19 crisis may have opened a "window of opportunity" to spur sustainability transformations. Still, evidence for such a dynamic is currently lacking. Here, we propose the linkage of "big data" and "thick data" methods for monitoring debates on transformation processes by following the COVID-19 discourse on ecological sustainability in Germany. We analysed variations in the topics discussed by applying text mining techniques to a corpus with 84,500 newspaper articles published during the first COVID-19 wave. This allowed us to attain a unique and previously inaccessible "bird's eye view" of how these topics evolved. To deepen our understanding of prominent frames, a qualitative content analysis was undertaken. Furthermore, we investigated public awareness by analysing online search behaviour. The findings show an underrepresentation of sustainability topics in the German news during the early stages of the crisis. Similarly, public awareness regarding climate change was found to be reduced. Nevertheless, by examining the newspaper data in detail, we found that the pandemic is often seen as a chance for sustainability transformations—but not without a set of challenges. Our mixed-methods approach enabled us to bridge knowledge gaps between qualitative and quantitative research by "thickening" and providing context to data-driven analyses. By monitoring whether or not the current crisis is seen as a chance for sustainability transformations, we provide insights for environmental policy in times of crisis.

Keywords: frames; SDG; green deal; content analysis; natural language processing; NLP

Citation: de Brito, M.M.; Otto, D.; Kuhlicke, C. Tracking Topics and Frames Regarding Sustainability Transformations during the Onset of the COVID-19 Crisis. *Sustainability* **2021**, *13*, 11095. https://doi.org/10.3390/su131911095

Academic Editors: Baojie He, Ayyoob Sharifi, Chi Feng and Jun Yang

Received: 18 September 2021
Accepted: 5 October 2021
Published: 7 October 2021

Publisher's Note: MDPI stays neutral with regard to jurisdictional claims in published maps and institutional affiliations.

1. Introduction

Historically, crises have motivated fundamental social, ecological, economic, and political transformations [1]. The evolving COVID-19 pandemic is no exception. In a matter of weeks, it cast a stark light on how changes can take place overnight. This includes, among others, short-term negative impacts such as an increased waste generation due to the lockdowns [2,3] but also positive environmental changes such as an improved air quality [4,5]. What remains to be understood is if fast-moving crises such as the COVID-19 pandemic can usher in lasting and far-reaching transformations [6].

Particularly at the beginning of the pandemic, an increasingly vocal debate on potential long-term changes with a rather optimistic view was initiated, with proponents arguing that the COVID-19 crisis may have opened a window of opportunity to spur on systemic ecological sustainability transformations [7]. This narrative was highlighted, for instance, by António Guterres: "We must turn the recovery from the pandemic into a real opportunity to build a better future" [8]. Similarly, multiple scholars have stated that they see an opportunity for accelerated sustainability transitions in the post-crisis period—mainly in the form of "restarting" society or by "rebuilding it better" [6,9,10]. The adherence to this narrative is echoed by both scientists and politicians. Yet, empirical evidence for such a dynamic currently lacking.

Previous experience suggests that whether or not crises result in an opportunity for initiating or accelerating sustainability transformations depends on factors such as the crisis severity, the public and media framing [11], socio-economic capacities, and political interests [7,12,13]. Hence, to understand whether or not the COVID-19 crisis can prompt sustainability transformations, it is necessary to monitor political, public, and media frames and their translation into policies. In our view, this calls for a systematic monitoring of discursive material and an in-depth study of how the relationship between the pandemic and ecological sustainability is established. This is crucial because the way environmental concerns are framed sets the stage for how they are addressed [14].

Since the beginning of the pandemic, a plethora of discursive material on COVID-19 has been produced. To varying degrees, this material has influenced the perceptions of and the reactions to the crisis. Indeed, it is widely acknowledged that traditional news media play an important role in shaping public awareness [15–17]. While social media and other forms of media have gained relevance, the analysis of newspapers is still a valuable tool for monitoring public knowledge [18]. Media coverage has been shown to influence public concern regarding climate change [19]; it can impact policy processes [20] and people's behavioural intentions [21,22]. The influence of the media in shaping opinion is even more evident in times of crisis [23]. This has been studied in relation to the Ebola outbreak in 2014 [24], the Brexit referendum [25], and the European drought of 2018 [26,27]. In this context, media news can serve as a proxy for understanding how public discourses on ongoing transformations evolve. Similarly, online search behaviour can provide insights into public awareness levels [28].

Given the multitude of discourse material available, new and variably applicable analytical approaches are needed to analyse these data, aiming to produce timely results that allow researchers to make cross-context generalisations. To this end, "big data" tools such as natural language processing (NLP) techniques [29,30] can be used to offer a "bird's eye view" of how discourses change. At the same time, context-specific and interpretative qualitative methods (i.e., "thick data" tools) are needed to elucidate the underlying patterns observed in data-driven analyses, as they allow for an in-depth analysis of prominent frames [31] (for a definition of discourse and frames see [32]).

Here, we propose the integration of NLP and qualitative content analysis on newspaper data to understand how sustainability aspects are discussed within news related to COVID-19. More specifically, we analyse how the reporting about topics underpinning sustainability has evolved in the media debate about COVID-19. The proposed approach is demonstrated by following the COVID-19 discourse on sustainability during the onset of the crisis in Germany. In a first step, we used topic modelling and quantitative content analysis to investigate the extent of the discussion on sustainability topics from March until June 2020. We also analysed how public awareness developed compared to the previous year by considering online search behaviour. In a second step, we conducted a qualitative content analysis on a smaller sample of articles, focusing on news published during the first COVID-19 wave. This allowed us to delve deeper into the discursive material and track specific frames. Emphasis was given to sustainability topics in line with the European Green Deal framework and the UN Sustainable Development Goals (SDGs) related to ecological sustainability.

2. Big Data and Thick Data: Methodological Considerations

Big data has been discussed as a central issue for empirical research. Some see it as a challenge for empirical social sciences [33,34]. Others weigh up the related threats and opportunities [35–38] or question whether the use of big data is truly innovative, arguing that traditional types of big data—such as administrative records and newspaper articles—have been studied for centuries [39]. Additionally, the definition of what constitutes big data has been contested, and it remains a "rather loose ontological framing" [40].

Overall, characteristics commonly associated with big data include large volumes of data that: are accessible for computational analyses, are rapidly created and shifting, have a complex and untidy structure, and are exhaustively captured instead of sampled [35,39–41]. Due to challenges arising from pure data-driven quantitative analyses, integrating qualitative components into the research design is often advised [38,39,41,42]. In this context, the goal of mixed-methods designs can be to: (a) provide additional coverage by adding more perspectives and data on a particular issue, (b) verify results by "convergent findings" of quantitative and qualitative methods or (c) achieve "sequential contributions", meaning that what is learned in one method serves as input for the other [43]. For big data, all three aspects are applicable.

Following this rationale, we argue that the monitoring of sustainability transformations needs both big and thick data. Thick data is understood here as qualitative and detailed data which support the understanding of the context in which a pattern occurs [44]. While big data tools can help to track and describe changes in discourses over time, it is difficult to understand these transformations and the reasons behind them based solely on data-driven approaches [31]. In this regard, thick data methods can be used to add a deep-diving and context-sensitive dimension to the data-driven outcomes, enabling more complex interpretations.

Here, we propose to "thicken" the data by reducing the number of data points while enhancing the thickness of their descriptions. Hence, instead of solely providing a trend analysis of the topics mentioned in the media, we analyse how they are framed. We use the metaphor of "thick data" in reference to Geertz's "thick description" [45], which describes the need for ethnographic descriptions to go beyond recording what people are doing, as this only provides a superficial account of actual situations. Taking golfing as an example, a "thin" description portrays a person "repeatedly hitting a little round white object with a club-like device" [46]. A "thick" description sensitively interprets the context of the behaviour. It adds the frames of a golf course, gaming rules, and equipment handling in an attempt to grasp the situation fully. Thick descriptions not only enable an understanding of cultural contexts beyond "thin" statements of observable behaviour, but they also allow a further understanding of patterns emerging in big data. Conversely, big data can support sampling strategies or highlight hotspots for thick data tools. Big data can also take on the role of thick data if, for instance, qualitative text analyses are complemented by the study of large databases.

3. Material and Methods

3.1. Newspapers Sample Selection

Newspaper data were collected from a news aggregator database (wiso-net.de). The articles in our sample were published between 1 March to 30 June 2020. By using the search string "Corona*" or "Covid" or "SARS-CoV-2", 459,129 articles were retrieved. Given that the aforementioned database does not allow web scraping, the articles had to be downloaded manually. Hence, to reduce the sample size to a manageable number while at the same time ensuring topic coverage and geographical equity [26], only the newspapers with the highest circulation rates in each German federal state were considered. Hence, 2 nationwide and 19 regional newspapers were used (Table S1). This reduced the sample to 84,587 articles. The Jaccard similarity coefficient [47] was used to identify duplicate articles. A threshold of 0.7 was applied, where 0 indicates no similarity and 1 indicates full similarity. The final data set consisted of 61,514 unique records (see Figure S1 for sampling flowchart).

3.2. Automated Text Analysis

Before the analysis was conducted, common NLP methods were applied to clean the data. First, we removed numbers and punctuation and converted the characters to lowercase. Then, the articles' sentences were tokenised into individual words. Stop words such as articles, pronouns and prepositions were removed from the corpus (see Table S2 for additional stop words). Finally, metadata relating to the date of publication and newspaper were extracted using regular expressions. All coding was carried out in R.

3.2.1. Topic Modelling

In the first step, topic modelling was conducted to provide an overview of the content of the articles and investigate if sustainability issues emerged as a relevant topic. Topic modelling is an unsupervised machine-learning algorithm, where patterns of word co occurrences are used to identify topics that describe the corpus [48]. Each topic contains a cluster of words that frequently occur together and refer to similar subjects [49].

The Latent Dirichlet allocation (LDA) was chosen to identify predominant topics in our dataset. The LDA model assumes that articles can have multiple topics. For instance, an article might be 40% about topic 1 and 60% about topic 2. The number of topics in our LDA model was defined as 70 after computing the coherence when varying the number of topics from 10 to 100. Additional stop words used are shown in Table S2. As an outcome of the LDA, a list of the most common words and the topic probabilities for each article were obtained. The topic titles were defined inductively based on the 20 most unique words from each topic (i.e., words that tend to occur mainly in a specific topic).

3.2.2. Quantitative Content Analysis

Pattern matching was used to classify the articles into different sustainability top ics. The articles were classified according to eight environmental topics drawn from the European Green Deal framework and the UN SDGs.

The keywords (Table S3) used to classify the articles within these topics were de fined based on two sustainability glossaries [50,51]. Additional terms were identified by analysing unigram frequencies and word co-occurrences (i.e., words that occur in the same sentence). Furthermore, ten experts in sustainability transformations were consulted to select unambiguous terms. Based on that, we tagged sentences where a keyword or combination of keywords occurred as related to a given topic.

Of the 61,514 unique articles, only 2343 mentioned at least one of the sustainability keywords. A random sample of 40% of the articles was read to validate the automatic classification and identify false positives and false negatives. Missclassifications found were manually corrected. In a final step, we plotted the results obtained for each of the eight topics against the German national SDG performance [52].

3.2.3. Bibliometric Analysis

To add a temporal perspective to the quantitative content analysis, we conducted a bibliometric analysis [53] on the whole wiso-net database, only considering German newspapers (>39 million articles, 240 newspapers). The same sustainability keywords (Table S3) were used, but with a different time frame: from August 2017 to July 2020. For these searches, the keywords related to COVID-19 were omitted. This allowed us to determine the frequencies of articles referring to sustainability over time and see if their number increased or declined during the first wave of the pandemic. The outputs were normalised according to the total number of articles per month. This reduced the bias due to the variation in the monthly number of news.

3.2.4. Co-Occurrence of Sustainability Topics

To analyse the topics that were reported simultaneously by the same article, a co occurrence analysis was conducted in the sustainability dataset (n = 2343 articles). Spear man correlation coefficients were computed to assess whether they co-occur by chance or follow a pattern. Circos plots [54] were used to visualise the interdependencies between the sustainability topics.

3.2.5. Word Frequencies

NLP procedures, such as term-document matrices, word frequency analyses and bigram analyses, were employed. Based on that, a network analysis graph was developed to visualise how the corpus' words are associated. In addition, word clouds were created to demonstrate how the terms' usage has changed over time and according to different sustainability topics. For the compilation of the comparison word clouds, all mentions of a word in the corpus were counted. Then, the proportional use of the word was calculated based on the total words used in all articles across the specified topic or month.

3.2.6. Public Awareness on Sustainability

In a final step, Google Trends (GT) data (trends.google.com) were used to measure public awareness on environmental topics. GT measures the search popularity in relative terms according to a random sample of all search terms used in queries within the investigated period. A detailed description of the GT measure can be found in Rousseau and Deschacht (2020). The GT database can be searched according to user-defined "search terms" and "topics". The user-defined keywords used here are shown in Table S3. Furthermore, the pre-defined "Environment" topic was considered.

To identify the effect of the COVID-19 crisis on the population's awareness regarding sustainability topics, we compared the GT search popularity indicator from March to June 2020 with data from the same period in 2019. The nonparametric Wilcoxon test was used to identify significant differences.

3.3. Qualitative Content Analysis

A qualitative content analysis was conducted to provide context and an in-depth analysis of the frames which connect COVID-19 and sustainability. Based on the previous steps, 461 articles were selected as a thick data sample. These articles discussed two or more of the sustainability topics considered, and hence, were more likely to add to a "window of opportunity" narrative for sustainability transformations.

Following Schreier [55] and Mayring [56], we applied a multi-step approach using a combination of inductive and deductive reasoning. First, we developed three broad categories: (a) COVID-19 impacts with implications for the environmental SDGs and Green Deal goals; (b) COVID-19 recovery measures and their relation to the environmental SDGs and Green Deal goals; and (c) COVID-19 as a window of opportunity for sustainability transformations. Two coders worked through all the articles to identify segments related to the above-mentioned categories. Disagreements were discussed in this process and code descriptions were refined.

After the first coding round and based on the discussion between the coders, sub-categories were added in vivo to create a final version of the coding scheme. In the end, eight prominent frames were identified. Once all the articles were analysed, we calculated the number of articles in each category. We also compared the frames with the automated text analysis outputs by developing a heatmap and computed the co-occurrence of frames and topics using the same method described in Section 3.2.4.

4. Results

4.1. Quantitative Analysis: Media Debate on Topics Regarding Sustainability during the Early Stages of the COVID-19 Crisis

4.1.1. Media Attention and Public Awareness on Sustainability

Overall, there was a sharp decline in the number of articles that mention sustainability-related keywords after the start of the COVID-19 pandemic in Germany (Figure 1). This could indicate that ecological concerns were brushed aside during the initial shock of COVID-19. Before the onset of the pandemic, clear peaks occurred during protests such as the Global Climate Strikes (GCS) (Figure 1).

With regard to the public awareness on sustainability, an adverse effect of the COVID-19 crisis on the online search behaviour for the topics climate change and environment was found (Figure 2). Conversely, there was an increase in searches regarding air quality, with more searches in 2020 compared to the same period in 2019.

Figure 1. Bibliometric analysis showing the frequency of articles that mention sustainability-related keywords (Table S3) in the German media according to the wiso-net database. The data were normalised against the total N of published articles per month.

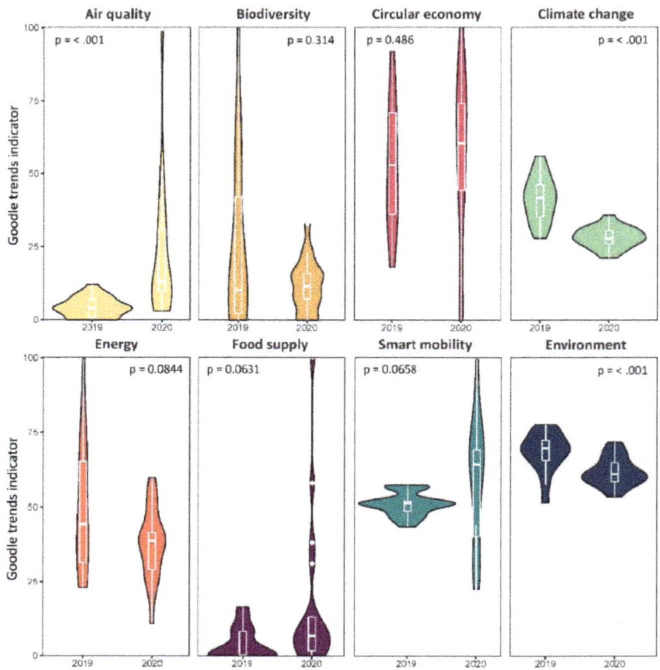

Figure 2. Search behaviour in 2019 and 2020 according to Google Trends data. Higher values indicate a higher amount of searches. For both years, data from March to June were considered. Wider sections represent a higher density of people searching for a term; the thinner sections represent a lower density of searches. *p*-values were obtained by using the nonparametric Wilcoxon test.

4.1.2. Media Statements about Sustainability (MSS) during the Onset of the COVID-Crisis

To analyse the relationship between sustainability and the COVID-19 crisis, only articles that mention the pandemic were considered (for sampling procedure, see Section 3.1). Topic modelling outcomes suggest that environmental problems are underrepresented in contrast to other topics (Table S4). The topics that dominate the corpus were related mainly to family and everyday life issues (3.5%), financial insecurity (3.4%), and quarantine restrictions (3.1%). Of the 70 topics, only one is directly related to environmental issues (1.5% of the corpus, n = 932). It includes news about outdoor recreation, gardening, and smart mobility (see topic 62 in Table S4). Hence, based solely on this unsupervised machine-learning algorithm, it was not possible to detect sustainability as a core issue in the COVID-19 discourse.

To further investigate the COVID-19 discourse on sustainability, all COVID-19-related articles were classified into eight sustainability topics following the scheme presented in Table S3. Based on that, 2935 media statements about sustainability (MSS) were identified. Most of the MSS refer to climate change aspects (39%, n = 1142), followed by food supply (22%, n = 637) (Figure 3). These numbers are low compared to the number of articles included in the analysis (n = 61,514). Indeed, for each week and topic, an average of only 0.6% of the COVID-19 articles reported about sustainability issues. This suggests that the early COVID-19 discourse in Germany was not centred on sustainability issues, which was expected, because issues that directly impact people tend to become a priority in an immediate crisis.

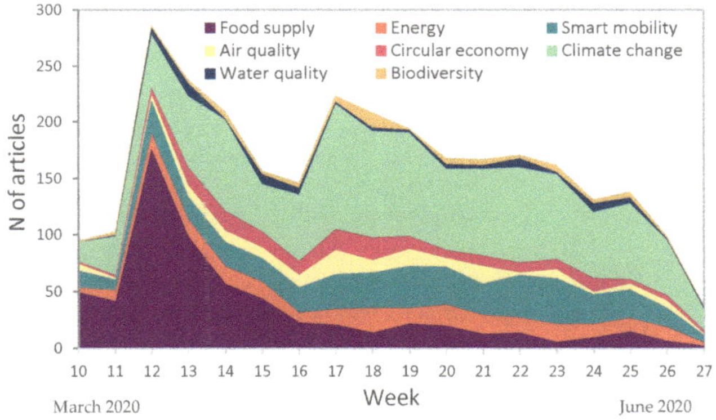

Figure 3. Temporal distribution of the MSS topics. The keywords used for the searches are provided in Table S3.

When considering the MSS frequency in relation to the total number of news articles, we see an increase in the proportion of climate change MSS (Figure 4). In March 2020, about 0.9% of all articles (n = 61,514) reported about climate change issues. This number increased to 2.7% in April 2020. This is also reflected in the usage of terms, which shifted drastically over time (Figure S2a). Indeed, the word stem *"Klima**"* (climate in english) appeared at a frequency of 0.0038 (words appeared 815 times across 210,311 words for this period) in April, whereas in March, the frequency was only 0.0016 (n = 411 out of 248,046 words). With regard to food supply MSS, there was higher interest in this topic in March 2020, with a rapid decline in the following months. In May, the discussion about smart mobility intensified, as reflected by the usage of the words "car", "Tesla", "railway", and "bicycles". Finally, in June, concerns regarding the "future" and the economy (e.g., "euro", "economic stimulus package") became more evident. In summary, there was a shift in the discourse, from a primary focus on food supply in March, to increasing references to a "chance" for climate change protection in April, followed by an intensified discussion on mobility in May and a focus on the restoring the economy in June.

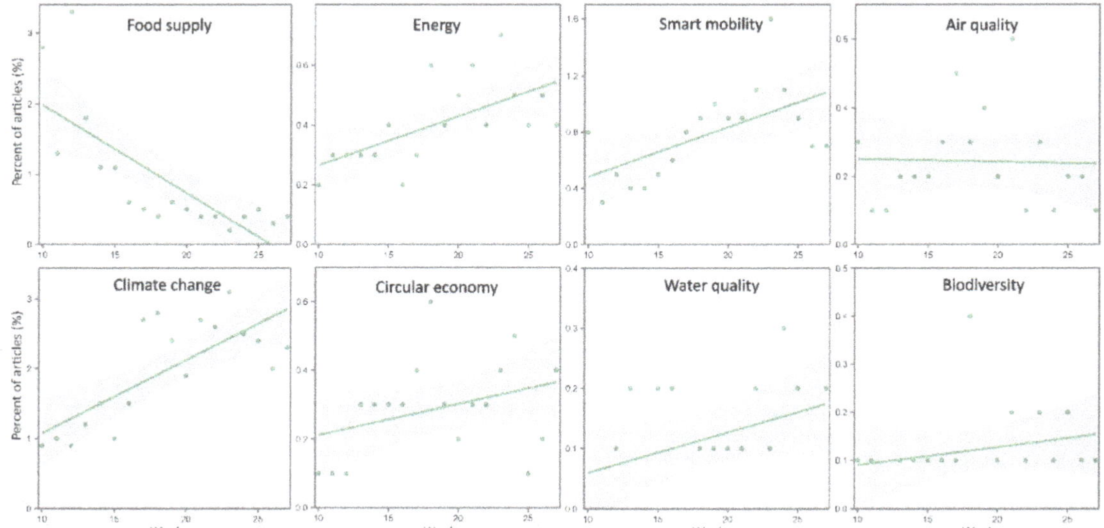

Figure 4. Temporal distribution of the sustainability topics with a 95% confidence interval. For each topic, the frequency of articles that contain the keywords in Table S3 in the text is shown for each week.

4.1.3. Co-Occurring Sustainability Topics and Keywords

To better understand the patterns behind the articles that mention multiple sustainability topics, a co-occurrence analysis was conducted. An average clustering coefficient of 0.91 was obtained, which indicates that most of the topics are connected at least in some of the articles (Figure 5). Still, the majority of the MSS were reported alone (64.1%, n = 1880). Climate change MSS were frequent drivers of network connections (present in 71.8% of all co-occurrences, n = 481). Indeed, energy, biodiversity, and air quality MSS were mentioned mainly together with climate change MSS. This is reflected in the strong correlation found between climate change and other topics (Table S5). Conversely, food supply, which was the second-most dominant MSS in our database, was primarily reported in isolation (n = 594), with few co-occurrences with climate change. This could indicate that these MSS are not directly related to sustainability but refer to individual food supply needs. This was verified when validating our classification system, where it became clear that the discourse on food supply is centred on panic buying and not on long-term food security.

To identify weakly integrated themes, a biography network was elaborated (Figure S3). It illustrates how the words in the corpus are associated. Terms related to the COVID-19 impacts cluster together (e.g., "consequences", "people", "million"). On the other hand, sustainability-related terms are side-lined (e.g., "climate change", "environment"). Several word pairs are disconnected (e.g., "Green Deal", "Fridays for Future", "renewable energy"). This indicates that there is little crossover between sustainability and the COVID-19 consequences as portrayed by the German media. It also shows that sustainability is not the central theme depicted in the corpus, even in the sample of articles that discuss at least two sustainability topics.

4.1.4. Results Validation

We evaluated the accuracy of the automatic classification system by conducting close reading and annotating 40% of the articles, as mentioned in Section 3.2.2. The automatic classification of the MSS was accurate in 96.1% of the cases, with a standard deviation of 2.2% (Table S6). The misclassifications correspond mostly to articles that mention sustainability-related keywords but outside of the COVID-19 context. Overall, MSS topics such as circular economy and food supply presented the highest levels of accuracy (98.9%

and 98.7%, respectively). Conversely, MSS about climate change were overestimated (accuracy of 92.3%).

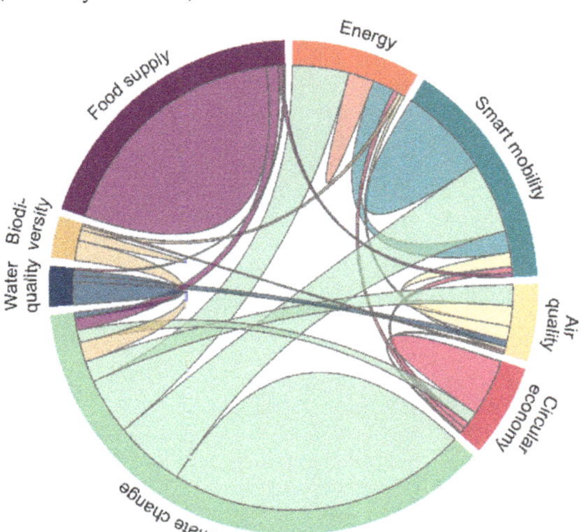

Figure 5. Co-occurring MSS topics. The ribbons connecting two MSS indicate copresence links, and their size is proportional to the number of links. Ribbons without connections indicate MSS that were reported alone. For clarity purposes, only co-occurrences >5 are shown. For an interactive visualisation see Supplementary Material S1.

As an additional validation measure, we built a comparison word cloud, which highlights the most unique terms in each topic (Figure S2b). The results indicate that the MSS are consistent with the sustainability topics they were assigned to. For instance, the word "water" was mentioned 168 times in the "water quality" MSS articles and only 13 times in the "biodiversity" MSS articles.

4.2. Qualitative Analysis: Frames Connecting COVID-19 and Sustainability

The big data bird's eye view (Figure 3) shows a decrease in coverage on sustainability since the beginning of the pandemic and only reveals a few articles that link COVID-19 and sustainability. In an in-depth interpretative content analysis, we differentiated these linkages further by analysing three bundles of frames, as shown in Table 1.

Of the 461 articles included in the qualitative analysis, 79.6% (n = 367) had substantive information related to the frames mentioned in Table 1. The following items discuss each of these frames and the main actors supporting them. Notably, these frames are interlinked and overlap, both in terms of occurrence over time (Figure S4) and content (Figure 6). Indeed, they do not occur as separate lines of discourse but instead supplement or counteract one another, suggesting that there is no dominant frame.

4.2.1. The COVID-19 Crisis Led to Observable Short-Term Positive Environmental Impacts (PEIs)

About a quarter of the articles (23.4%, n = 108) highlight that the pandemic has given nature time to recover. These were connected to early scientific reports on the environmental effects of the first lockdowns. It is stated that there was an improvement in air and water quality, as well as biodiversity, due to reduced economic activities and limited human impacts on the environment. The clean canals of Venice, less nitrogen dioxide emissions in China and India, and increased sightings of wildlife are frequent points of reference for this frame. However, this frame rarely ends with this simple description. Most of the articles reference researchers and emphasise that these effects are temporary and are not enough to impact overall climate developments [7]. They might be enough to enable Germany to accomplish its 2020 greenhouse gas emission reduction goals as some commentators note, but they do not present a solution for the climate crisis. This line of argument is accompanied by warnings of potential rebound effects. Parallels are drawn between post-pandemic scenarios and the fast return to pre-crisis emission levels that accompanied the economic recovery after the Global Financial Crisis (2008/09). This frame is closely linked to air and water quality MSS (Figure 7).

Table 1. Overview of sustainability frames and percent of articles (n = 367) that mention them.

Category	COVID-19 Sustainability Frame	%
COVID-19 impacts with implications for the SDGs and Green Deal goals	The COVID-19 crisis led to observable short-term Positive Environmental Impacts (PEIs)	23.4
	The COVID-19 crisis spurred Behavioural Changes that have an impact on sustainable development (BC)	24.3
COVID-19 recovery measures and their relation to the SDGs and Green Deal goals	The COVID-19 recovery programme and Measures Address or should address environmental Sustainability goals (MAS)	27.3
	The COVID-19 recovery Measures are Not Doing Enough to achieve sustainability goals (MNDE)	7.4
	The COVID-19 recovery Measures Focus on the Economy First (MFEF)	4.6
COVID-19 as a window of opportunity for sustainability transformations	The COVID-19 crisis is an Opportunity for Change towards more Sustainable development (OCS)	25.5
	The COVID-19 crisis poses Challenges for Sustainability Transformations (CST)	21.9
	The COVID-19 crisis as an Analogy for the Climate Crisis (ACC)	6.3

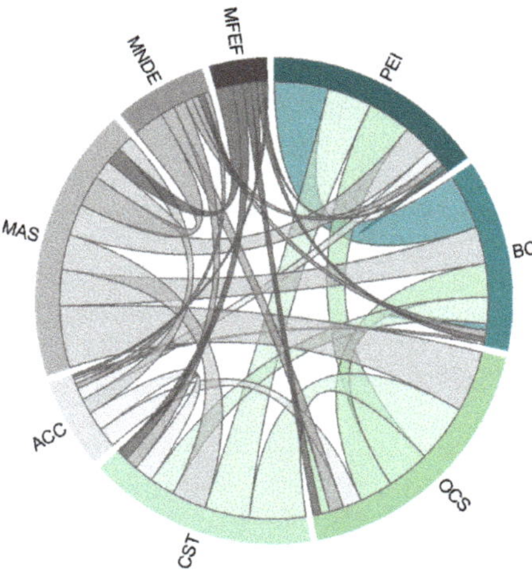

Figure 6. Relationships between the frames. For the acronyms, see Table 1. The ribbons connecting two frames indicate copresence of links and their size is proportional to the number of links. Ribbons without connections indicate frames that were reported alone. For an interactive visualisation, see Supplementary Material S2.

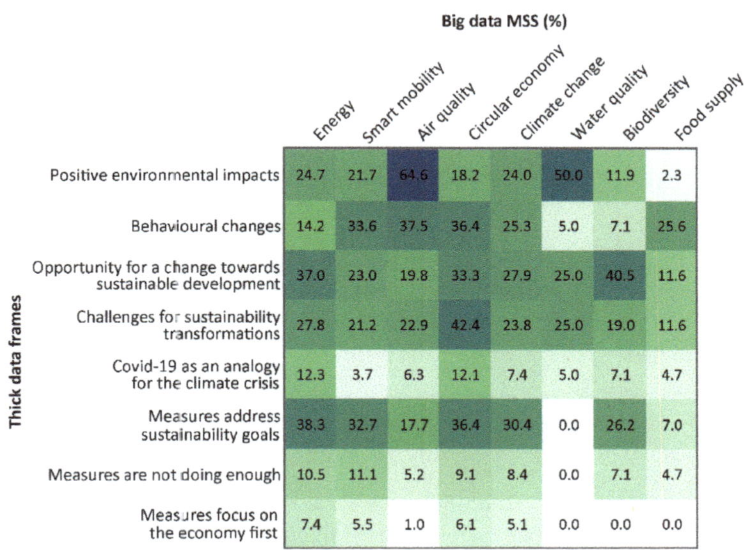

Thick data frames	Energy	Smart mobility	Air quality	Circular economy	Climate change	Water quality	Biodiversity	Food supply
Positive environmental impacts	24.7	21.7	64.6	18.2	24.0	50.0	11.9	2.3
Behavioural changes	14.2	33.6	37.5	36.4	25.3	5.0	7.1	25.6
Opportunity for a change towards sustainable development	37.0	23.0	19.8	33.3	27.9	25.0	40.5	11.6
Challenges for sustainability transformations	27.8	21.2	22.9	42.4	23.8	25.0	19.0	11.6
Covid-19 as an analogy for the climate crisis	12.3	3.7	6.3	12.1	7.4	5.0	7.1	4.7
Measures address sustainability goals	38.3	32.7	17.7	36.4	30.4	0.0	26.2	7.0
Measures are not doing enough	10.5	11.1	5.2	9.1	8.4	0.0	7.1	4.7
Measures focus on the economy first	7.4	5.5	1.0	6.1	5.1	0.0	0.0	0.0

Figure 7. Relationships between the MSS and the frames identified in the thick data analysis. The numbers represent the percentage of articles in the subsample (n = 367) that are tagged according to an MSS and present a specific frame. For instance, 64.6% of the air quality articles in the subsample were also classified as having the positive environmental impacts (PEIs) frame.

4.2.2. The COVID-19 Crisis Spurred Behavioural Changes That Have an Impact on Sustainable Development (BC)

A total of 112 articles (24.3%) emphasised that the COVID-19 crisis has led to multiple behavioural changes that affect the environment either positively or negatively. These include, for instance, increased use of bicycles, mobility reduction, and a decline in oil demand. The pandemic has also affected consumption patterns (e.g., panic buying) and waste generation, as indicated by the high linkage between the BC frame and circular economy and food supply MSS (Figure 7). Similar to the positive environmental impacts (PEIs) frame, it is stated that these changes may be reversed after the crisis is over. Still, it is argued that the digitalisation of work will remain a feature of the post-COVID world, thus reducing mobility needs and overall fossil fuel consumption. Nevertheless, countervailing negative behavioural changes are also expected, such as a preference for using private cars and the avoidance of public transportation. Furthermore, experts highlight that recreational air travel is not likely to decrease after the pandemic.

4.2.3. The COVID-19 Recovery Programme and Measures Address or Should Address Environmental Sustainability Goals (MAS)

This is the most common frame in our sample (27.3%, n = 126). It is reasoned that the recovery plan should boost the economy while at the same time focusing on long-term strategies for advancing the climate agenda. Experts and politicians argue that the UN SDGs may act as a normative guide for selecting the targets for future investments. Overall, the ecological modernisation of transport and the car industry is a central point of this frame, as evidenced by the linkage between this frame and MSS on smart mobility (Figure 7). Of the measures discussed, the most prominent (and controversial) is the environmental bonus programme for the purchase or lease of electric or plug-in hybrid vehicles. We also found broader arguments that called for recovery programmes to focus on sustainability or include "climate-friendly" technologies and energy. From a long-term perspective, it is argued that the breakdowns of normality caused by the pandemic should be used to assess the possibilities for a more sustainable recovery (i.e., innovative, digital, low CO2 emissions). It is noteworthy that the water quality MSS are not linked to any of the measures frames (MAS, MNDE, and MFEF) (Figure 6). This could indicate that water issues may be overlooked in the news media reporting.

4.2.4. The COVID-19 Recovery Measures Are Not Doing Enough to Achieve Sustainability Goals (MNDE)

This frame (7.4%, n = 34) criticises the recovery measures for not including regulations or practices that would accelerate sustainability efforts or even for hampering those efforts. Politicians and environmental NGOs question the use of environmental bonus programmes for transforming the mobility sector. Furthermore, they are critical of government rescue programmes for airline companies. They argue that investments in environmental protection projects are too low to achieve the climate goals of the Paris Agreement. This frame is usually mentioned together with the MAS frame (COVID-19 recovery programme and measures address or should address environmental sustainability goals) by pointing out existing challenges (Figure 6).

4.2.5. The COVID-19 Recovery Measures Focus on the Economy First—Rollback for Sustainability (MFEF)

The MFEF frame emphasises a quick economic recovery that should not be hampered by ecological concerns. This frame is consistent with policy discourses that propose halting or postponing environmental standards and goals [57,58]. Only a few newspapers (4.6%, n = 21) report on this position. Nevertheless, this number should be interpreted cautiously as only articles that reported on at least two sustainability topics were included in the qualitative content analysis (Figure S1). This frame is supported mainly by politicians from federal states with extensive vehicle production facilities or industry sector representatives. It focuses on the situation of large industries (especially the car industry) during the pandemic and suggests rollbacks of various climate regulations (such as CO2 taxes or emission levels). The articles also mention the pressure to loosen up regulations regarding the circular economy and to postpone the adoption of environmental measures for the energy and car industries (Figure 7).

4.2.6. The COVID-19 Crisis Is an Opportunity for Change towards More Sustainable Development (OCS)

These articles (25.2%, n = 116) highlight that the government should consistently pursue socio-ecological transformations to deal with the aftermath of the pandemic. A central challenge mentioned is the transformation of transport, energy, and agriculture in an economically sensible, socially acceptable, and sustainable way. At the heart of the debate is the Green New Deal. Both politicians and experts emphasise that the COVID-19 crisis and the climate crisis must be tackled together. Similarly, economic institutes argue that the ongoing crisis is a good time to make longer-term investments in climate protection. Another point often mentioned is that the COVID-19 crisis has shown how society can be mobilised and how quickly changes can be made. Indeed, structures, institutions, and behaviours that were considered unchangeable were put to the test. Hence, the COVID-19 crisis is seen as an opportunity to break up ossified structures to trigger a wave of green recovery in Germany. Articles in this class also point out that the COVID-19 crisis is an opportunity for the renewable energy industry because it is accelerating the coal phase-out and, as some put it, "ending the oil age".

4.2.7. The COVID-19 Crisis Poses Challenges for Sustainability Transformations (CST)

Various factors that might challenge sustainability transformations are connected in this frame (21.9%, n = 101). From an economic perspective, the financial feasibility of climate objectives is questioned due to the costs of recovery programmes and the strain that COVID-19 regulations are putting on industry and the service sector. Articles that are sceptical about the successful implementation of climate protection measures or regulations in light of the pandemic tend to reference industry actors and (conservative) politicians in particular.

In addition to these economic challenges, the articles also describe a shift of attention away from climate issues. As society is focused on the immediate COVID-19 response, there are limited capacities to address the climate crisis with its uncertain timeframes. Furthermore, this loss of attention is linked to secondary effects on climate activism and (international) climate diplomacy. Focal points of this frame are the postponement of major climate conferences and changes in how climate activists stage protests (e.g., virtual formats for Fridays for Future). Both issues are seen as challenging for the sustainability efforts, as the lack of demonstrations reduces the attention given to environmental protection and the postponement of significant international meetings limits climate policy. Similar delays and shifts are reported for environmental research, technology development, and legislation processes. Interestingly, all MSS were well-represented in this frame, indicating that the challenges encompass a broad range of sustainability aspects (Figure 7).

4.2.8. The COVID-19 Crisis as an Analogy for the Climate Crisis (ACC)

Articles mentioning this frame (6.3%, n = 29) indicate that the COVID-19 outbreak may provide an illustrative analogy for sustainability challenges. This frame exemplifies how the COVID-19 and climate change crises require an analogous response even though they occur on different temporal scales. The analogy for the climate crisis frame is mainly supported by scientists who argue that similar to the COVID-19 pandemic, innovative and far-reaching interventions are needed to address the climate crisis [59]. These may include, for instance, the creation of new institutions and radical changes in behavioural patterns and policy governance. Along with this frame, it is stated that to curb the spread of coronavirus, ingrained societal behaviours were put to the test, including citizens´ freedom of movement. Likewise, deeply entrenched behaviours related to resource consumption and travel need to be challenged to tackle climate change. Energy and circular economy were the MSS most mentioned within this frame (Figure 7). Furthermore, even though these classes were not prominent in the MSS database (8% and 6% of the MSS, respectively), together with climate change MSS, they were the ones that were represented more strongly in the investigated frames.

5. Discussion

Monitoring how the media portrays different topics and frames on sustainability transformations is challenging due to the complexity of the social systems and processes involved [60] and the large amounts of discursive material available. Against this background, we proposed a mixed-methods approach [61] for following the COVID-19 discourse on sustainability in Germany. In this section, we discuss the empirical and methodological contributions of the article and the limitations of the proposed approach.

5.1. Empirical Findings: Evidence for a "Window of Opportunity" Narrative

Our empirical results show that the frequency of articles that mention sustainability topics in the early stages of the pandemic is limited (Figure 1). Indeed, an average of 0.6% of the articles in the corpus discussed sustainability topics each week (Figure 4). Topic modelling results indicate that (Table S4) only 1.5% of the articles were connected to environmental issues. Furthermore, google searches on sustainability on topics such as climate change, and environment were significantly lower between March and June 2020 compared to the same period in 2019 (Figure 2). This is not surprising, as during the first months of the crisis, media attention focused on the pandemic's social, economic, and epidemiological aspects rather than the climate crisis.

More broadly, these results support empirical studies that show that environmental sustainability is being overshadowed by COVID-19 [62]. Findings by [63] show that environmental issues have become less of a priority for international organisations since the COVID-19 crisis. Recent assessments indicate that the pandemic will likely undermine progress towards 12 of the 17 SDG goals [64,65]. Indeed, it is estimated that two-thirds of the 169 SDG targets are under threat due to the pandemic [66].

To discuss how our results relate to the SDGs, we linked our findings to Germany's national SDG performance [52] (Figure 8). We found that two SDGs in which Germany is not performing well were widely reported by the German news media: the SDG 13 (climate action) and the SDG 2 (food security and sustainable agriculture). A possible interpretation is that the COVID-19 crisis put the existing fragilities into the spotlight. Conversely, it could be argued that SDG 13 is intrinsically linked to all other SDGs, and the high number of SDG 2 articles is linked to panic buying during lockdowns and not necessarily connected to long-term food security. The analysis also showed that SDG 12 (sustainable consumption and production) is underrepresented in the COVID-19 media discourse, even though it is another area in which Germany is not performing well [52]. Still, the evidence indicates that COVID-19 has aggravated consumerism and the generation of waste [2,3]. This was identified in the challenge for sustainability frame (CST), where challenges for the circular economy were frequently pointed out.

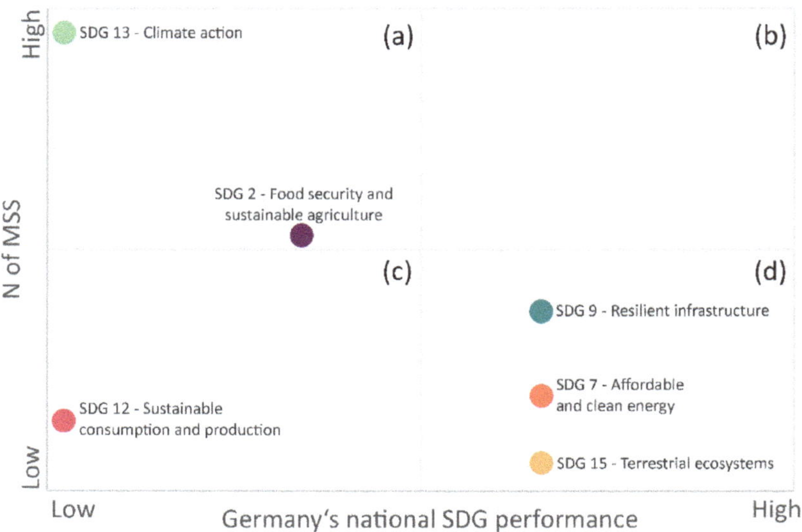

Figure 8. Matrix comparing Germany's national SDG performance [52] and the number of MSS. The matrix shows four patterns: (**a**) high number of MSS and low performance; (**b**) high number of MSS and high performance; (**c**) low number of MSS and low performance; and (**d**) low number of MSS and high performance.

By looking closely at the qualitative data, we found that frames supporting a more sustainable green COVID-19 recovery programme (MAS, n = 126) are more frequent than those promoting economic recovery without advocating for climate action (MFEF, n = 21). Nevertheless, this cannot be translated into unequivocal support of a "window of opportunity" frame. It became apparent in our qualitative content analysis that, even in articles that see opportunities for accelerated sustainability transformations, such acceleration does not come on its own. Both the environmental impacts (PEIs) and challenge for sustainability transformation (CST) frames point out that the pandemic could also lead to an increased environmental burden.

In sum, there certainly are thoughts about rebuilding better and initialising sustainability transformations with COVID-19 recovery measures. There is, however, no dominant frame that sees the pandemic as a chance for sustainability. In fact, the frames are closely interwoven (Figure 6). News media give contradictory representations of the relationship between COVID-19 sustainability that range from stories on environmental recreation during lockdowns to accounts that prioritise an economic recovery and see sustainability regulations as an obstacle. Depending on the actors involved, the pandemic is seen as a chance to rebuild more sustainably, or it is perceived as an economic threat that side-lines sustainability because more urgent issues are at hand. Therefore, it is necessary to continue monitoring the discourses as the pandemic unfolds since policy priorities and opinions are expected to change based on the severity of further waves of infection.

5.2. Methodological Considerations: Thickening Big Data

By linking big data and thick data tools, we were able to deal with large volumes of text data; a volume that surpasses researchers' ability to conduct a close reading [67], yet the analysis remains true to the principles of traditional qualitative approaches. The use of NLP tools allowed us to scrutinise huge collections of documents across 21 news outlets and identify thematic clusters as a means of following discursive changes. Therefore, our results can be more effectively generalised. To delve deeper into the data, qualitative content analysis was employed to track specific frames.

The use of this approach allowed us to bridge knowledge gaps between qualitative and quantitative research by providing context to the "big data" through rigorous empirical qualitative research. This offers both a broad overview on relevant topics and a close analysis of particular frames. As such, the approach can yield new insights not easily achievable through traditional text mining or qualitative social science methods alone. On the one hand, NLP techniques can provide an overview of the topics that are being discussed in large-scale textual corpora through time. On the other hand, qualitative tools can yield a background to these findings by adding context. Therefore, here, we do not consider data-driven quantitative tools as an "end in itself", nor do we see traditional qualitative tools as a mere "supplement". Instead, we promote the use of both approaches alongside each other.

5.3. Limitations and Further Research

Despite its advancements, the approach is not without limitations. First, we relied on a stratified sample of articles as the articles need to be downloaded manually, given that the database used does not allow for web scraping (see Section 3.1). Hence, less popular newspapers were ignored. Similarly, the thick data analysis focused on articles that mentioned at least two sustainability topics (Table S3, Figure S1, n = 461). As a result, they may be skewed towards the "window of opportunity" frame.

A further drawback is that we only considered the first wave of the COVID-19 pandemic, which limits the extrapolation of the findings. With this in mind, we recommend that future studies consider how the frames on sustainability have changed during different COVID-19 waves. This will allow one to assess whether or not the pandemic has had a discernible lasting impact on the media frames used to discuss sustainability.

Future work should more deeply examine how the media debate around sustainability within the COVID-19 crisis is shaped by both the actors involved and the power relationships of the organisations that promote these discourses. Additionally, the results of the qualitative analysis can also be retrofitted to the big data analysis, to capture the extent to which the identified frames and actors were mentioned in the broad sample.

6. Conclusions

In this study, we examined the COVID-19-related discourse regarding sustainability transformations during the early stages of the crisis in Germany. The results show that sustainability topics were not prominent in contrast to other topics. When we looked deeper into the data, we found that there is no dominant "window of opportunity" frame that sees the pandemic as an unequivocal chance for sustainability. Some frames support a green COVID-19 recovery programme, and they appear more often than articles that support economic recovery without advocating for climate action. Still, they only represent a small fraction of the published articles. Furthermore, even in articles that perceive an opportunity for accelerated sustainability transformations, such acceleration is not taken

for granted. In fact, the presence of a window of opportunity does not necessarily impl
that this opportunity will actually be taken [7].

Pragmatically, these findings are significant because reduced coverage about sustain
ability could dampen political will to act on the climate crisis. Indeed, as stated by Webe
and Stern [17], the news media can influence people's thoughts and actions regardless c
whether they are accurate or not. Even if the articles do not influence the public's opinion
the attention they bring to the topic can still be relevant as they can influence future new
coverage. Thus, monitoring the immediate and long-term impacts of different COVID-1!
waves on the sustainability discourse and their translation into policies remains a task fo
future investigation.

The mixed-methods approach described here can assist with this task by equippin;
scientists with a reliable toolkit for investigating discourses on rapidly evolving transforma
tion processes. Researchers can use it to explore ongoing media discourses on sustainabilit
transformations and detect trends in large-scale textual corpora.

Supplementary Materials: The following are available online at https://www.mdpi.com/article
10.3390/su131911095/s1, Table S1. Selected newspapers, number of articles per month, duplicate
and articles included in the analysis, Table S2. Additional stop words for the word cloud, networl
and topic modelling, Table S3. Sustainability transformation topics, their related SDGs and Gree
Deal goals, and keywords used for coding the newspaper articles and for the google trends searcl
Table S4. Topics obtained by applying the LDA model to the 61,514 COVID-19-related articles. Eacl
topic consists of a series of keywords, a topic name (assigned by the authors based on manua
inspection of keywords and articles), the topic coherence and prevalence. Following LDA, one articl
can be assigned to more than one topic, Table S5. Correlation matrix showing Pearson correlation
between the MSS, where 0 indicates a low correlation and 1 a strong correlation, Table S6. Automati
classification accuracy (%) for different MSS topics. Of the 2343 articles, 40% were randomly selecte
and read to determine the accuracy of the classification system, Figure S1. Sampling flowchar
Figure S2. Comparison word cloud of the changing discourse in the news media in Germany. Th
size of the word is proportional to its frequency. It compares the predominance of specific terms fo
each (a) month and (b) sustainability topic. For instance, the word "Klimawandel" (climate change
was used 158 times in April 2020 (frequency = 0.0007) and 91 times in May 2020 (frequency = 0.0004
Because this word was used more frequently in April, it only appears in the April section of th
comparison cloud, Figure S3. Network analysis graph with word relationships in the MSS corpus. I
portrays words that are strongly vs. weakly integrated. Each node represents a word, which mus
have at least 30 occurrences in the entire corpus to be present in the graph. The size and colour of th
node represents the word's frequency. The darkness of the connecting lines depicts the strength c
the word pair's relationship, Figure S4. Distribution of articles (n = 367) according to different frame
over time. The size of the bubbles varies according to the number of articles per week.

Author Contributions: M.M.d.B.: Conceptualisation, methodology, software, validation, formal ana
ysis, visualisation, data curation, writing—original draft, writing—review and editing. D.O.: Cor
ceptualisation, methodology, validation, formal analysis, writing—original draft, writing—reviev
and editing. C.K.: Conceptualisation, supervision, writing—review and editing. All authors hav
read and agreed to the published version of the manuscript.

Funding: This research received no external funding.

Institutional Review Board Statement: Not applicable.

Data Availability Statement: Data used for the analyses are available upon request.

Acknowledgments: Christian Kahmann and Tim Golla are thanked for their valuable research assistanc

Conflicts of Interest: The authors declare no conflict of interest.

References

1. Walby, S. *Crisis*; Polity Press: Cambridge, UK, 2015; ISBN 978-0745647616.
2. Aldaco, R.; Hoehn, D.; Laso, J.; Margallo, M.; Ruiz-Salmón, J.; Cristobal, J.; Kahhat, R.; Villanueva-Rey, P.; Bala, A.; Batlle-Baye
   L.; et al. Food waste management during the COVID-19 outbreak: A holistic climate, economic and nutritional approach. *Sc
   Total Environ.* **2020**, *742*, 140524. [CrossRef]

3. You, S.; Sonne, C.; Ok, Y.S. COVID-19's unsustainable waste management. *Science* **2020**, *368*, 1438. [CrossRef] [PubMed]
4. He, G.; Pan, Y.; Tanaka, T. The short-term impacts of COVID-19 lockdown on urban air pollution in China. *Nat. Sustain.* **2020**, *3*, 1005–1011. [CrossRef]
5. Ju, M.J.; Oh, J.; Choi, Y.-H. Changes in air pollution levels after COVID-19 outbreak in Korea. *Sci. Total Environ.* **2021**, *750*, 141521. [CrossRef] [PubMed]
6. Lehmann, P.; Beck, S.; de Brito, M.M.; Gawel, E.; Groß, M.; Haase, A.; Lepenies, R.; Otto, D.; Schiller, J.; Strunz, S.; et al. Environmental Sustainability Post-COVID-19: Scrutinizing Popular Hypotheses from a Social Science Perspective. *Sustainability* **2021**, *13*, 8679. [CrossRef]
7. Lehmann, P.; de Brito, M.M.; Gawel, E.; Groß, M.; Haase, A.; Lepenies, R.; Otto, D.; Schiller, J.; Strunz, S.; Thrän, D. Making the COVID-19 crisis a real opportunity for environmental sustainability. *Sustain. Sci.* **2021**. [CrossRef]
8. UNESCO United in Science Report: Climate Change Has Not Stopped for COVID-19. Available online: https://en.unesco.org/news/united-science-report-climate-change-has-not-stopped-covid19 (accessed on 15 October 2020).
9. Sarkis, J.; Cohen, M.J.; Dewick, P.; Schröder, P. A brave new world: Lessons from the COVID-19 pandemic for transitioning to sustainable supply and production. *Resour. Conserv. Recycl.* **2020**, *159*, 104894. [CrossRef]
10. Bodenheimer, M.; Leidenberger, J. COVID-19 as a window of opportunity for sustainability transitions? Narratives and communication strategies beyond the pandemic. *Sustain. Sci. Pract. Policy* **2020**, *16*, 61–66. [CrossRef]
11. Hay, C. Crisis and the Structural Transformation of the State: Interrogating the Process of Change. *Br. J. Polit. Int. Relations* **1999**, *1*, 317–344. [CrossRef]
12. Geels, F.W. The impact of the financial–economic crisis on sustainability transitions: Financial investment, governance and public discourse. *Environ. Innov. Soc. Transit.* **2013**, *6*, 67–95. [CrossRef]
13. Haase, A.; Bedtke, N.; Begg, C.; Gawel, E.; Rink, D.; Wolff, M. On the Connection Between Urban Sustainability Transformations and Multiple Societal Crises. In *Urban Transformations*; Springer: Cham, Switzerland, 2018; pp. 61–76.
14. Swim, J.K.; Vescio, T.K.; Dahl, J.L.; Zawadzki, S.J. Gendered discourse about climate change policies. *Glob. Environ. Chang.* **2018**, *48*, 216–225. [CrossRef]
15. Moser, S.C. Communicating climate change: History, challenges, process and future directions. *Wiley Interdiscip. Rev. Clim. Chang.* **2010**, *1*, 31–53. [CrossRef]
16. Pérez-González, L. 'Is climate science taking over the science?': A corpus-based study of competing stances on bias, dogma and expertise in the blogosphere. *Humanit. Soc. Sci. Commun.* **2020**, *7*, 92. [CrossRef]
17. Weber, E.U.; Stern, P.C. Public Understanding of Climate Change in the United States. *Am. Psychol.* **2011**, *66*, 315–328. [CrossRef] [PubMed]
18. Peterson, E. Not Dead Yet: Political Learning from Newspapers in a Changing Media Landscape. *Polit. Behav.* **2021**, *43*, 339–361. [CrossRef]
19. Carmichael, J.T.; Brulle, R.J. Elite cues, media coverage, and public concern: An integrated path analysis of public opinion on climate change, 2001–2013. *Env. Polit.* **2017**, *26*, 232–252. [CrossRef]
20. Blair, B.; Zimny-Schmitt, D.; Rudd, M.A. U.S. News Media Coverage of Pharmaceutical Pollution in the Aquatic Environment: A Content Analysis of the Problems and Solutions Presented by Actors. *Environ. Manag.* **2017**, *60*, 314–322. [CrossRef] [PubMed]
21. Arlt, D.; Hoppe, I.; Wolling, J. Climate change and media usage: Effects on problem awareness and behavioural intentions. *Int. Commun. Gaz.* **2011**, *73*, 45–63. [CrossRef]
22. Quesnel, K.J.; Ajami, N.K. Changes in water consumption linked to heavy news media coverage of extreme climatic events. *Sci. Adv.* **2017**, *3*, e1700784. [CrossRef]
23. Kim, Y. Understanding publics' perception and behaviors in crisis communication: Effects of crisis news framing and publics' acquisition, selection, and transmission of information in crisis situations. *J. Public Relations Res.* **2016**, *28*, 35–50. [CrossRef]
24. Moodley, P.; Lesage, S.S. A discourse analysis of Ebola in South African newspapers (2014–2015). *S. Afr. J. Psychol.* **2020**, *50*, 158–169. [CrossRef]
25. Krzyżanowski, M. Brexit and the imaginary of 'crisis': A discourse-conceptual analysis of European news media. *Crit. Discourse Stud.* **2019**, *16*, 465–490. [CrossRef]
26. De Brito, M.M.; Kuhlicke, C.; Marx, A. Near-real-time drought impact assessment: A text mining approach on the 2018/19 drought in Germany. *Environ. Res. Lett.* **2020**, *15*, 1040a9. [CrossRef]
27. De Brito, M.M. Compound and cascading drought impacts do not happen by chance: A proposal to quantify their relationships. *Sci. Total Environ.* **2021**, *778*, 146236. [CrossRef] [PubMed]
28. Rousseau, S.; Deschacht, N. Public Awareness of Nature and the Environment During the COVID-19 Crisis. *Environ. Resour. Econ.* **2020**, *76*, 1149–1159. [CrossRef]
29. Ulibarri, N.; Scott, T.A. Environmental hazards, rigid institutions, and transformative change: How drought affects the consideration of water and climate impacts in infrastructure management. *Glob. Environ. Chang.* **2019**, *59*, 102005. [CrossRef]
30. De Fortuny, E.J.; de Smedt, T.; Martens, D.; Daelemans, W. Media coverage in times of political crisis: A text mining approach. *Expert Syst. Appl.* **2012**, *39*, 11616–11622. [CrossRef]
31. Smets, A.; Lievens, B. Human Sensemaking in the Smart City: A Research Approach Merging Big and Thick Data. *Ethnogr. Prax. Ind. Conf. Proc.* **2018**, *2018*, 179–194. [CrossRef]

32. Hajer, M.A.; Pelzer, P. 2050—An Energetic Odyssey: Understanding 'Techniques of Futuring' in the transition towards renewable energy. *Energy Res. Soc. Sci.* **2018**, *44*, 222–231. [CrossRef]
33. Burrows, R.; Savage, M. After the crisis? Big Data and the methodological challenges of empirical sociology. *Big Data Soc.* **2014**, *1*, 205395171454028. [CrossRef]
34. Strong, C. The challenge of "Big Data": What does it mean for the qualitative research industry? *Qual. Mark. Res. An Int. J.* **2014**, *17*, 336–342. [CrossRef]
35. Bail, C.A. The cultural environment: Measuring culture with big data. *Theory Soc.* **2014**, *43*, 465–482. [CrossRef]
36. Borgman, C.L. *Big data, little data, no data: Scholarship in the networked world*; The MIT Press: Cambridge, MA, USA, 2015; ISBN 978-0-262-02856-1.
37. Kitchin, R. Big data and human geography. *Dialogues Hum. Geogr.* **2013**, *3*, 262–267. [CrossRef]
38. Mills, K.A. What are the threats and potentials of big data for qualitative research? *Qual. Res.* **2017**, *18*, 591–603. [CrossRef]
39. Baur, N.; Graeff, P.; Braunisch, L.; Schweia, M. The Quality of Big Data. Development, Problems, and Possibilities of Use of Process-generated Data in the Digital Age. *Hist. Soc. Res./Hist. Sozialforsch* **2020**, *45*, 209–243. [CrossRef]
40. Kitchin, R.; McArdle, G. What makes Big Data, Big Data? Exploring the ontological characteristics of 26 datasets. *Big Data Soc.* **2016**, *3*, 205395171663113. [CrossRef]
41. Hesse, A.; Glenna, L.; Hinrichs, C.; Chiles, R.; Sachs, C. Qualitative Research Ethics in the Big Data Era. *Am. Behav. Sci.* **2019**, *63*, 560–583. [CrossRef]
42. Felt, M. Social media and the social sciences: How researchers employ Big Data analytics. *Big Data Soc.* **2016**, *3*, 205395171664582. [CrossRef]
43. Morgan, D.L. *Integrating Qualitative and Quantitative Methods: A Pragmatic Approach*; SAGE Publications, Inc.: London, UK, 2014; ISBN 9780761915232.
44. Bornakke, T.; Due, B.L. Big–Thick Blending: A method for mixing analytical insights from big and thick data sources. *Big Data Soc.* **2018**, *5*, 2053951718765026. [CrossRef]
45. Geertz, C. *The Interpretation of Cultures*; Basic Books: New York, NY, USA, 1973.
46. Ponterotto, J. Brief Note on the Origins, Evolution, and Meaning of the Qualitative Research Concept Thick Description. *Qual. Rep.* **2006**, *11*, 538–549.
47. Levandowsky, M.; Winter, D. Distance between Sets. *Nature* **1971**, *234*, 34–35. [CrossRef]
48. Callaghan, M.W.; Minx, J.C.; Forster, P.M. A topography of climate change research. *Nat. Clim. Chang.* **2020**, *10*, 118–123. [CrossRef]
49. Luiz, O.J.; Olden, J.D.; Kennard, M.J.; Crook, D.A.; Douglas, M.M.; Saunders, T.M.; King, A.J. Trait-based ecology of fishes: A quantitative assessment of literature trends and knowledge gaps using topic modelling. *Fish Fish.* **2019**, *20*, 1100–1110. [CrossRef]
50. Aachener Stiftung Lexikon der Nachhaltigkeit. Available online: https://www.nachhaltigkeit.info/suche/a-z/b/index.htm (accessed on 4 October 2021).
51. GreenFacts Glosar Nachhaltigkeit. Available online: https://www.greenfacts.org/de/glossar/abc/index.htm (accessed on 4 October 2021).
52. Sachs, J.; Schmidt-Traub, G.; Kroll, C.; Lafortune, G.; Fuller, G. Sustainable Development Report 2019. 2019. Available online: https://www.sdgindex.org/reports/sdg-index-and-dashboards-2018/ (accessed on 4 October 2021).
53. Youngblood, M.; Lahti, D. A bibliometric analysis of the interdisciplinary field of cultural evolution. *Palgrave Commun.* **2018**, *4*, 120. [CrossRef]
54. Krzywinski, M.; Schein, J.; Birol, I.; Connors, J.; Gascoyne, R.; Horsman, D.; Jones, S.J.; Marra, M.A. Circos: An information aesthetic for comparative genomics. *Genome Res.* **2009**, *19*, 1639–1645. [CrossRef]
55. Schreier, M. *Qualitative Content Analysis in Practice*; SAGE: Los Angeles, CA, USA, 2012; ISBN 978-1-84920-592-4, ISBN 978-1-84920-593-1.
56. Mayring, P. Qualitative Content Analysis. *Forum Qual. Sozialforsch. Forum Qual. Soc. Res.* **2000**, *1*. [CrossRef]
57. Topham, G.; Harvey, F. Carmakers Accused of Trying to Use Crisis to Avert Emissions Crackdown. 2020. Available online: https://www.theguardian.com/business/2020/mar/27/carmakers-accused-of-using-covid-19-weaken-environmental-laws (accessed on 4 October 2021).
58. Simon, F. Green Deal Facing Delays Due to Coronavirus, EU Admits. 2020. Available online: https://www.euractiv.com/section/energy-environment/news/green-deal-facing-delays-due-to-coronavirus-eu-admits/ (accessed on 4 October 2021).
59. Engler, J.-O.; Abson, D.J.; von Wehrden, H. The coronavirus pandemic as an analogy for future sustainability challenges. *Sustain. Sci.* **2021**, *16*, 317–319. [CrossRef]
60. Williams, S.; Robinson, J. Measuring sustainability: An evaluation framework for sustainability transition experiments. *Environ. Sci. Policy* **2020**, *103*, 58–66. [CrossRef]
61. Almoradie, A.; Brito, M.M.; Evers, M.; Bossa, A.; Lumor, M.; Norman, C.; Yacouba, Y.; Hounkpe, J. Current flood risk management practices in Ghana: Gaps and opportunities for improving resilience. *J. Flood Risk Manag.* **2020**, *13*, e12664. [CrossRef]
62. Zhang, D.; Hao, M.; Morse, S. Is Environmental Sustainability Taking a Backseat in China after COVID-19? The Perspective of Business Managers. *Sustainability* **2020**, *12*, 10369. [CrossRef]
63. Barreiro-Gen, M.; Lozano, R.; Zafar, A. Changes in Sustainability Priorities in Organisations due to the COVID-19 Outbreak: Averting Environmental Rebound Effects on Society. *Sustainability* **2020**, *12*, 5031. [CrossRef]
64. Barbier, E.B.; Burgess, J.C. Sustainability and development after COVID-19. *World Dev.* **2020**, *135*, 105082. [CrossRef] [PubMed]

5. Nature. Time to revise the Sustainable Development Goals. *Nature* **2020**, *583*, 331–332. [CrossRef] [PubMed]
6. Naidoo, R.; Fisher, B. Reset Sustainable Development Goals for a pandemic world. *Nature* **2020**, *583*, 198–201. [CrossRef] [PubMed]
7. Davidson, E.; Edwards, R.; Jamieson, L.; Weller, S. Big data, qualitative style: A breadth-and-depth method for working with large amounts of secondary qualitative data. *Qual. Quant.* **2019**, *53*, 363–376. [CrossRef] [PubMed]

sustainability

Article

Investigating the Drivers of Supply Chain Resilience in the Wake of the COVID-19 Pandemic: Empirical Evidence from an Emerging Economy

Mohammad Ali Yamin

Department of Human Resources Management, Collage of Business, University of Jeddah, Jeddah 23454, Saudi Arabia; mayameen@uj.edu.sa

Abstract: The COVID-19 pandemic has disrupted supply chain operations globally. Nevertheless, resilient firms have the capacity to combat an unprecedented situation with the right strategic approach. The current research has developed an integrated research model that combines factors such as supply chain intelligence, supply chain communication, leadership commitment, risk management orientation, supply chain capability and network complexity to investigate supply chain resilience. The research model of this study was empirically tested with 309 responses collected from supply chain managers. Results revealed that supply chain resilience is measured with supply chain intelligence, supply chain communication, leadership commitment, risk management orientation, supply chain capability and network complexity and demonstrated a substantial variance R^2 of 0.548% towards supply chain resilience. Practically, this study suggests that supply chain managers should focus on factors such as big data analytics, risk management orientation,1 supply chain communication and leadership commitment to enhance supply chain resilience and sustainable supply chain performance.

Keywords: supply chain resilience; supply chain intelligence; communication; leadership commitment; big data analytics; sustainable supply chain performance

Citation: Yamin, M.A. Investigating the Drivers of Supply Chain Resilience in the Wake of the COVID-19 Pandemic: Empirical Evidence from an Emerging Economy. *Sustainability* **2021**, *13*, 11939. https://doi.org/10.3390/su132111939

Academic Editors: Marc A. Rosen, Gaojie He, Ayyoob Sharifi, Chi Feng and Jun Yang

Received: 4 September 2021
Accepted: 21 October 2021
Published: 28 October 2021

Publisher's Note: MDPI stays neutral with regard to jurisdictional claims in published maps and institutional affiliations.

1. Introduction

In this dynamic environment, the major challenge for supply chain practitioners is to deal with supply chain upheavals, disruption, and unforeseen events [1]. If a firm faces upheavals in supply chain operations and continue to perform, that situation is characterized by resilience [2]. Supply chain resilience is defined as operational capacity of a firm to return to its initial state after being disrupted and be stronger than before in a supply chain process [3]. The importance of supply chain resilience is highlighted in earlier studies [1,3–5]. According to Brandon-Jones et al. [1] firms are facing more disruption due to natural and manmade events and therefore resilience phenomenon should be investigated to understand how resilience help firms to recover quickly after disruption. In current situation wherein COVID-19 pandemic has left devastated impact global economy and badly affected supply chain operations and therefore organizations are now seeking resilient kind of strategies to confront unforeseen events [6,7].

Concerning supply chain disruption, the literature indicates that the pandemic has destabilized supply chain operations and negatively impacted customer needs requirements and satisfaction [5,8,9]. Therefore, the current study fills a research gap and develops an integrated supply chain resilience model with a combination of factors such as supply chain intelligence, supply chain communication, leadership commitment, risk management orientation, supply chain capability, network complexity and big data analytics to investigate supply chain resilience during the COVID-19 pandemic and sustainability supply chain performance in a post-pandemic context. To enhance supply chain resilience, the researcher has paid attention to supply chain intelligence and communication strategies. Supply chain intelligence is a process of integrating knowledge that is derived from suppliers, customers and competitors and using that knowledge to manage supply chain operations [10,11]. Therefore, communication is the extent to which a firm facilitates supply chain partners through communication, messages, and communication networks [12]. The impact of leadership commitment is found to be positive in measuring supply chain resilience. For instance, authors such as Speier, Whipple, Closs, and Voss [13] postulated that leadership commitment has been used as a foundation in designing and implementing supply chain strategies. Similarly, prior research conducted by Wieland and Marcus Wallenburg [14] confirmed that risk orientation reduces the failure chances in the supply chain process. Furthermore, supply chain capability and network complexity have shown a positive influence on measuring supply chain resilience [15,16].

The research model as shown in Figure 1 is extended with the moderating role of big data analytics. Big data analytics is identified as a combination of technologies, processes and techniques that enable organizations to collect, organize, visualize and analyze data and bring swiftness into supply chain operations [17]. The moderating effect of big data analytics is examined between supply chain resilience and sustainable supply chain performance. To the best of the authors knowledge, this study is the first that integrates factors such as leadership, technology and network factors altogether to investigate supply chain resilience and sustainable supply chain performance. This study is significant as it investigates the role of supply chain resilience and sustainable supply chain performance during the COVID-19 pandemic. In addition to that, the results of this research disclose several useful findings for the manufacturing industry to understand how to bring resilience in supply chain operations whenever they confront unforeseen events. The remaining part of the research is included in the literature review, methodology, data analysis, discussion and the conclusion of this study.

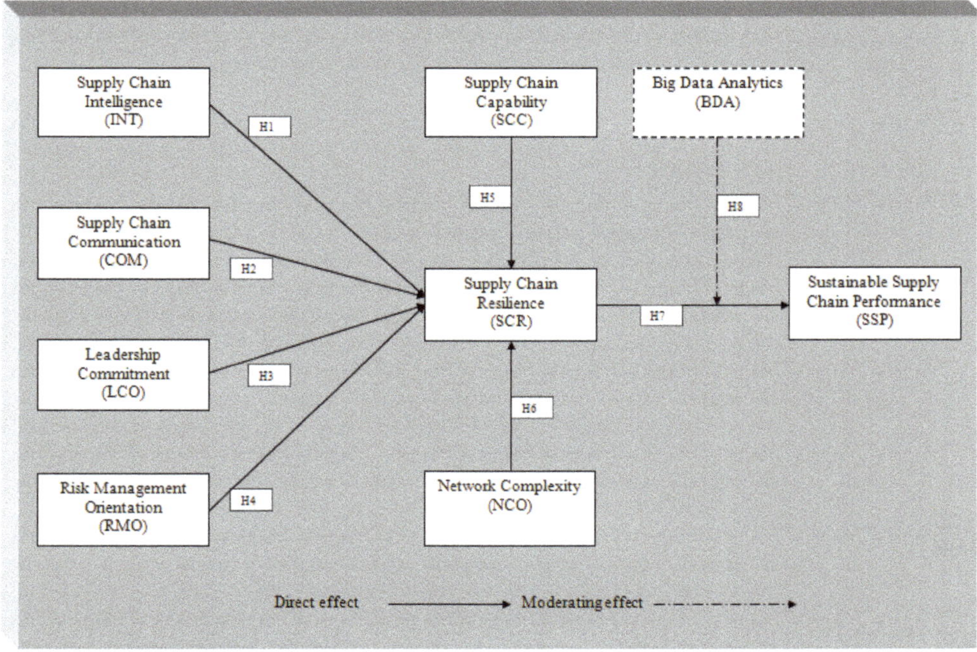

Figure 1. Research framework.

2. Literature Review

2.1. Supply Chain Intelligence and Communication

The concept of supply chain intelligence is extracted from knowledge-based resources and is explained as the extent to which a firm integrates knowledge that is derived from suppliers, customers and competitors and uses this to manage supply chain operations [11]. Supply chain intelligence is a process of knowledge integration among supply chain partners [11,18]. The use of a supply chain intelligence strategy gives a holistic view of the supply chain process and reduces disruption in supply chain operations [18–20]. Earlier studies have established that intelligence characteristics enable organizations to forecast accurately, reduce risk and make firms more resilient in response to supply chain uncertainty [18,19,21]. Therefore, communication is the extent to which a firm facilitates supply chain partners through communication, messages and communication networks [12] The literature has shown that intelligence and communication bring supply chain integration, responsiveness, information exchange and leverage to a higher supply chain performance [12,21,22]. A recent study conducted by Asamoah et al. [22] has confirmed that both communication and supply chain intelligence positively impact supply chain resilience. Therefore, the following hypotheses are proposed:

Hypothesis 1 (H1). *Supply chain intelligence significantly impacts SC resilience.*

Hypothesis 2 (H2). *Supply chain communication significantly impacts SC resilience.*

2.2. The Role of Leadership Commitment and Risk Management Orientation

The importance of leadership commitment is vital in planning, designing and implementing a supply chain strategy [15]. As suggested by Speier et al. [13] leadership commitment is considered as a foundation in the supply chain process. The literature has confirmed that a leader's commitment motivates employees, brings pro-activeness and ensures that resources are being used adequately [13,23–25]. It is established that leadership commitment assists supply chain managers in implementing decisions transparently to avoid supply chain disruption, which, in turn, enhances resilience in operations [26,27]. Concerning risk orientation, earlier studies have confirmed that risk orientation reduces the failure chances in a supply chain process [14]. Organizations can handle uncertainty in a supply chain process through risk orientation, which ultimately boosts supply chain resilience [28–30]. Therefore, the following hypotheses are proposed:

Hypothesis 3 (H3). *Leadership commitment significantly impacts SC resilience.*

Hypothesis 4 (H4). *Risk management orientation significantly impacts SC resilience.*

2.3. Supply Chain Capability and Network Complexity

Supply chain capability denotes to the ability of a firm to identify, manage, and synchronize information that facilitates a firms internal and external supply chain operations [16]. The literature provides abundant evidence that the supply chain capability improves a firm's operational and financial performance by reducing cost and enhancing resilience in supply chain operations [2,16,31]. Therefore, in the current research, we centered our attention toward the supply chain capability and resilience. According to Nishat Faisal, Banwet, and Shankar [32] strong coordination is required among supply chain partners to manage, forecast and replenish inventory. Authors such as Bhamra et al. [2] asserted that supply chain capability plays a vital role in the flow of goods, reduces lead time, brings transparency in supply chain operations, speeds up the payment cycle, reduces inventory and avoids over production. The network complexity is identified as the degree of connectivity between the network length and the number of nodes [15]. The complexity of the network is increased with an increase in the number of nodes and network length [4,13]. Thus, the researcher assumed that supply chain resilience may enhance with a decrease in the numbers of nodes and the network length. Therefore, the following hypotheses are proposed:

Hypothesis 5 (H5). *Supply chain capability significantly impacts supply chain resilience.*

Hypothesis 6 (H6). *Decrease in network complexity significantly impacts supply chain resilience*

2.4. Big Data Analytics

Big data analytics is an emerging concept and is being acknowledged as a process that enables organizations to collect, process, store and analyze data further to obtain useful insights [17]. In a supply chain setting, big data analytics is conceptualized as a combination of technologies, processes and techniques that enable organizations to collect, organize, visualize and analyze data and bring swiftness to their supply chain operations [17]. The extant literature has confirmed the significant influence of big data in reducing SC disruption while increasing SC resilience [33–38]. According to Janssen et al. [34] data analytics is significantly related to the supply chain innovation process. Another study conducted by Dubey et al. [17] has confirmed that big data analytics positively impacts the predication of supply chain resilience. Following the above arguments and supported by [17]. the current research tests the moderating effect of BDA on the relationship of supply chain resilience and sustainable supply chain performance. Thus, the following hypotheses are proposed:

Hypothesis 7 (H7). *Resilience significantly impacts sustainable SC performance.*

Hypothesis 8 (H8). *Big data analytics has a moderating impact on SC resilience and sustainable SC performance.*

3. Research Methods

3.1. Designing Questionnaire and Instrument Development

In line with research objectives, the research framework of this study is developed under a quantitative and cross-sectional research approach. The researcher has followed a positivist paradigm to design the research in line with [39]. In addition to this, data are collected through research questionnaires. The research questionnaire in this study comprises construct items and demographic characteristics of the respondents. Construct items were developed by reviewing the literature and previously established scales. Scale items for network complexity were adopted from [15]. Risk management orientation scale is adopted from Wieland and Marcus Wallenburg [14] and then slightly adapted into the current research context. Supply chain resilience scale items were adopted from previously developed scale by Brandon-Jones et al. [1] then slightly adapted. Scale items for the construct big data analytics were adopted from [17]. Supply chain communication scale items were adopted from [21]. Similarly, supply chain intelligence items were adopted from [22]. Therefore, supply chain capability scale was adapted from [16,22]. Scale items for leadership commitment were adapted from [7]. Construct items for sustainable supply chain performance were adopted from [40,41]. The scale items are measured using a 7-point Likert scale, where 1 denotes strongly disagree and 7 denotes strongly agree [42].

3.2. Sampling and Data Collection

This study uses an empirically tested research framework to examine supply chain resilience and sustainable supply chain performance. Therefore, manufacturing firms are selected as the unit of analysis in this research. Earlier studies have emphasized that manufacturing companies provide in depth understanding about supply chain operations and organizational performance [2,16,31]. The population of this research comprises supply chain managers working in manufacturing companies in Saudi Arabia. As suggested by M. Yamin [43] and consistent with prior power analysis, a sample of 300 respondents was selected for data analysis. Research data of this study were collected through an online research survey. Data were collected during the month of May 2021. The research survey was designed and conducted during COVID-19 pandemic; consequently, the researcher used online tools such as social media platforms, Google forms and email for data collection. Concerning respondent selection, the researcher selected respondents through a convenience sampling approach. For data collection, 750 questionnaires were forwarded to supply chain managers with a request to fill out an online research questionnaire according to their knowledge about supply chain resilience and sustainable supply chain performance. In response to the online research survey, 317 questionnaires were retrieved from respondents. Nonetheless, 8 questionnaires were discarded due to inadequate and inappropriate filling. Thus, a total of 309 responses were used for inferential analysis. Descriptive analyses were conducted with SPSS software. The results of the descriptive analysis revealed that the data set comprised 198 males and 111 females among 309 respondents. Similarly, the respondents' ages were considered. The results indicate that the majority of the respondents 143 were found to be between 21 and 30 years. Next to this, 35 respondents were found to be between 41 and 50 years old. Furthermore, 131 respondents were aged between 31 and 40 years. Finally, data were estimated with a structural equation modeling (SEM) approach.

4. Data Analysis

4.1. Common Method Bias

The common method variance bias (CMV) needed to be examined before the inferential analysis to ensure that the research data had no bias [44]. Since the data were collected from a single source, a CMV issue may occur in this research [45,46]. Referring to the CMV issue, the literature has suggested two well-known methods, namely, procedural and statistical remedies [45,47–49]. Following the procedural method, the questionnaire items were jumbled up, consistent with [48]. Therefore, CMV is statistically confirmed with Harman's single factor solution. Harman's single factor solution suggests that the variance explained by the first factor must not be higher than 50% [48]. The results, as depicted in Table 1, indicate that the variance explained by the first factor is less than the threshold value (50%), hence confirming that CMV is not a potential issue in this study.

Table 1. Harman's single factor analysis.

	Total Variance Explained					
Factors	Eigenvalues			Extraction Sums of Squared Loadings		
	Total	% of Variance	Cumulative %	Total	% of Variance	Cumulative %
1	9.422	23.737	23.737	9.422	23.737	23.737

4.2. Structural Equation Modeling

The theoretical framework of this study was tested with a structural equation modeling (SEM) approach [42]. The researcher opted for a two-stage approach for the structural equation modeling computation, which included the assessment of a measurement model and a structural model [43,50]. The reliability and validity of the constructs was tested with a measurement model [43]. Therefore, the hypothesized relationship was confirmed with a structural model [51]. The data were evaluated with Smart PLS software using a partial least square (PLS) approach [52].

4.2.1. Assessing Measurement Model

The measurement model was examined to establish the construct's validity and reliability. The convergent validity of the constructs was achieved with average variance extracted (AVE) following the criterion that that AVE values must be higher than 0.50. Nevertheless, the construct's reliability was achieved following the values of (α) and composite reliability in line with Mohammad Ali Yousef Yamin and Sweiss [53] who suggested that the CR and Cronbach's alpha values should be higher than 0.70, reflecting adequate construct reliability and validity. The findings, as depicted in Table 2, revealed the adequate reliability, validity and convergent validity of the constructs.

Table 2. Measurement model.

Indicator	Loadings	(α)	CR	AVE
Big Data Analytics (BDA)				
BDA1: In this firm, advance data analytics tools are used to take decisions.	0.824	0.757	0.861	0.673
BDA2: In this firm, information is extracted using big data analytics to take decision.	0.843			
BDA3: For data visualization, this firm use dashboard display to assist supply chain managers.	0.793			
Supply Chain Communication (COM)				
COM1: This firm has multiple communication channels to facilitate supply chain operations.	0.826	0.854	0.902	0.696
COM2: This firm uses an integrated organizational system to communicate with stakeholders.	0.841			
COM3: This firm uses the latest integrated communication tools for channel communication.	0.825			
COM4: The use of frequent communication among supply chain partners enhances firm resilience.	0.845			
Supply Chain Intelligence (INT)				
INT1: This firm is able to search, retrieve and store business information to boost supply chain operation.	0.717	0.700	0.834	0.628
INT3: This firm is ability to understand sales trends and customer preferences using integrated supply chain tools.	0.869			
INT4: This firm uses integrated information retrieved from past events to deal with any kind of unprecedented situation.	0.784			
Leadership Commitment (LCO)				
LCO1: The leadership of this organization is committed to handling all kinds of profit and loss.	0.752	0.785	0.860	0.606
LCO2: Leaders of this organization take responsibility for all the departments to tackle with unprecedented situation.	0.798			
LCO3: Leaders of this organization support long term organizational goals.	0.752			
LCO4: Leadership of this organization shows pro-activeness to recover business operations.	0.809			
Network Complexity (NCO)				
NCO2: This organization invests heavily on infrastructure to reduce network complexity.	0.802	0.788	0.876	0.702
NCO3: In this organization, network complexity occurred due to unexpected changes in supply chain operations.	0.853			
NCO4: This organization has a strategic plan to deal with supply chain nodes that reduce network complexity.	0.858			
Risk Management Orientation (RMO)				
RMO2: Risks in this organization are monitored continuously and managed proactively.	0.748	0.736	0.849	0.652
RMO3: This organization has the ability to identify the source of disruption in a systematic way.	0.820			
RMO4: This organization is efficient in assessing own risk, customer risk and supplier risk.	0.852			
Supply Chain Capability (SCC)				
SCC1: In this firm, the information flow is more effective between the firm and supply chain partners.	0.796	0.848	0.898	0.687
SCC2: This firm has the capacity to handle follow-up activities proactively.	0.836			
SCC3: This firm has strong coordination with stake holders for planning and forecasting.	0.850			
SCC4: This firm has the competency to respond quickly to changing customer needs and demands.	0.832			
Supply Chain Resilience (SCR)				
SCR1: This firm has the competency to recover supply chain operations quickly.	0.845	0.876	0.915	0.729
SCR2: In this firm, inventory flow would not take long to restore.	0.863			
SCR3: This firm is able to restore operating performance.	0.867			
SCR4: This firm has the capacity to deal with all kinds of supply chain disruption without any delay.	0.840			
Sustainable Supply Chain Performance (SSP)				
SSP1: This firm has reduced buffer stock in the supply chain process.	0.882	0.851	0.910	0.770
SSP2: This firm is following all environmental standards according to customer requirements.	0.889			
SSP3: This firm has controlled the supply chain wastage significantly.	0.862			

The measurement model had established the construct's convergent validity and reliability. Therefore, Fornell and Larcker analysis was incorporated for the assessment of the construct's discriminant validity [24,54]. The Fornell and Larcker analysis suggests that the square root values of the average variance extracted must be higher than the other constructs correlations [24]. The findings indicate that the square root of AVE is higher when compared with other constructs correlations, thus establishing the discriminant validity of the measure. The findings of the Fornell and Larcker analysis are tabulated in Table 3.

Table 3. Discriminant validity.

	BDA	COM	INT	LCO	NCO	RMO	SCC	SCR	SSP
BDA	0.820								
COM	0.299	0.834							
INT	0.465	0.244	0.792						
LCO	0.400	0.323	0.386	0.778					
NCO	0.595	0.277	0.513	0.396	0.838				
RMO	0.390	0.315	0.384	0.897	0.352	0.808			
SCC	0.271	0.706	0.262	0.318	0.279	0.325	0.829		
SCR	0.407	0.470	0.408	0.653	0.367	0.662	0.455	0.854	
SSP	0.398	0.500	0.346	0.543	0.335	0.519	0.471	0.719	0.878

Although the Fornell and Larcker analysis was used extensively, it has some deficiencies in computation [54–56]. The cross-loading method was used in this study as an alternative method and consistent with earlier studies by [54]. The cross-loading method suggests that the indicator loading of the construct must be higher than corresponding constructs loading [57]. The results of the cross-loading method revealed that construct loadings are satisfactory when compared with the corresponding constructs loading and, therefore, establish the discriminant validity of the constructs. The results of the cross loading are exhibited in Table 4.

Table 4. Cross loadings.

	BDA	COM	INT	LCO	NCO	RMO	SCC	SCR	SSP
BDA1	0.824	0.232	0.388	0.360	0.564	0.371	0.257	0.365	0.352
BDA2	0.843	0.273	0.332	0.298	0.442	0.284	0.229	0.306	0.311
BDA3	0.793	0.233	0.422	0.322	0.450	0.298	0.177	0.325	0.312
COM1	0.212	0.826	0.200	0.320	0.218	0.288	0.562	0.391	0.432
COM2	0.242	0.841	0.186	0.264	0.195	0.266	0.533	0.415	0.409
COM3	0.295	0.825	0.221	0.293	0.254	0.298	0.619	0.377	0.427
COM4	0.252	0.845	0.210	0.202	0.259	0.199	0.646	0.384	0.400
INT1	0.342	0.187	0.717	0.268	0.327	0.263	0.233	0.287	0.270
INT3	0.409	0.190	0.869	0.328	0.433	0.332	0.187	0.352	0.282
INT4	0.351	0.205	0.784	0.317	0.451	0.313	0.210	0.327	0.274
LCO1	0.309	0.258	0.296	0.752	0.361	0.540	0.245	0.433	0.385
LCO2	0.345	0.262	0.329	0.798	0.340	0.700	0.233	0.464	0.415
LCO3	0.256	0.208	0.239	0.752	0.219	0.725	0.186	0.497	0.415
LCO4	0.335	0.276	0.333	0.809	0.323	0.793	0.312	0.608	0.464
NCO2	0.391	0.228	0.439	0.292	0.802	0.287	0.245	0.288	0.299
NCO3	0.491	0.195	0.445	0.316	0.853	0.274	0.214	0.290	0.244
NCO4	0.599	0.267	0.410	0.381	0.858	0.321	0.242	0.340	0.297
RMO2	0.310	0.257	0.324	0.711	0.291	0.748	0.237	0.429	0.388
RMO3	0.304	0.233	0.278	0.721	0.240	0.820	0.241	0.531	0.408
RMO4	0.333	0.275	0.334	0.748	0.323	0.852	0.302	0.621	0.456
SCC1	0.245	0.570	0.233	0.275	0.211	0.271	0.796	0.380	0.384
SCC2	0.219	0.619	0.154	0.221	0.216	0.241	0.836	0.351	0.382
SCC3	0.208	0.582	0.246	0.255	0.229	0.285	0.850	0.369	0.369
SCC4	0.225	0.569	0.230	0.295	0.263	0.277	0.832	0.403	0.423
SCR1	0.346	0.360	0.301	0.610	0.294	0.633	0.383	0.845	0.554
SCR2	0.331	0.331	0.342	0.532	0.283	0.559	0.369	0.863	0.535
SCR3	0.355	0.435	0.324	0.498	0.352	0.502	0.404	0.867	0.616
SCR4	0.354	0.465	0.416	0.582	0.320	0.566	0.393	0.840	0.726
SSP1	0.335	0.445	0.280	0.466	0.318	0.457	0.420	0.633	0.882
SSP2	0.362	0.455	0.382	0.527	0.306	0.476	0.423	0.676	0.889
SSP3	0.351	0.414	0.240	0.430	0.254	0.431	0.396	0.578	0.862

Prior studies have proposed Heterotrait-monotrait (HTMT) analysis for the testing of discriminant validity. HTMT analysis was introduced by Gold, Malhotra, and Segars [55] and suggests that the HTMT ratios must be lower than 0.85 or 0.90 to indicate the adequate discriminant validity of the constructs [55,56]. The results of the HTMT ratio analysis show that the HTMT values are less than 0.90 and hence, confirm the adequate discriminant validity of the variables. The values of HTMT are presented in Table 5.

Table 5. Heterotrait-Monotrait analysis.

	BDA	COM	INT	LCO	NCO	RMO	SCC	SCR	SSP
BDA									
COM	0.374								
INT	0.637	0.318							
LCO	0.516	0.394	0.518						
NCO	0.757	0.336	0.690	0.503					
RMO	0.519	0.398	0.536	0.170	0.460				
SCC	0.335	0.832	0.343	0.381	0.339	0.406			
SCR	0.496	0.537	0.516	0.771	0.437	0.810	0.525		
SSP	0.494	0.585	0.446	0.656	0.406	0.651	0.553	0.821	

4.2.2. Assessing Structural Model

The structural model tests the hypothesized relationship between variables. In order to mitigate the normality issue, data were bootstrapped with dummy data of 4000, as suggested by [58]. The results of the structural model are given in Table 6 comprising path value, standard error, t-statistics and the significance of the hypotheses.

Table 6. Results of the hypotheses.

Hypothesis	Relationship	Path Coefficient	STDEV	T-Statistics	Significance	Decision
H1	INT → SCR	0.113	0.043	20.634	0.005	Accepted
H2	COM → SCR	0.183	0.064	20.859	0.003	Accepted
H3	LCO → SCR	0.210	0.080	20.635	0.005	Accepted
H4	RMO → SCR	0.326	0.081	40.012	0.000	Accepted
H5	SCC → SCR	0.116	0.060	10.933	0.028	Accepted
H6	NCO → SCR	0.028	0.042	0.679	0.249	Not Accepted
H7	SCR → SSP	0.667	0.032	20.678	0.000	Accepted

The results of the structural model indicate that although the exogenous variables have a significant impact on supply chain resilience, the relationship between network capacity and supply chain resilience is insignificant. The following results indicate that supply chain intelligence significantly impacts supply chain resilience and support H1: $\beta = 0.113$ path; significance, $p < 0.001$ and t-statistics, 2.634. Supply chain communication has a positive impact on supply chain resilience and confirms H2: $\beta = 0.183$ path; significance, $p < 0.001$ and t-statistics, 2.859. Similarly, leadership commitment has revealed a positive impact on measuring supply chain resilience and statistically confirms H3: $\beta = 0.210$ path; significance, $p < 0.001$ and t-statistics, 2.635. The statistics show that both risk management orientation and supply chain capability have a positive impact on supply chain resilience and are supported by $\beta = 0.326$ path; significance, $p < 0.001$ and t-statistics, 4.012; $\beta = 0.116$ path; significance, $p < 0.05$ and t-statistics, 1.933, hence establishing H4 and H5. Contrary to the researcher, the relationship between the expected network complexity and supply chain resilience was found to be insignificant ($\beta = 0.028$ path; significance, $p > 0.05$ and t-statistics, 0.679) and, therefore, H6 is rejected. Finally, the supply chain resilience has shown a positive impact toward sustainable supply chain performance and confirms H7: $\beta = 0.667$ path; significance, $p < 0.001$ and t-statistics, 2.678, the results of hypotheses including path coefficient and significant level are shown in Appendix A.

4.2.3. Assessing Effect Size, Predictive Power and Coefficient of Determination

The effect size of the variables is examined through f^2 analysis [50,51]. According to Samar Rahi et al. [58] the coefficient of determination, R^2, reveals the collective impact of all the exogenous variables towards the criterion variable. Nevertheless, variable size at a single factor should be assessed with effect size analysis. The results demonstrate that in measuring the supply chain sustainable performance, the effect size of supply chain resilience is substantial and, therefore, a potential construct for managerial implication.

Concerning the coefficient of determination, the results of the structural model revealed that exogenous variables have a substantial variance R^2 of 0.548% on supply chain resilience. Similarly, the findings indicate a substantial variance R^2 of 0.550% in measuring sustainable supply chain performance, which was predicted by supply chain resilience and big data analytics. Aside from the substantial coefficient of determination for supply chain resilience and sustainable supply chain performance, the predictive power Q^2 of the framework was computed. The result of the blindfolding method has established that the research model has substantial power to predict supply chain resilience and sustainable supply chain performance. The results of the effect size, predictive power and coefficient of determination are shown in Table 7.

Table 7. Measuring R^2, f^2 and Q^2.

Supply Chain Resilience				
Constructs	R^2	Q^2	f^2	Findings
Supply Chain Resilience	54.8%	37.4%		
Supply chain communication			0.036	Small
Supply chain intelligence			0.019	Small
Leadership commitment			0.018	Small
Network complexity			0.001	Small
Risk management orientation			0.045	Small
Supply chain capability			0.014	Small
Sustainable Supply Chain Performance				
Constructs	R^2	Q^2	f^2	Findings
Sustainable Supply Chain Performance	55.0%	40.0%		
Big data analytics			0.030	Small
Supply chain resilience			0.826	Substantial

4.2.4. Importance and Performance Analysis

The current research model has integrated numerous factors to determine sustainable supply chain performance. Therefore, the importance and performance of the variable are tested with importance performance matrix analysis (IPMA) consistent with earlier studies by [42,59]. Before applying IPMA analysis, it was important to choose an outcome variable. In this study, the researcher selected supply chain sustainable performance as the outcome variable. The findings of the importance performance analysis indicate that supply chain resilience has the highest importance/total effect in measuring sustainable supply chain performance. Therefore, the importance of big data analytics, risk management orientation, supply chain communication and leadership commitment show an intermediate level of importance. The results of the IPMA analysis are exhibited in Table 8 with the importance and performance indexes.

The importance of the constructs is observed through an IPMA map. The IPMA map exhibits that supply chain capability, intelligence and network complexity have less importance in measuring sustainable supply chain performance. Therefore, the importance of big data analytics, risk management orientation, supply chain communication, leadership commitment and supply chain resilience is considerable. Thus, managers and policy makers should focus on factors such as the importance of big data analytics, risk management orientation, supply chain communication, leadership commitment and supply chain resilience to enhance the sustainable supply chain performance in their organization. The IPMA map exhibited in Figure 2 clearly shows the performance of the constructs on the Y-axis and importance on the X-axis.

Table 8. Results of IPMA.

Constructs	Total Effects of Constructs	Total Performance of the Constructs
Big data analytics	0.157	7.890
Supply chain communication	0.128	73.379
Supply chain intelligence	0.095	69.381
Leadership commitment	0.167	67.870
Network complexity	0.023	72.166
Risk management orientation	0.261	68.990
Supply chain capability	0.086	71.953
Supply chain resilience	0.690	66.904

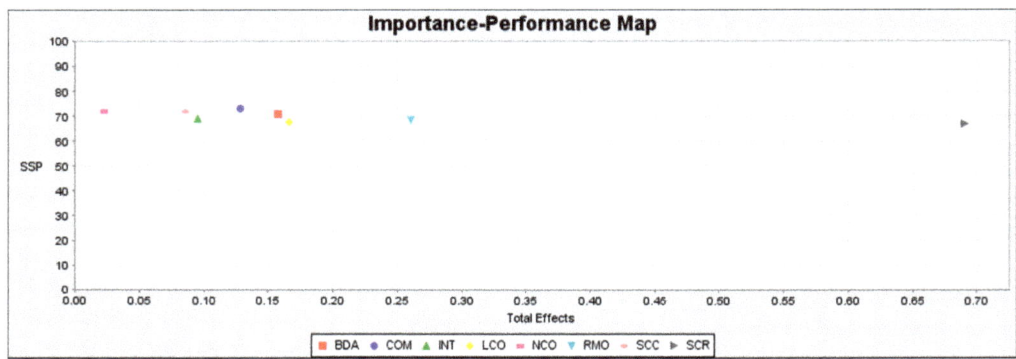

Figure 2. IPMA analysis map.

4.3. Moderating Effect of Big Data Analytics

The impact of technology is substantial in achieving the strategic goals of a firm [60]. In line with the above argument, this study has outlined big data analytics in a research model as moderating the variables between supply chain resilience and sustainable supply chain performance. The moderating effect of big data analytics was tested through a product indicator approach consistent with earlier studies by [53]. For statistical computation, the orthogonalization method was selected. The findings indicate that BD analytics moderates the relationship outlined between supply chain resilience and sustainable supply chain performance and statistically confirms H8 ($\beta = 0.138$; significant at $p < 0.001$; t-statistics 4.094). Figure 3 exhibits the findings of the moderating effect.

The moderating effect was further analyzed with simple slope analysis to test the strength of the moderating effect. According to S. Rahi [61] simple slope analysis reveals the trend of the moderating effect i.e., whether it moderates strongly or weakly. Nevertheless, the result of the simple slope analysis shows an upward BDA trend of +1SD on the green line compared to the negative BDA at −SD on the red line, which means that simple slope analysis has a declining trend. Nonetheless, the blue line indicates a neutral impact between the highest and lowest moderating effect. Hence, the result of the moderating effect establishes that the higher use of big data analytics in a supply chain operation will increase the supply chain resilience and sustainable supply chain performance. The simple slope analysis graph is displayed in Figure 4.

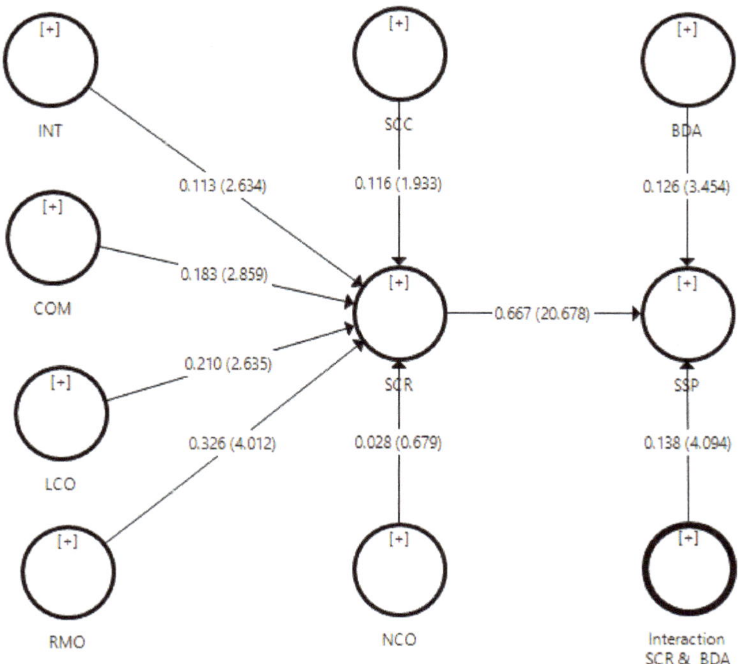

Figure 3. Moderating analysis.

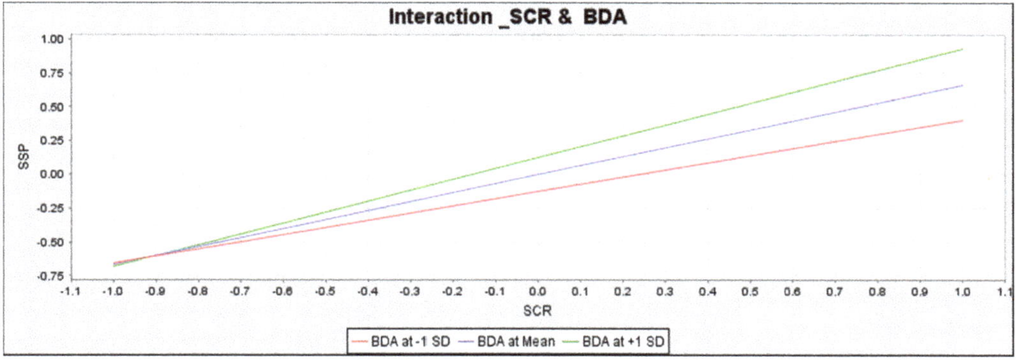

Figure 4. Output of simple slope analysis.

5. Discussion

The results of this research unfold interesting facts for both academic researchers and practitioners. The current research has synthesized the literature into two main streams. At the first stage, supply chain resilience is determined by supply chain intelligence, supply chain communication, leadership commitment, risk management orientation, supply chain capability, network complexity and the substantial variance R^2 of 54.8%. Therefore, the second stream of the literature focused on sustainable supply chain performance with supply chain resilience and big data analytics and revealed a considerable variance R^2 of 55.0% in sustainable supply chain performance. These findings established the theoretical validity of the research model in determining supply chain resilience and sustainability supply chain performance. Similarly, the literature has confirmed the moderating role of big data analytics between the relationship of supply chain resilience and sustainable supply chain performance. The moderating effect of big data analytics indicates that the higher use of BD analytics in supply chain operations will raise supply chain resilience and sustainable supply chain performance.

Concerning the hypothesized relationships, the findings of the structural model indicate that supply chain intelligence positively influences supply chain resilience, which is consistent with earlier studies by [11,18]. Similarly, supply chain communication has shown a positive impact on supply chain resilience, which is in line with [12,22,38]. Leadership commitment had revealed a positive impact on measuring supply chain resilience, which supports prior studies by [13,15]. Pointing to risk management orientation and supply chain capability, both factors have shown a positive impact on supply chain resilience which was supported by earlier studies by [28–30]. Nevertheless, network complexity has shown an insignificant influence supply chain resilience that was beyond our expectation. Therefore, supply chain resilience has shown a positive impact toward sustainable supply chain performance, which is consistent with an earlier study by [17]. Following IPMA and effect size analysis, this research suggests that during the COVID-19 pandemic, supply chain resilience and sustainable supply chain process could be enhanced by focusing on factors such as big data analytics, risk management orientation, supply chain communication and leadership commitment.

Research Contribution to Theory and Practice

The findings of this research have contributed to theory and practice. For instance, the current research shows that factors such as supply chain intelligence, communication, leadership commitment, risk orientation, supply chain capability and network complexity positively relate to supply chain resilience. These findings clearly indicate that policy makers and supply chain managers should focus on the outlined factors to enhance supply chain resilience in supply chain operations. Another aspect of this research is to shed light on sustainable supply chain performance. The results of this study reveal that a sustainable supply chain is determined by supply chain resilience and big data analytics. Therefore, supply chain managers should improve supply chain resilience in order to improve sustainable supply chain performance. Aside from industry utilization, this research contributes to the academic literature by developing an integrated supply chain model that comprises market-oriented factors to determine supply chain resilience and sustainable supply chain performance. In addition to that, the variance explained by exogenous variables was substantial for supply chain resilience and sustainable supply chain performance, which in turn, confirms the validity of the research model. Furthermore, this study contributes to information system literature by testing the moderating effect of big data analytics between supply chain resilience and sustainable supply chain performance. The findings establish that a higher use of big data analytics in supply chain operations will increase supply chain resilience and sustainable supply chain performance. Therefore, managers and policy makers should introduce big data analytics tools in supply chain operations to boost supply chain resilience and sustainable supply chain performance.

6. Conclusions

The aim of this study was to investigate factors that influence supply chain resilience and sustainable supply chain performance during the COVID-19 pandemic. The current study developed an integrated research model that combined factors such as supply chain intelligence, supply chain communication, leadership commitment, risk management orientation, supply chain capability and network complexity to investigate supply chain resilience. On the other side, the research model was extended with the moderating effect of big data analytics between the relationship of supply chain resilience and sustainable supply chain performance. The research model of this study was tested using structural equation modeling. The results of the structural model computation revealed that supply chain intelligence, supply chain communication, leadership commitment, risk management orientation, supply chain capability and network complexity have shown a substantial variance R^2 of 0.548% toward supply chain resilience. Therefore, supply chain resilience and big data analytics have an explanation to the variance R^2 of 0.550% in measuring sustainable supply chain performance. In addition to that, the validity of the research model was tested with predictive analysis Q^2 using a blind folding procedure. The results of the predictive analysis Q^2 revealed that the research model has large predictive power to predict supply chain resilience and sustainable supply chain performance 37.4% and 40.0%, respectively. This research contributes to the theory by developing an integrated research model toward supply chain resilience and sustainable supply chain performance. Therefore, for practical implications, this study suggests that supply chain managers should focus on factors such as big data analytics, risk management orientation, supply chain communication and leadership commitment to enhance supply chain resilience and sustainable supply chain performance. In addition to that, the moderating effect of big data analytics is confirmed between the relationship of supply chain resilience and sustainable supply chain performance. These findings established that the use of big data analytics in a supply chain operation will increase supply chain resilience and sustainable supply chain performance. Thus, supply chain managers and policy makers should incorporate big data analytics in supply chain operations to enhance supply chain resilience and sustainable supply chain performance.

Research Limitations and Future Direction

This research has some limitations that reveal a potential area of future research. First, this study was developed as an integrated supply chain research model that combined factors such as supply chain intelligence, supply chain communication, leadership commitment, risk management orientation, supply chain capability, network complexity and big data analytics to investigate supply chain resilience and sustainable supply chain performance phenomenon. Nevertheless, there are some other factors that could impact supply chain resilience such as supplier relationship, uncertainty and inter departmental coordination. Thus, extending the current research model with some additional factors could reveal interesting findings. Another limitation of this research is related to research design. The research design was based on a cross-sectional design and, therefore, respondents were restrained to participate at once in the research survey. It is expected that results may differ in a longitudinal research design. Therefore, the supply chain resilience phenomenon should be investigated with a longitudinal research design. The data were collected through a convenience sampling approach, which is a non-probability sampling approach. It is suggested that future researchers collect data through a probability sampling approach to mitigate any kind of sampling risk. The research model was designed from a developing countries perspective and tested the resilience behavior of a Saudi manufacturer. However, a future researcher may test the current research model in the context of other developing countries to enhance the generalizability of the research model.

Funding: This work is funded by the University of Jeddah, Jeddah, Saudi Arabia, under grant No. (UJ-21-DR-46). The authors, therefore, acknowledge with thanks the University of Jeddah technical and financial support.

Institutional Review Board Statement: Not Applicable.

Informed Consent Statement: Not Applicable.

Data Availability Statement: Delivered on demand.

Acknowledgments: Author would like to thank anonymous reviewers for the critical and constructive suggestions which in turn give an opportunity to rethink and reconstruct manuscript. In addition to that, author is grateful to the editor-in-chief and assistant editor for their humbleness during the revision process.

Conflicts of Interest: The author declares no conflict of interest. The authors declare that they have no known competing financial interests or personal relationships that could have appeared to influence the work reported in this paper.

Appendix A. Path Coefficient and Significance Level

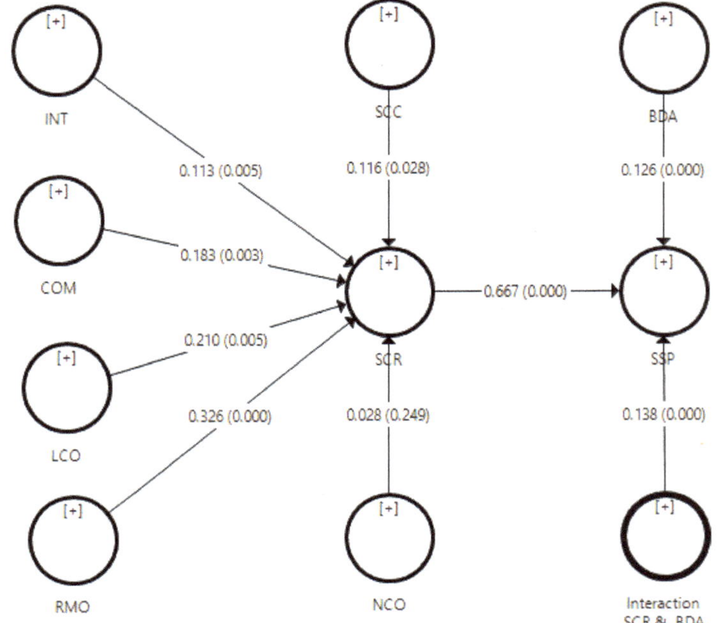

References

1. Brandon-Jones, E.; Squire, B.; Autry, C.W.; Petersen, K. A Contingent Resource-Based Perspective of Supply Chain Resilience and Robustness. *J. Supply Chain Manag.* **2014**, *50*, 55–73. [CrossRef]
2. Bhamra, R.S.; Dani, S.; Burnard, K.J. Resilience: The concept, a literature review and future directions. *Int. J. Prod. Res.* **2011**, *49*, 5375–5393. [CrossRef]
3. Bode, C.; Wagner, S.; Petersen, K.; Ellram, L.M. Understanding Responses to Supply Chain Disruptions: Insights from Information Processing and Resource Dependence Perspectives. *Acad. Manag. J.* **2011**, *54*, 833–856. [CrossRef]
4. Blackhurst, J.; Dunn, K.S.; Craighead, C.W. An Empirically Derived Framework of Global Supply Resiliency. *J. Bus. Logist.* **2011**, *32*, 374–391. [CrossRef]
5. Karmaker, C.L.; Ahmed, T.; Ahmed, S.; Ali, S.M.; Moktadir, M.A.; Kabir, G. Improving supply chain sustainability in the context of COVID-19 pandemic in an emerging economy: Exploring drivers using an integrated model. *Sustain. Prod. Consum.* **2021**, *26*, 411–427. [CrossRef]
6. Belhadi, A.; Kamble, S.; Jabbour, C.J.C.; Gunasekaran, A.; Ndubisi, N.O.; Venkatesh, M. Manufacturing and service supply chain resilience to the COVID-19 outbreak: Lessons learned from the automobile and airline industries. *Technol. Forecast. Soc. Chang.* **2021**, *163*, 120447. [CrossRef] [PubMed]
7. Kaynak, H. The relationship between total quality management practices and their effects on firm performance. *J. Oper. Manag.* **2003**, *21*, 405–435. [CrossRef]
8. Butt, A.S. Strategies to mitigate the impact of COVID-19 on supply chain disruptions: A multiple case analysis of buyers and distributors. *Int. J. Logist. Manag.* **2021**, ahead-of-print. [CrossRef]
9. Jabbour, A.B.L.D.S.; Jabbour, C.J.C.; Hingley, M.; Vilalta-Perdomo, E.; Ramsden, G.; Twigg, D. Sustainability of supply chains in the wake of the coronavirus (COVID-19/SARS-CoV-2) pandemic: Lessons and trends. *Mod. Supply Chain Res. Appl.* **2020**, *2*, 117–122. [CrossRef]
10. Paul, S.K.; Chowdhury, P.; Moktadir, A.; Lau, K.H. Supply chain recovery challenges in the wake of COVID-19 pandemic. *J. Bus. Res.* **2021**, *136*, 316–329. [CrossRef]
11. Yang, J.; Xie, H.; Yu, G.; Liu, M. Achieving a just–in–time supply chain: The role of supply chain intelligence. *Int. J. Prod. Econ.* **2020**, *231*, 107878. [CrossRef]

2. Barki, H.; Pinsonneault, A. A Model of Organizational Integration, Implementation Effort, and Performance. *Organ. Sci.* **2005**, *16*, 165–179. [CrossRef]
3. Speier, C.; Whipple, J.M.; Closs, D.J.; Voss, M.D. Global supply chain design considerations: Mitigating product safety and security risks. *J. Oper. Manag.* **2011**, *29*, 721–736. [CrossRef]
4. Wieland, A.; Wallenburg, C.M. Dealing with supply chain risks. *Int. J. Phys. Distrib. Logist. Manag.* **2012**, *42*, 887–905. [CrossRef]
5. Durach, C.F.; Wieland, A.; Machuca, J.A. Antecedents and dimensions of supply chain robustness: A systematic literature review. *Int. J. Phys. Distrib. Logist. Manag.* **2015**, *45*, 118–137. [CrossRef]
6. Wu, F.; Yeniyurt, S.; Kim, D.; Cavusgil, S.T. The impact of information technology on supply chain capabilities and firm performance: A resource-based view. *Ind. Mark. Manag.* **2006**, *35*, 493–504. [CrossRef]
7. Dubey, R.; Gunasekaran, A.; Childe, S.J.; Wamba, S.F.; Roubaud, D.; Foropon, C. Empirical investigation of data analytics capability and organizational flexibility as complements to supply chain resilience. *Int. J. Prod. Res.* **2021**, *59*, 110–128. [CrossRef]
8. Schoenherr, T.; Swink, M. The Roles of Supply Chain Intelligence and Adaptability in New Product Launch Success. *Decis. Sci.* **2015**, *46*, 901–936. [CrossRef]
9. Handfield, R. *Supply Market Intelligence: A Managerial Handbook for Building Sourcing Strategies*, 1st ed.; CRC Press: New York, NY, USA, 2006.
10. Yamin, M.A.Y.; Mahasneh, M.S. The Impact of Values-based Leadership on Ethical Loyalty in Saudi Arabian Health Organizations. *Int. Rev. Manag. Mark.* **2018**, *8*, 6–13.
11. Zhang, Q.; Cao, M. Exploring antecedents of supply chain collaboration: Effects of culture and interorganizational system appropriation. *Int. J. Prod. Econ.* **2018**, *195*, 146–157. [CrossRef]
12. Asamoah, D.; Agyei-Owusu, B.; Andoh-Baidoo, F.; Ayaburi, E. Inter-organizational systems use and supply chain performance: Mediating role of supply chain management capabilities. *Int. J. Inf. Manag.* **2021**, *58*, 102195. [CrossRef]
13. Benedek, Z.; Fertő, I.; Marreiros, C.G.; de Aguiar, P.M.; Pocol, C.B.; Čechura, L.; Põder, A.; Pääso, P.; Bakucs, Z. Farm diversification as a potential success factor for small-scale farmers constrained by COVID-related lockdown. Contributions from a survey conducted in four European countries during the first wave of COVID-19. *PLoS ONE* **2021**, *16*, e0251715. [CrossRef]
14. Fornell, C.; Larcker, D.F. Structural Equation Models with Unobservable Variables and Measurement Error: Algebra and Statistics. *J. Mark. Res.* **1981**, *18*, 382–388. [CrossRef]
15. Yamin, M.A.Y. Examining the effect of organisational innovation on employee creativity and firm performance: Moderating role of knowledge sharing between employee creativity and employee performance. *Int. J. Bus. Innov. Res.* **2020**, *22*, 447–467. [CrossRef]
16. Grötsch, V.M.; Blome, C.; Schleper, M.C. Antecedents of proactive supply chain risk management—A contingency theory perspective. *Int. J. Prod. Res.* **2013**, *51*, 2842–2867. [CrossRef]
17. Hall, D.J.; Skipper, J.B.; Hazen, B.; Hanna, J.B. Inter-organizational IT use, cooperative attitude, and inter-organizational collaboration as antecedents to contingency planning effectiveness. *Int. J. Logist. Manag.* **2012**, *23*, 50–76. [CrossRef]
18. Jüttner, U.; Maklan, S. Supply chain resilience in the global financial crisis: An empirical study. *Supply Chain Manag. Int. J.* **2011**, *16*, 246–259. [CrossRef]
19. Lin, C.-C.; Wang, T.-H. Build-to-order supply chain network design under supply and demand uncertainties. *Transp. Res. Part B Methodol.* **2011**, *45*, 1162–1176. [CrossRef]
20. Schmitt, A.J. Strategies for customer service level protection under multi-echelon supply chain disruption risk. *Transp. Res. Part B Methodol.* **2011**, *45*, 1266–1283. [CrossRef]
21. Brusset, X.; Teller, C. Supply chain capabilities, risks, and resilience. *Int. J. Prod. Econ.* **2017**, *184*, 59–68. [CrossRef]
22. Faisal, M.N.; Banwet, D.K.; Shankar, R. Information risks management in supply chains: An assessment and mitigation framework. *J. Enterp. Inf. Manag.* **2007**, *20*, 677–699. [CrossRef]
23. Akter, S.; Wamba, S.F.; Gunasekaran, A.; Dubey, R.; Childe, S.J. How to improve firm performance using big data analytics capability and business strategy alignment? *Int. J. Prod. Econ.* **2016**, *182*, 113–131. [CrossRef]
24. Janssen, M.; van der Voort, H.; Wahyudi, A. Factors influencing big data decision-making quality. *J. Bus. Res.* **2017**, *70*, 338–345. [CrossRef]
25. Lai, Y.; Sun, H.; Ren, J. Understanding the determinants of big data analytics (BDA) adoption in logistics and supply chain management. *Int. J. Logist. Manag.* **2018**, *29*, 676–703. [CrossRef]
26. Snapp, S. Is Big Data the Silver Bullet for Supply-Chain Forecasting? In *Business Forecasting: The Emerging Role of Artificial Intelligence and Machine Learning*; Gilliland, M., Tashman, L., Sglavo, U., Eds.; Wiley: Hoboken, NJ, USA, 2021; pp. 136–141.
27. Wamba, S.F.; Akter, S.; Edwards, A.; Chopin, G.; Gnanzou, D. How 'big data' can make big impact: Findings from a systematic review and a longitudinal case study. *Int. J. Prod. Econ.* **2015**, *165*, 234–246. [CrossRef]
28. Yu, W.; Zhao, G.; Liu, Q.; Song, Y. Role of big data analytics capability in developing integrated hospital supply chains and operational flexibility: An organizational information processing theory perspective. *Technol. Forecast. Soc. Chang.* **2021**, *163*, 120417. [CrossRef]
29. Rowley, J. Designing and using research questionnaires. *Manag. Res. Rev.* **2014**, *37*, 308–330. [CrossRef]
30. Gunasekaran, A.; Papadopoulos, T.; Dubey, R.; Wamba, S.F.; Childe, S.J.; Hazen, B.; Akter, S. Big data and predictive analytics for supply chain and organizational performance. *J. Bus. Res.* **2017**, *70*, 308–317. [CrossRef]

41. Bag, S.; Wood, L.C.; Xu, L.; Dhamija, P.; Kayikci, Y. Big data analytics as an operational excellence approach to enhance sustainable supply chain performance. *Resour. Conserv. Recycl.* **2020**, *153*, 104559. [CrossRef]
42. Hair, J.F., Jr.; Hult, G.T.M.; Ringle, C.; Sarstedt, M. *A Primer on Partial Least Squares Structural Equation Modeling (PLS-SEM)*, 2nd ed.; Sage Publications: Thousand Oaks, CA, USA, 2016.
43. Yamin, M. Examining the role of transformational leadership and entrepreneurial orientation on employee retention with moderating role of competitive advantage. *Manag. Sci. Lett.* **2020**, *10*, 313–326. [CrossRef]
44. Rahi, S.; Mansour, M.M.O.; Alharafsheh, M.; Alghizzawi, M. The post-adoption behavior of internet banking users through the eyes of self-determination theory and expectation confirmation model. *J. Enterp. Inf. Manag.* **2021**, ahead-of-print. [CrossRef]
45. Podsakoff, P.M.; Organ, D.W. Self-Reports in Organizational Research: Problems and Prospects. *J. Manag.* **1986**, *12*, 531–544. [CrossRef]
46. Sweiss, M.I.K.; Yamin, M.A.Y. The influence of organisational and individual factors on organisational innovation with moderating role of innovation orientation. *Int. J. Bus. Innov. Res.* **2020**, *23*, 103. [CrossRef]
47. Podsakoff, P.M.; Bommer, W.H.; Podsakoff, N.P.; MacKenzie, S.B. Relationships between leader reward and punishment behavior and subordinate attitudes, perceptions, and behaviors: A meta-analytic review of existing and new research. *Organ. Behav. Hum. Decis. Process.* **2006**, *99*, 113–142. [CrossRef]
48. Podsakoff, P.M.; MacKenzie, S.B.; Lee, J.-Y.; Podsakoff, N.P. Common method biases in behavioral research: A critical review of the literature and recommended remedies. *J. Appl. Psychol.* **2003**, *88*, 879–903. [CrossRef] [PubMed]
49. Rahi, S.; Khan, M.M.; Alghizzawi, M. Extension of technology continuance theory (TCT) with task technology fit (TTF) in the context of Internet banking user continuance intention. *Int. J. Qual. Reliab. Manag.* **2020**, *38*, 986–1004. [CrossRef]
50. Rahi, S.; Mansour, M.M.O.; Alghizzawi, M.; Alnaser, F.M. Integration of UTAUT model in internet banking adoption context. *J. Res. Interact. Mark.* **2019**, *13*, 411–435. [CrossRef]
51. Yamin, M.A.Y. The relationship between right ethical behavior perspective, demographic factors, and best ethical performance. *Int. Rev. Manag. Mark.* **2020**, *10*, 27–39. [CrossRef]
52. Ringle, C.M.; Wende, S.; Becker, J.-M. SmartPLS 3. Boenningstedt: SmartPLS. Available online: https://www.smartpls.com (accessed on 22 May 2021).
53. Yamin, M.A.Y.; Sweiss, M.I.K. Investigating employee creative performance with integration of DeLone and McLean information system success model and technology acceptance model: The moderating role of creative self-efficacy. *Int. J. Bus. Excel.* **2020**, *2*, 396. [CrossRef]
54. Yamin, M.A.Y.; Alyoubi, B.A. Adoption of telemedicine applications among Saudi citizens during COVID-19 pandemic: An alternative health delivery system. *J. Infect. Public Health* **2020**, *13*, 1845–1855. [CrossRef] [PubMed]
55. Gold, A.H.; Malhotra, A.; Segars, A.H. Knowledge Management: An Organizational Capabilities Perspective. *J. Manag. Inf. Syst.* **2001**, *18*, 185–214. [CrossRef]
56. Kline, R.B. *Principles and Practice of Structural Equation Modeling*, 3rd ed.; The Guilford Press: New York, NY, USA, 2011.
57. Rahi, S.; Ghani, M.A.; Ngah, A.H. Factors propelling the adoption of internet banking: The role of e-customer service, website design, brand image and customer satisfaction. *Int. J. Bus. Inf. Syst.* **2020**, *33*, 549. [CrossRef]
58. Rahi, S.; Ghani, M.A. Integration of DeLone and McLean and self-determination theory in internet banking continuance intention context. *Int. J. Account. Inf. Manag.* **2019**, *27*, 512–528. [CrossRef]
59. Hair, J.F.; Ringle, C.M.; Sarstedt, M. Partial least squares structural equation modeling: Rigorous applications, better results and higher acceptance. *SSRN J.* **2013**, *46*, 1–12. [CrossRef]
60. Ranjan, J.; Foropon, C. Big Data Analytics in Building the Competitive Intelligence of Organizations. *Int. J. Inf. Manag.* **2021**, *56*, 102231. [CrossRef]
61. Rahi, S. *Structural Equation Modeling Using SmartPLS*, 1st ed.; CreateSpace Independent Publishing Platform: Scotts Valley, CA, USA, 2017.

Article

Evolution of the Global Scientific Research on the Environmental Impact of Food Production from 1970 to 2020

Alessio Cimini

Department for Innovation in the Biological, Agrofood and Forestry Systems, University of Tuscia, Via S.C. de Lellis, 01100 Viterbo, Italy; a.cimini@unitus.it

Abstract: Food production and consumption account for a significant share of the impact of various pressing and important environmental concerns such as climate change, eutrophication, and loss of biodiversity. In this work, a bibliometric analysis of the last 50 years of research papers, written in English and indexed on Scopus database, was carried out to highlight the evolution of the global scientific research in the environmental assessment of food production (EAFP). The research papers in EAFP started to significantly increase from 2005, being most frequently published by the *Journal of Cleaner Production* and *International Journal of Life Cycle Assessment*. The United States of America was the first publishing country, followed by China, the United Kingdom, and Italy. Wheat, rice, fish, maize, and milk were the food items mainly studied, with different importance depending on the authors' publishing country. *Life Cycle Analysis, Carbon Footprint*, and *Water Footprint* were the first three standard methods used to assess *climate change, energy consumption*, and *environmental impact*. The *Wageningen University, Chinese Academy of Sciences* and *Research Centre*, and *China Agricultural University* were the main publishing research centers. All the papers published worldwide received 18.1 citations per paper, the UK and Chinese papers being those mostly and minimally cited, respectively. Over the last five years, this research field largely aimed to managing the agricultural practices, mitigating global warming and water use, assuring food security and sustainable food consumption, while minimizing food waste formation. Such an objective evaluation of this research topic might help guide researchers on where to address their future research work.

Keywords: environmental analysis; food production; bibliometric study; citation number; authors; affiliations; Scopus

Citation: Cimini, A. Evolution of the Global Scientific Research on the Environmental Impact of Food Production from 1970 to 2020. *Sustainability* **2021**, *13*, 11633. https://doi.org/10.3390/su13 2111633

Academic Editors: Baojie He, Ayyoob Sharifi, Chi Feng, Jun Yang and Marc A. Rosen

Received: 30 August 2021
Accepted: 10 October 2021
Published: 21 October 2021

Publisher's Note: MDPI stays neutral with regard to jurisdictional claims in published maps and institutional affiliations.

1. Introduction

Earth's environment is affected by human activities. Since the late 20th century, numerous studies and published papers have highlighted how pollution, burning fossil fuels, deforestation and land exploitation may directly or indirectly cause damage to the environment [1,2]. Today, the increasing human overpopulation and its resulting adverse impacts affect the Earth's environment through ocean acidification, global warming, biodiversity loss, soil erosion, air pollution and undrinkable water. Every choice made can exert multifaceted environmental effects, either negative or positive, on four subsystems [3] like renewable and non-renewable materials; water consumed; land used for agricultural, forest, and grazing areas; raw material extraction or private housing [4]; and emissions such as greenhouse gases (GHG) and several other pollutants, namely SO_X, NO_X, and O_3.

Food production accounts for a significant share of the total impact of several important environmental categories, such as climate change, eutrophication and loss of biodiversity. Between 22% and 37% of global anthropogenic emissions may be attributed to the overall food system [5]. Agricultural production involves the manufacture of fertilizers, pesticides, equipment and energy, as well as land-use change, and is responsible for the great majority (72–82%) of the above GHG emissions. The post-production (2.4 Tg

CO$_{2e}$) and post-sale (1.0 Tg CO$_{2e}$) steps produce results of only a minor magnitude, while emissions embedded in food wasted at the consumption phase (~1.6 Tg CO$_{2e}$) are no way insignificant. About 63% of food-related GHG emissions derived from the production and consumption of animal-based products, except fish and fisheries, 8.5 ± 2.4 Tg CO$_{2e}$ in 2010 [5].

Food production systems are often complex and involve biological systems, which are difficult to control and measure. Some examples are the evaluation of *land occupation*, generally resulting in different crop outputs; *soil quality*, this depending on the water content, presence of organic matter, nutrients, heavy metals, living organisms and soil texture and structure [6]; *carbon storage in soils* and *standing biomass*; *yield variability*, between years owing to weather conditions and other factors; and *consumer behavior* towards food consumption, portion sizes, packaging, as well as wastage,. Given the many factors involved, it is quite a difficult task to assess the environmental impact of food products and production systems.

Life Cycle Assessment (LCA) dates to the 1960s, when the energy analyses of industrial systems started to be carried out to account for the oil crises of the early 1970s. The rise of environmental awareness in the late 1980s had the effect of increasing the attention paid to LCA as a potentially valuable environmental management tool. The International Standards Organization (ISO) released four ISO LCA standards (ISO 14040 to 14043) from 1997 to 2000. Following this, the LCA became a decision-support tool for several food companies. The Coca Cola Co. had started to account for the environmental impact of its packaging since 1969. Then, other companies (e.g., Tetrapak, Nestlé, Unilever, Arla Foods) and many beer companies engaged in assessing and improving their environmental sustainability [7,8].

Over the last decade there has been significant advances in terms of the environmental analysis of food products (EAFP). A bibliometric analysis of the research studies performed so far might be used as a basis for the comprehensive understanding of current research on the EAFP and, thus, highlight some potential future research directions.

Several bibliometric studies have been recently published to assess the scientific productivity related to climate change [9–11], food security [12], as well as food security in the context of climate change [13] or food waste [14]. Other scientometric reviews analyzed the effect of climate change on carbon sink [15], water quality [16], and human health [13]. However, no bibliometric study has so far attempted to analyze the scientific literature regarding the environmental assessment of food production and consumption.

Different multidisciplinary citation indices are available online, namely ISI Web of Science, Scopus, Google Scholar, these allowing the extraction of bibliometric indicators which can be classified as *quantity, quality,* and *structural indicators* [17]. *Primary bibliometric indicators* are elementary measures. They consist of the simple count of publications (*quantity indicators*) and/or citations produced/received (*performance indicator*) by a single author, research group, or journal in a certain time interval and represent the starting point of many bibliometric analyses, even if such measures are usually not very suitable for representing the complexity in the various contexts of application. With regards to the citation count, there is a general agreement about its capacity to represent the impact of an article in the scientific community, but the citation behavior might be influenced by its publication date and language, the author's reputation, membership of an important scientific Institution, as well as the journal characteristics, such as the degree of internationalization, accessibility, and so on. At the same time, the citation data extracted can be used to highlight any connectivity between scientific fields, research groups, as well as authors.

The combinations of the primary indicators with other data (i.e., time interval, average number, etc.) allow the definition of a set of *secondary bibliometric indicators*, such as the *Impact* Factor (IF), and *H-Index* [17].

The construction of the *landscape of science* [18] is now considered an established and prolific field of research and application of bibliometrics. Thanks to natural language processing techniques and a linguistic filter employed by the elaboration software, terms occurring in titles and abstracts of a set of identical elements extracted from publications presents in a database of selected literature are represented as circles in a two-dimensional map [19]. The representation of a research field as a *term map*, or *co-word map*, allows strongly related terms to be located close to each other, while the greater their distance the weaker their relationship will be.

The main aim of this work was to trace global research trends and scientific evolution in the EAFP research from 1970 to 2020, by resorting to a bibliometric, textual, and map analysis to provide a basis for the comprehensive understanding of current research and to highlight the countries, institutions, authors, and journals more productive and influent in this research field.

2. Materials and Methods

2.1. Bibliometric Analysis of Scientific Literature

Elsevier Scopus database (that is, one of the largest abstract and citation database of peer-reviewed literature) was consulted in March 2021 to retrieve bibliographic records related to the environmental impact and sustainability of food production from peer-reviewed articles that were published from 1970 to 2020 (included), this time period having been also subdivided into two time intervals (e.g., 1970–2015, and 2016–2020). This choice implied the exclusion of even important non-peer reviewed articles, some proceedings, communications, and patents, but they were out of the scope of this work. The outcome obtained from a database research analysis can be heavily conditioned by the query string. Indeed, the latter represents the most important tool to extract reliable results for a bibliometric analysis. The words present in the query string were selected as the most important related terms in the field of the environmental assessment of food production. To overcome the numerous limitations related to this kind of exercise, the database queries included two different categories of selected terms related to the topic, as detailed in the electronic supplement (Table S1). Namely, category 1 (C1) was composed of two thematic groups. The first one (A) was related to the *Standard Methods* usable for measuring the environmental impact of food production and consumption, and included 17 words; while the second one (B) included nine generic terms often used in research papers. Category 2 (C2) was composed of words identifying different *food products*. Owing to the numerous foods available, the FAO database (https://www.fao.org/faostat/en/#data, accessed on 18 October 2021) was consulted to identify the foods most relevant worldwide. To this end, all the words contained in the domain *Production* were extracted and included in a few categories, such as *Crop production* (Table S2), and *Crop processed, Live Animals, Livestock Primary,* and *Livestock Processed* (Table S3), by eliminating redundancies and similarity. Moreover, Table S4 shows both several *Generic* terms and other ones characterizing main foods and beverages as derived from the so-called FoodEx2 catalog, that is the *standardized* system recognized by the European Food Safety Agency (EFSA) (https://www.efsa.europa.eu/it/data/data-standardisation, accessed on 18 October 2021) for classifying and describing *foods* and *beverages*. Thus, a total number of 275 words were listed in Tables S2–S4.

The database queries consisted of a string obtained from all the possible binary combinations of all the terms included in the two categories C1 and C2, thus the 24 words in C1 were combined with each one of the 275 words in C2 using the 'AND' Scopus operator. In this exercise, only documents containing simultaneously the two terms in the *Title* were extracted. For an accurate description of the Scopus logical operators refer to the Scopus Search Guide (https://dev.elsevier.com/tips/ScopusSearchTips.htm, accessed on 18 October 2021). Obviously, it is almost impossible to produce a *perfect query string*, and background noise will be always produced. To improve the performance of this study, the Scopus search was performed by choosing the only *Title* (T) as research field (excluding Abstract and keywords) on the assumption that this was the most accurate strategy in relation to terms and research selection efficiency [20]. Bibliometric analysis is usually applied at three levels, the so-called macro level referring to national systems, the meso level to institutions, including individual Universities, and finally the micro on to research groups or individual researchers [21].

The resulting database including the title, citation count, year, author, affiliation, country, and abstract was saved in CSV format and elaborated in the following different ways.

1. *A simple data collection* to show quantitative and qualitative bibliometric analysis in tables and graphs reporting the most publishing countries, affiliations, authors, journals, and founding sponsors, as well as the citation number-per-paper index (CPPI).

2. *Abstract textual analysis* to highlight the most studied food and beverage items and used standard methods by resorting to a custom-made Python 3 script. This operation analyzed all the words present in the Abstracts extracted from the database, and gave rise to a txt file to obtain the frequency (occurrence) of the words listed in both categories C2 and C1, as well as other terms related to the main environmental *impact categories, production phases*, and *packaging materials* used (Table S5) to give a broader overview of the subject studied.

3. *Map analysis* by using the bibliometric mapping and clustering approach. Thus, world publication maps were plotted using color intensities proportional to the number of publications by means of the VOSviewer v. 1.6.5.0 software (freely available at www.vosviewer.com, accessed on 18 October 2021). This software was specifically developed for creating, visualizing, and exploring scientific bibliometric maps [22,23]. In such a visual map, strongly or weakly related terms are contiguous or distant from each other, respectively. Only terms occurring at least 50 times were extracted from the publications retrieved. The next step was to identify clusters of related terms by means of a software applying the clustering technique [19]. The assignment of terms to the same cluster depended on their co-occurrences in the title and abstract of the publications retrieved, terms often co-occurring were strongly related to each other, and were automatically assigned to the same cluster. On the contrary, terms with a low co-occurrence, or no-occurrences at all, were assigned to different clusters. A cluster made up of terms characterized by the same color represented a research theme in which one or more research topics were identified. A thesaurus file was also used to ensure consistency for different term spelling, or synonyms. For instance, the expression *wheat productivity* or *wheat production* was termed *wheat yield*, while terms considered not relevant to the search (i.e., names of cities or countries) were omitted.

The search was restricted to publications written in English because it was almost impossible to translate all the scientific terms and keywords in English for the elaboration and analysis. Thus, many authors that wrote papers in their native language were excluded. Even if the database obtained did not include all the papers published in the EAFP field, the data collected allowed a general picture of the world scientific research in EAFP topic, and were not intended to draw up a ranking among countries, affiliations, or authors. In general, it was possible to assume a 10–15% underestimation of the data retrieved, as approximately evaluated by counting manually the number of articles published by several authors and automatically those extracted from the database using the search keys. Figure 1 shows a schematic diagram to highlight the bibliometric procedure used in this work.

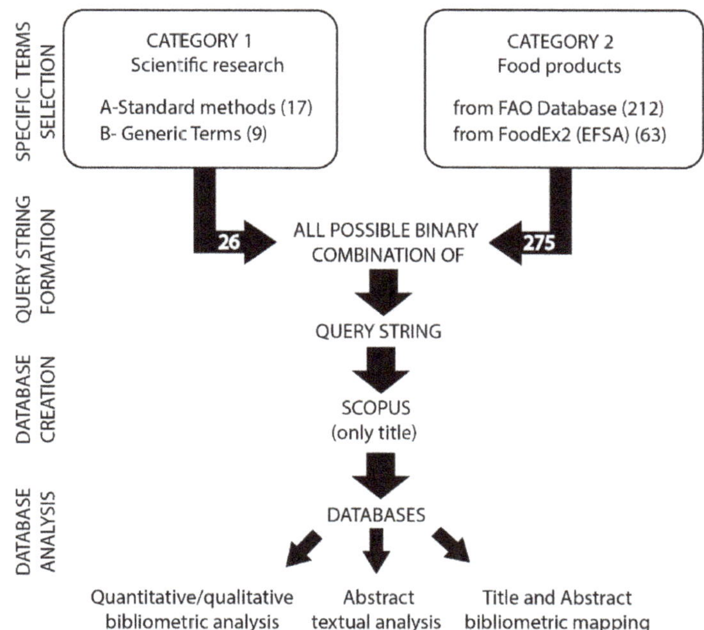

Figure 1. A schematic diagram to highlight the bibliometric procedure used in this work.

2.2. Time Horizon

Since the Scopus search had been conducted in March 2021, publications relative to the year 2020 were included in the analysis. Thus, the publications relative to the year 2021 were underestimated, their overall number being not yet completely indexed by the Scopus database. For the same reason, the citation count regarding the papers published in the year 2021 was excluded.

The selected bibliography was used to extract information regarding the number of publications, countries, main affiliations, authors, citations, and journals. The resulting information was also segmented for the five countries with greater publishing rate in this research topic. The Scimago database (https://www.scimagojr.com/, accessed on 18 October 2021) was used to extract journal info.

3. Results and Discussion

3.1. Bibliometric Analysis

3.1.1. Publication Trends from 1970 to 2020

An overall number of 4186 scientific research works resulted from the research performed here, their annual distribution being shown in Figure 2.

Such a trend was in line with that resulting from a research article that analyzed food security in the context of climate change from 1980 to 2019 [13] and retrieved as many as 5960 documents. The first paper appeared in the years 1968–1970 and until 1990 the overall number of papers published remained almost insignificant; then, two increasing trends were clearly identified. The first one covered the latest decade of the 20th century. Such a period represented a turning point of awareness of the scientific community and public opinion. First, the Antarctic ozone hole was discovered in the 1985. Within two years the United States and more than 100 other countries pledged to phase out the use of ozone-depleting compounds. In 1988, the Intergovernmental Panel on Climate Change (IPCC) was established by the World Meteorological Organization (WMO) and the United Nations Environment Program (UNEP) in order to provide the world with a clear and scientifically based view about the current state of knowledge on climate change and its potential environmental and socio-economic impacts. According to [24], the first Life Cycle Assessment (LCA) study on food products was performed at the beginning of the 1990s and in 1996 was hold the first *LCA food* conference dealing with the environmental impacts analysis of the agri-food sector [25]. A second trend extending from 2000 to 2020 exhibited a definitively steeper growth rate of about 800 papers per year consequently not only to the recent attention to the environmental consequences of food production and consumption, but also to the increasing number of affiliations/authors publishing in this research field.

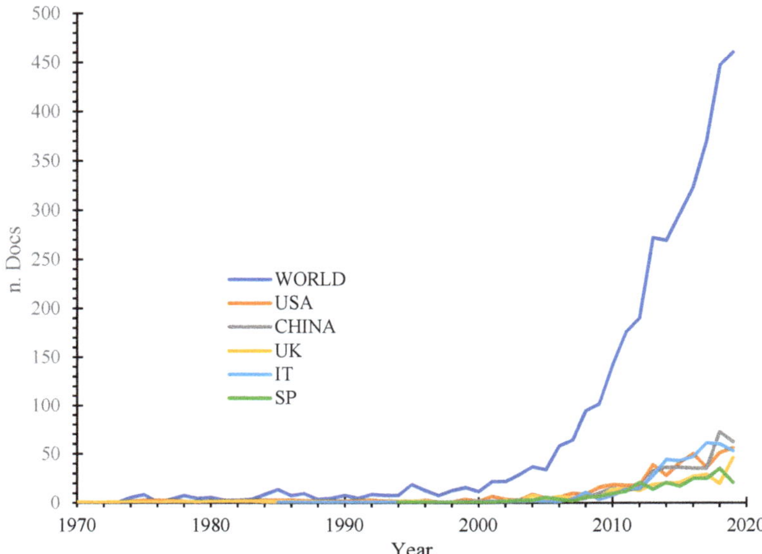

Figure 2. Annual distribution of the overall number of scientific research papers retrieved from the Scopus database from 1970 to 2020.

Table 1 lists the first 20 countries most publishing in this topic, as well as the total number of papers published together with their relative and cumulative percentages. Even if any document might be attributed to multiple countries, the total number of papers retrieved was higher than 4186 (Table 2)

The first 10 nations covered about 53% of the total documents produced, seven of these being European countries. The USA, China, and Italy were the three most productive countries with 548, 465, and 464 documents, respectively, even if the Climate Change Performance Index (CCPI) ranked their efforts to combat climate change at the 61st, 33rd and 27th position, respectively [26]. In Europe, Italy (464), the UK (328), and Spain (283) were the first three publishing nations with their CCPI at the 27th, 5th, and 4ht position respectively [26]. Surprisingly, they produced more scientific publications than The Netherlands (170), Denmark (141), and Sweden (161), but, in all, it is likely this achievement is linked to the overall number of inhabitants. Even in this case, in 2021 the CCPI of such countries inversely ranked at the 29th, 6th, and 4th position, respectively [27]. In 2019 the world's largest CO_2 emitters were in descending order China (30.4% of total CO_2 emissions), the United States (13.43%), India (6.83%), the EU27 + UK (8.69%), Russia (4.71%) and Japan (3.03%), these together accounting for 67% of total global fossil CO_2 emitted [27]. As shown in Table 1, China, the USA, and EU28 carried out an intense research activity in this sector (as related to food), showing a certain attention to the problem, differently from Japan and Russia, that ranked at the 17th place and by far lower than the 20th place respectively. Over the last five years, China, Italy, and USA were the main publishing countries. From 2016 to 2020 the number of publications was higher than that produced from 1970 to 2015 for most of the countries listed in Table 1. A different behavior was noted for Denmark, Sweden, the Netherlands, Australia, and Canada, probably because their research interest switched to other topics.

Table 1. Overall number of publications by country, as well as their relative and cumulative frequencies, as indexed by the Scopus database in different time periods (i.e., 2020–1970; 2020–2016; 2015–1970).

No.	2020–1970				2020–2016				2015–1970			
	Country	No. Docs	Relative [%]	Cumulative [%]	Country	No. Docs	Relative [%]	Cumulative [%]	Country	No. Docs	Relative [%]	Cumulative [%]
1	USA	547	9.7	9.7	China	287	9.3	9.3	USA	273	10.7	10.7
2	China	465	8.2	17.9	Italy	283	9.1	18.4	Italy	181	7.1	17.8
3	Italy	464	8.2	26.1	USA	274	8.8	27.2	China	178	7.0	24.8
4	UK	328	5.8	31.9	UK	169	5.5	32.7	UK	159	6.2	31.0
5	Spain	283	5.0	36.9	Spain	154	5.0	37.7	Spain	129	5.1	36.1
6	Germany	194	3.4	40.4	Germany	106	3.4	41.1	France	97	3.8	39.9
7	France	189	3.3	43.7	Brazil	97	3.1	44.2	Sweden	96	3.8	43.6
8	Australia	173	3.1	46.8	India	97	3.1	47.4	Netherlands	95	3.7	47.3
9	Netherlands	170	3.0	49.8	France	92	3.0	50.3	Australia	94	3.7	51.0
10	Sweden	161	2.9	52.6	Iran	80	2.6	52.9	Germany	88	3.4	54.5
11	India	160	2.8	55.5	Australia	79	2.6	55.5	Canada	79	3.1	57.6
12	Brazil	157	2.8	58.3	Netherlands	75	2.4	57.9	Denmark	74	2.9	60.5
13	Denmark	141	2.5	60.8	Malaysia	71	2.3	60.2	India	63	2.5	62.9
14	Iran	140	2.5	63.2	Thailand	68	2.2	62.4	Brazil	60	2.4	65.3
15	Canada	138	2.4	65.7	Denmark	67	2.2	64.5	Iran	60	2.4	67.6
16	Malaysia	114	2.0	67.7	Sweden	65	2.1	66.6	Belgium	47	1.8	69.5
17	Thailand	107	1.9	69.6	Indonesia	63	2.0	68.7	Japan	47	1.8	71.3
18	Belgium	86	1.5	71.1	Canada	59	1.9	70.6	Malaysia	43	1.7	73.0
19	Switzerland	86	1.5	72.6	Switzerland	49	1.6	72.1	Thailand	39	1.5	74.5
20	Japan	85	1.5	74.1	Turkey	45	1.5	73.6	Switzerland	37	1.4	76.0

Table 2. Top 5 nations ranked by the number of documents published over the 1970–2020 period together with the overall number of citations and average citations per paper index (CPPI).

Time Period	Country	No. Doc.s	No. Citations	CPPI
1972–2020	WORLD	4186	84917	20.3
1973–2020	USA	548	13604	24.8
2002–2020	China	465	6782	14.6
1970–2020	Italy	464	9749	21.0
1985–2020	UK	328	9699	29.6
1994–2020	Spain	283	6673	23.6

From the bibliometric analysis carried out here, it was possible to extract an overview of the number of citations per each article. This assessment was just restricted to the year 2020, because many of the articles published in the 2021 have not been cited yet. Table 2 show the first five countries ranked by the number of documents published during the time periods accounted for, together with total citation count and citation per paper index (CPPI). All the scientific papers retrieved from Scopus database received an average number of 20.3 citations per paper. Interestingly, Table 2 displays that CPPI ranged from 16.6 to 29.6 for Chinese and UK papers, respectively. For some countries (i.e., the UK, USA, Spain, and Italy) CPPI was greater than the world average value, whereas for the second publishing country (i.e., China), such an index was smaller than 20.3. Thus, the CPPI might be regarded as an index measuring the scientific relevance of a paper that should no way be confused with its scientific quality.

3.1.2. Affiliations

The affiliation search returned a list of the institutions involved in the environmental analysis of food production. Table 3 shows the number of papers published by these affiliations from 1970 to 2020.

The *Wageningen University & Research* (NL) resulted to be the first affiliation in terms of number of publications, followed by the *Chinese Academy of Sciences*, which includes other affiliations separately named in Scopus, such as the *Institute of Geographical Sciences* and *Natural Resources Research Chinese Academy of Sciences*, *University of Chinese*

Academy of Sciences. Moreover, other Chinese Institutions ranked at the 3rd, 12th, and 16th places, whereas the *Aarhus Universitet* (DK) together with other five European affiliations is in the Top 10 world ones. The first 20 affiliations altogether represented about 20% of the world publications over the years 1970–2020. Thus, there is a widely distributed interest on this scientific research field. Generally, in some countries, and particularly in the USA, the papers retrieved were produced by numerous institutions. Inversely in other countries, like Sweden, Denmark, and Iran, such papers were performed by specialized research centers.

3.1.3. Authors

A search dedicated to the papers' authors returned the list of the most publishing authors, shown in Table 4, together with the overall number of papers released by such authors in their own country over the time intervals examined. Prof. Adisa Azapagic at the University of Manchester (UK) resulted to be the most publishing author with 31 papers, followed by Prof. María Teresa Moreira (27) at the Universidad de Santiago de Compostela (Spain), and Ian Vázquez-Rowe (25) at the Pontificia Universidad Catolica de Peru (Peru), and Feijoo Gumersindo (24) at the Universidad de Santiago de Compostela (Spain). It can be also noted that one author from Iran and another one from Thailand are included in the first Top 10, indicating a high specialization activity in both countries. The research string used did not allow to number all the papers published by the authors listed in Table 4, being the underestimation of the order of 10–15%, as reported above.

3.1.4. Journals

A journal-based search returned the list of the peer-reviewed journals publishing more frequently papers related to the topic of concern. Table 5 shows the number of publications hosted by such journals, their ranking in quartiles (Q), Scimago Scientific Journal Ranking (SJR), and H-index, as extracted from https://www.scimagojr.com/, accessed on 18 October 2021. In particular, the quartile ranking classifies scientific journals on the basis of their impact factor or impact index.

Table 3. Number of publications for affiliation, as referred to different time periods (2020–1970; 2020–2016; 2015–1970).

	2020–1970		2020–2016		2015–1970	
No.	Affiliation	No. Docs	Affiliation	No. Docs	Affiliation	No. Docs
1	Wageningen Univ. & Res.	79	Chinese Academy of Sciences	50	Wageningen Univ. & Res.	51
2	Chinese Academy of Sciences	78	China Agricultural Univ.	41	Swedish Ins.e for Food and Biotech.	41
3	China Agricultural Univ.	70	Univ. degli Studi di Milano	32	Aarhus Universitet	35
4	Aarhus Universitet	55	M. of Agric. of the People's Rep. China	32	China Agricultural Univ.	29
5	Univer. de Santiago de Compostela	47	The Univ. of Manchester	31	Chinese Academy of Sciences	28
6	Univ. of Tehran	45	Danmarks Tekniske Universitet	29	Agriculture et Agroalimentaire Canada	25
7	Univ. degli Studi di Milano	45	Wageningen Univ. & Res.	28	Univ. de Santiago de Compostela	25
8	Centre INRAE Bretagne-Normandie	44	Ministry of Education China	26	Centre INRAE Bretagne-Normandie	24
9	Swedish Ins.e for Food and Biotecg.	41	Univer. of Chinese Acad. of Sciences	25	Sveriges lantbruksuniversitet	24
10	The Univ. of Manchester	41	Northwest A&F Univ.	25	Univ. of Tehran	23
11	Sveriges lantbruksuniversitet	38	Univ. of Tehran	22	Nanjing Agricultural Univ.	19
12	Ministry of Education China	37	Univer. de Santiago de Compostela	22	Agrocampus Ouest	18
13	Agriculture et Agroalimentaire Canada	37	Chinese Acad. of Agricultural Sciences	21	Ins. de Recerca I Technologia Agroal.	18
14	Univ. of Chinese Acad. of Sciences	37	Centre INRAE Bretagne-Normandie	20	Universiteit Gent	18
15	Danmarks Tekniske Universitet	36	Aarhus Universitet	20	Univ. of California. Davis	16
16	M. of Agricul. of the People's Rep. China	36	Univ. degli Studi di Bari	19	CIRAD Centre de Rech. de Montpellier	16
17	Agrocampus Ouest	35	CIRAD Centre de Rech. de Montpellier	19	Chalmers Univ. of Technology	15
18	CIRAD Centre de Rech. de Montpellier	35	INRAE	19	Malaysian Palm Oil Board	14
19	Northwest A&F Univ.	33	Univ. degli Studi della Tuscia, Viterbo	18	Teagasc-Irish Agric. and Food Dev. Aut.	14
20	Univ. of California, Davis	32	Pontificia Univ. Catolica del Peru	18	Natural Resources Ins.e Finland Luke	14

Table 4. Number of publications by the first 20 highly publishing authors in the world, as referred to different time periods (i.e., 2020–1970; 2020–2016; 2015–1970).

No.	2020–1970			2020–2016			2015–1970		
	Name	No. Docs	Country	Name	No. Docs	Country	Name	No. Docs	Country
1	Azapagic A.	31	UK	Azapagic A.	25	UK	Sonesson U.	14	Sweden
2	Moreira M.T.	27	Spain	Bacenetti J.	17	Italy	Feijoo G.	13	Spain
3	Vázquez-Rowe I.	25	Peru	Moreira M.T.	15	Spain	Rafiee S.	13	Iran
4	Feijoo G.	24	Spain	Vázquez-Rowe I.	13	Peru	Hermansen J.E.	12	Denmark
5	Bacenetti J.	22	Italy	Gheewala S.H.	12	Thailand	Moreira M.T.	12	Spain
6	Rafiee S.	22	Iran	Holden N.M.	12	Ireland	Vázquez-Rowe I.	12	Peru
7	Gheewala S.H.	21	Thailand	Vignali G.	12	Italy	Antón A.	10	Spain
8	Sonesson U.	21	Sweden	Feijoo G.	11	Spain	Cederberg C.	10	Sweden
9	Knudsen M.T.	19	Denmark	González-García S.	10	Spain	Hospido A.	10	Spain
10	Holden N.M.	18	Ireland	Ingrao C.	10	Italy	Subramaniam V.	10	Malaysia
11	González-García S.	17	Spain	Knudsen M.T.	10	Denmark	Van Der Werf H.M.G.	10	France
12	Ingrao C.	16	Italy	Cimini A.	9	Italy	Dewulf J.	9	Belgium
13	Vignali G.	16	Italy	De Marco I.	9	Italy	Flysjö A.	9	Denmark
14	Basset-Mens C.	15	France	Jeswani H.K.	9	UK	Gheewala S.H.	9	Thailand
15	Subramaniam V.	15	Malaysia	Moresi M.	9	Italy	Knudsen M.T.	9	Denmark
16	Thoma G.	15	USA	Rafiee S.	9	Iran	May C.Y.	9	Malaysia
17	Bava L.	14	Italy	Bava L.	8	Italy	Andersson K.	8	Sweden
18	Nemecek T.	14	Switzerland	Birkved M.	8	Denmark	Basset-Mens C.	8	France
19	Zucali M.	14	Italy	Iannone R.	8	Italy	Berlin J.	8	Sweden
20	Hermansen J.E.	13	Denmark	Rosentrater K.A.	8	USA	Corson M.S.	8	France

Table 5. Top 20 world journals ranked by the number of EAFP-related publications for the periods examined (1970–2020).

No.	Journal	No. Docs	Q	SJR	H-Index
1	Journal of Cleaner Production	499	Q1	1.81	173
2	International Journal of Life Cycle Assessment	179	Q1	1.60	98
3	Science of the Total Environment	116	Q1	1.66	224
4	Sustainability Switzerland	106	Q2	0.58	68
5	Acta Horticulturae	67	Q4	0.18	54
6	Resources Conservation and Recycling	50	Q1	2.22	119
7	Journal of Environmental Management	46	Q1	1.31	161
8	Agricultural Systems	42	Q1	1.51	101
9	Agriculture Ecosystems and Environment	34	Q1	1.72	163
10	Energy	34	Q1	2.17	173
11	Iop Conference Series Earth and Environmental Science	33			
12	Journal of Food Engineering	32			
13	Environmental Science and Pollution Res.	30			
14	Ecological Indicators	29			
15	Journal of Dairy Science	28			
16	Water Switzerland	26			
17	Bioresource Technology	25			
18	Environmental Science and Technology	25			
19	Energies	23			
20	Animal	21			

The journals mainly specialized in the EAFP topic were those appearing in the first positions, namely in descending order the *Journal of Cleaner Production* with 499 papers published in the 1970–2020 interval, *International Journal of Life Cycle Assessment* with 179 papers, and *Science of the Total Environment* with 116 papers. Other journals, not exclusively publishing papers dealing with food production, firstly dealt with agricultural and environmental science aspects, and, then, with their climate ones. The great majority of these journals were first-quartile journals with SJR citation index in the top 25% of journals for at least one of its classified subdisciplines.

According to Table 6 that ranks all the above 20 journals with respect to the average citation per paper index (CPPI), the *International Journal of Life Cycle Assessment* yielded the greatest CPPI (35.8), followed by the *Journal of Cleaner Production* (32.4), *Science of the Total Environment* (23.5), and *Sustainability Switzerland* (7.8), the latter being the only open-access journal.

3.1.5. Founding Sponsors

Table 7 reports the data extracted from the Scopus database when accounting for the founding sponsors. Over the 1970–2020 period, the *European Commission* was the first sponsor in terms of publications produced, followed by the *National Natural Science Foundation of China*, and *Ministry of Science and Technology of the People's Republic of China*. Such a wide presence of Chinese papers again shows a certain sensitivity of China towards the environmental sustainability of food production, probably because this nation is currently the first CO_2 emitter in the world [25].

Table 6. Top 5 journals ranked by the overall number of papers published from 1970–2020 together with the overall number of citations, total citations per paper index (CPPI), and citation per paper index (10%-CPPI) and percentage of citations (10%-CG) generated by the first 10% of published articles in order of citation received.

No.	Journal	No. Doc.s	No. Cit.s	CPPI	10%-CPPI	10%-CG
1	Journal of Cleaner Production	499	16158	32.4	123.6	37.5
2	International Journal of Life Cycle Assessment	179	6412	35.8	140.1	37.1
3	Science of the Total Environment	116	2730	23.5	91.4	36.8
4	Sustainability Switzerland	106	824	7.8	29.7	36.0
5	Acta Horticulturae	67	279	4.2	23.8	51.3

Table 7. Founding Sponsors cited in the papers retrieved from the Scopus database over the 1970–2020 period.

No.	Sponsor	No. Docs
1	European Commission	162
2	National Natural Science Foundation of China	138
3	Ministry of Science and Technology of the People's Republic of China	74
4	UK Res. and Innovation	68
5	European Regional Development Fund	50
6	Engineering and Physical Sciences Res. Council	40
7	National Key Res. and Development Program of China	40
8	Seventh Framework Programme	38
9	National Science Foundation	37
10	Ministry of Education of the People's Republic of China	33
11	Conselho Nacional de Desenvolvimento Científico e Tecnológico	30
12	Horizon 2020 Framework Programme	30
13	Coordenação de Aperfeiçoamento de Pessoal de Nível Superior	27
14	National Basic Res. Program of China (973 Program)	26
15	Ministerio de Economía y Competitividad	25
16	Ministério da Ciência, Tecnologia, Inovações	25
17	Fundamental Res. Funds for the Central Universities	24
18	Government of Canada	24
19	U.S. Department of Agriculture	24
20	Ministry of Agriculture of the People's Republic of China	23

3.2. Textual Analysis

3.2.1. The Most Cited Food-Related Terms

Table 8 shows the mostly used terms (occurrence) and their relative frequency (%) in the database constructed using the Title of the papers produced worldwide or in the five mostly publishing countries (Table 1). The term *milk* was that most cited worldwide: the same conclusion can be drawn for other terms, such as *rice, maize, fish,* and *wheat.* Such terms were those most used also for the five most publishing countries, but in different order. In China, *rice* resulted to be the first term used, owing of course to the high amount of rice produced and consumed (http://www.fao.org/faostat/en/#data/QC/visualize, accessed on 18 October 2021). Italy and Spain differentiate from the other countries for their characteristic terms were related to their diet and/or production. Both countries had *wine* as the third and fourth term most frequently studied, respectively, followed by *wheat* and *milk.* The term *olive oil* was at 6th place, while the term *pasta* was present in Italian publications only. It is worth noting that worldwide the term *palm oil* resulted to be more studied than the terms *cattle, coffee,* or *cheese.* The term *beer* was only present in Chinese (just less cited than *tea*) and Italian papers at the 13th and 14h place, respectively.

Table 8. Ranking of the occurrence of the most food-related terms in the Abstract/Title of the papers indexed in the Scopus database, as referred to the world and five most publishing countries.

World			USA			China			Italy			UK			Spain		
Terms	Doc.s	%	Terms	Doc.s	%	Terms	Doc.s	%	Terms	Doc.s	%	Terms	Doc.s	%	Terms	Doc.s	%
milk	1383	9.6	dairy	228	13.0	rice	497	28.6	milk	208	13.5	dairy	96	8.2	milk	96	11.5
rice	1320	9.1	milk	205	11.7	wheat	219	12.6	dairy	120	7.8	rice	80	6.8	dairy	75	9.0
dairy	1125	7.8	rice	152	8.7	maize	188	10.8	wine	115	7.5	milk	79	6.8	fish	70	8.4
wheat	867	6.0	fish	105	6.0	vegetable	88	5.1	food waste	69	4.5	meat	78	6.7	wine	48	5.7
fish	599	4.1	cattle	77	4.4	food waste	84	4.8	wheat	67	4.3	food waste	73	6.2	meat	39	4.7
meat	593	4.1	meat	69	3.9	cotton	81	4.7	olive oil	53	3.4	fish	55	4.7	rice	35	4.2
maize	491	3.4	wheat	64	3.7	fish	55	3.2	rice	51	3.3	sugar	46	3.9	maize	29	3.5
food waste	479	3.3	maize	62	3.5	milk	45	2.6	pasta	51	3.3	wheat	43	3.7	cheese	27	3.2
sugar	465	3.2	food waste	52	3.0	dairy	35	2.0	cheese	43	2.8	meal	30	2.6	wheat	23	2.7
palm oil	410	2.8	cotton	42	2.4	rubber	27	1.6	meat	43	2.8	maize	29	2.5	rapeseed	21	2.5
cotton	323	2.2	meal	39	2.2	meat	27	1.6	maize	41	2.7	sheep	26	2.2	legumes	21	2.5
vegetable	277	1.9	wine	38	2.2	tea	26	1.5	fruits	39	2.5	wool	25	2.1	olive oil	18	2.2
wine	273	1.9	rubber	36	2.1	beer	24	1.4	vegetable	37	2.4	legumes	25	2.1	food waste	18	2.2
cattle	270	1.9	vegetable	32	1.8	sugar	23	1.3	beer	32	2.1	wine	24	2.1	vegetable	17	2.0
oil palm	266	1.8	palm oil	31	1.8	cassava	21	1.2	fish	26	1.7	rapeseed	23	2.0	cotton	16	1.9
coffee	257	1.8	sugar	28	1.6	wool	21	1.2	fat	26	1.7	vegetable	19	1.6	seafood	16	1.9
rubber	250	1.7	rye	26	1.5	legumes	20	1.2	cattle	22	1.4	tea	18	1.5	tomatoes	15	1.8
meal	219	1.5	legumes	25	1.4	sorghum	18	1.0	bread	22	1.4	palm oil	18	1.5	cattle	13	1.6
cheese	200	1.4	lettuce	24	1.4	seafood	18	1.0	sorghum	21	1.4	rapeseed oil	18	1.5	wool	11	1.3
seed	180	1.2	tea	24	1.4	barley	16	0.9	barley	20	1.3	bread	18	1.5	sugar	11	1.3
rapeseed	164	1.1	coffee	23	1.3	fruits	16	0.9	whey	19	1.2	chocolate	17	1.5	bread	10	1.2
tea	164	1.1	seafood	22	1.3	tobacco	15	0.9	meal	18	1.2	cotton	15	1.3	barley	9	1.1
bread	149	1.0	cheese	21	1.2	soybeans	13	0.7	tea	16	1.0	butter	15	1.3	yoghurt	9	1.1

3.2.2. The Most Cited Environmental Impact Categories and Standard Methods

Table 9 shows the environmental impact categories and standard methods most frequently used and their relative frequency, as searched in the Title or Abstract of published papers. *Life Cycle Assessment* (LCA), *Carbon Footprint* (CF), and *Water Footprint* (WF) were the first three standard methods mostly used. However, the LCA procedure does not a priori give any clue about the impact categories effectively considered. In any case, the most used methods relied on just a single environmental issue (i.e., CF including the Publicly Available Specification 2050, WF, *Ecological Footprint*, and *Cumulative Energy Demand*), followed by the standard methods that analyze the environmental impact with mid- (i.e., *Environmental Product Declaration, TRACI*) or end- (i.e., *IMPACT 2002+*) point impact categories, and more recently by the *Product Environmental Footprint* (PEF) method where as many as 16 mid-point categories are normalized with respect to the impacts caused by one person living in the world during one year and weighed excluding or including three toxicity related impact categories (i.e., human toxicity cancer, human toxicity non-cancer and freshwater eco-toxicity) to yield a single weighted score. Further details were summarized by Moresi et al. 2021, [28].

Table 9. Occurrence of the environmental impact assessment methods in the abstract of the papers published in the 1970–2020 period.

No.	1970–2020	Occurrence	1970–2015	Occurrence	2016–2020	Occurrence
1	LCA	2128	LCA	1183	LCA	945
2	Life Cycle Assessment	1372	Life Cycle Assessment	660	Life Cycle Assessment	712
3	Carbon Footprint	978	Carbon Footprint	474	Carbon Footprint	504
4	Water Footprint	804	Water Footprint	340	Water Footprint	464
5	Ecological Footprint	148	Ecological Footprint	102	CED	52
6	CED	84	Ecoindicator	37	Ecological Footprint	46
7	Eco-indicator	42	CED	32	EPD	8
8	Impact 2002	17	PAS 2050	13	PEF	8
9	PAS 2050	14	Impact 2002	11	Impact 2002	6
10	EPD	13	EPD	5	TRACI	5
11	PEF	11	TRACI	5	Eco-indicator	5
12	TRACI	10	PEF	3	PAS 2050	1
13	EPS 2000	1	AWCC	0	EPS 2000	1
14	AWCC	0	CML 2002	0	AWCC	0
15	CML 2002	0	EDIP 2003	0	CML 2002	0
16	EDIP 2003	0	EPS 2000	0	EDIP 2003	0
17	Eco Scarcity	0	Eco Scarcity	0	Eco Scarcity	0
	Total	5622		2865		2757

AWCC—Australian Wine Carbon Calculator; CED—Cumulative Energy Demand; EPD—Environmental Product. Declaration; PEF—Product Environmental Footprint.

3.3. Map Analysis

Figure 3 shows the term map referred to the 1970–2020 period. The 299 terms displayed on the map were grouped in six clusters, which appeared well separated and were identified by different colors.

Among the 116 terms marked in red, the three main terms were *development, consumer, challenge* and *food production*, these cluster terms were clearly linked to issues related to environmental policies, decision-making about the mitigation measures to be undertaken to address and resolve the environmental impacts of food production, and communication of environmental and sustainability aspects to different stakeholders. The term *diet* was closely related to another cluster (marked in yellow), and in particular to *meat*. Such a cluster represents an important sector of study regarding the *livestock* impact on food production, *milk* production and *land use*.

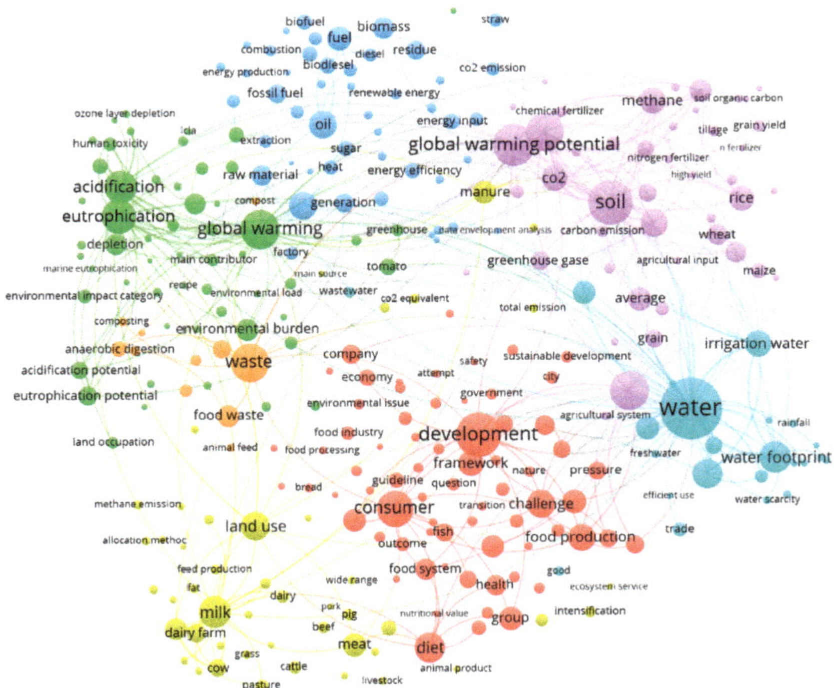

Figure 3. The term map referred to the 1970–2020 period.

The green-marked cluster included 55 items, regarding the impact categories mostly cited, the terms *global warming*, *eutrophication*, and *acidification* being the prevailing ones. This cluster is near the blue-marked one, where the biggest spot was represented by the terms related to *fossil fuels*, *oil* and *combustion* probably for transportation and *heat* generation. Clearly, the *fossil fuel* problems were emphasized by other terms, such as *biofuel*, *biodiesel*, *ethanol*, and *sugar*, these underlining the use of an edible raw material as energy source.

The viola-marked cluster was strictly related to *agriculture* and raw material production (in particular *wheat* and *rice*). *Nitrogen fertilizer*, *methane*, CO_2 resulted to be linked to the *soil* and *global warming potential* spots. One of the biggest spots included the term *water* with connection to all the other clusters and the smallest cluster regarding *waste*, *food waste*, *composting*, and *anaerobic digestion*.

About 70% of the global water consumed by humans is used in agriculture [29], primarily for irrigation. Thus, in food products coming from irrigated land, irrigation will dominate the water use.

Figures 4 and 5 elucidate the main links (i.e., co-occurrence between terms), as referred to the 1970–2016 and 2017–2020 periods. One of the main topics of the first period concerned several terms, namely *sustainability*, *milk*, *fertilizer*, *global warming*, *eutrophication*, and *land use*. On the contrary, in the following four years the main research topics diverted to *water*, *diet*, *consumer* and *sustainable food consumption*, *food waste*, and *waste disposal*, this including *incineration* and *recycling*, such as *composting* and *anaerobic digestion*.

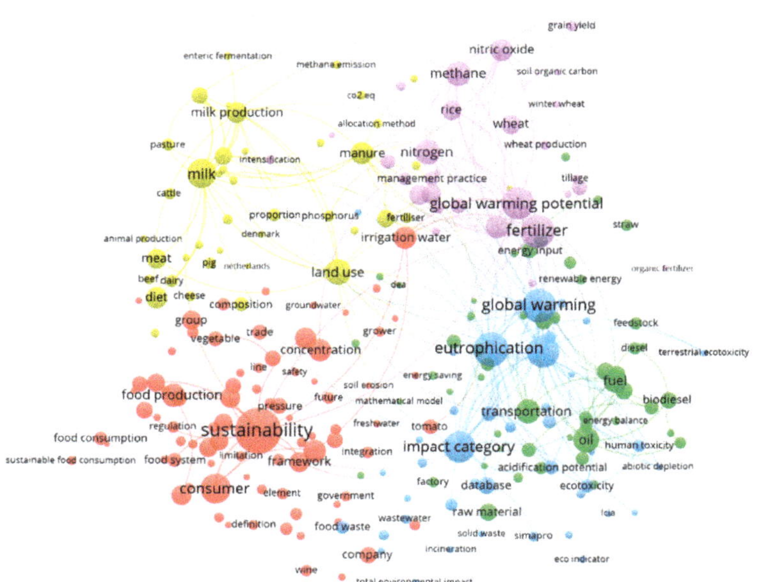

Figure 4. The term map referred to the 1970–2015 period.

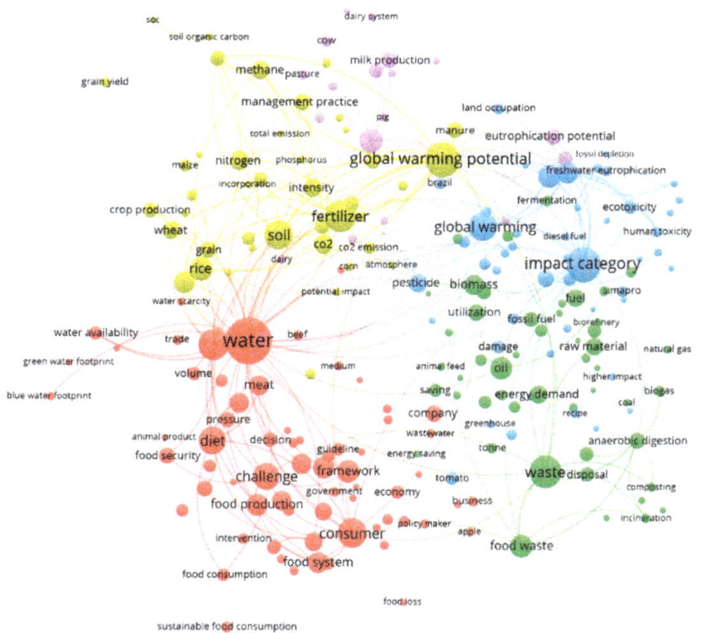

Figure 5. The term map referred to the 2016–2020 period.

3.4. Firstly Indexed and Mostly Cited Papers

Table S6 in the electronic supplement shows the first 10 published papers concerning the EAFP, these having been indexed by the Scopus database since 1968, as well as those mostly cited ones over the latest 50 years.

The term *environmental impact* firstly appeared in 1974 and was related to the disposal of sulfur removal sludges generated by the slaked lime-wet scrubber process, and use of antibacterial drugs in food animals. A paper published in 2006 explored the consumer attitude towards a more sustainable food consumption and resulted to be the most cited paper in this research sector, gathering more than one thousand citations.

4. Discussion of Results

The present bibliometric review highlighted the presence of some novel aspects regarding the actual EAFP sector. The first and most evident conclusion of this study pointed out that the production of the scientific community in terms of number of publications is still relatively small despite the social and scientific relevance of the environmental impact of food production and consumption in terms especially of food security, air and soil quality, diet and health. In all probability, the limited number of publications retrieved depended on the evolution of this sector that has just a recent origin. On the other hand, the data analysis at the macro level highlighted that two of the most GHG emitting countries, namely China and the USA (http://www.globalcarbonatlas.org/en/CO2-emissions, accessed on 18 October 2021) were the first ones in terms of the number of publications on this topic. Moreover, such nations displayed a long and consistent record of investment in research and development (R&D) (https://www.rdworldonline.com/2021-global-rd-funding-forecast-released/), accessed on 18 October 2021). Nevertheless, if their prolific scientific production in terms of total world scientific publications was related to their population and governmental investment in R&D (see Table S7 in the Supplementary Material), the number of scientific papers per each billion USD invested o per million citizens was just of the order of unity, while it roughly increased by a factor of 10 in the case of the most publishing European countries, such as Italy, the UK, and Spain.

Not by chance, the scientific research in the EAFP field resulted to be mainly carried out in Europe, since 10 out of the first 20 nations and 12 out of the first 20 Institutions/Research Centers, as well as the most publishing authors, were European.

The publications resulted to be concentrated in few journals (e.g., the *Journal of Cleaner Production*, *International Journal of Life Cycle Assessment*, and *Science of the Total Environment*), even if the EAFP topic has interested other magazines not always directly related to this research area as measured by a higher number of citations per paper with respect to the world average one.

The textual analysis showed that a great number of studies just focused on a single environmental issue, mainly the climate impact measured via the carbon footprint or water footprint methodology. This choice was probably encouraged by a wide-ranging public debate on this issue, a relatively important share of the overall GHG emissions by the food sector, and a good availability of data, even if such impact categories are regarded to be insufficient to describe the full range of environmental impacts deriving from the food systems. Less attention was up to now given to other standard methods, such as EPD® and PEF, likely for their relatively recent introduction. Since the characterization of the whole environmental profile of a single food or drink product is costly [30,31], and the climate change impact category is quite more reliable than all the other ones used in the aforementioned standard methods [32], the assessment of the product carbon footprint might be regarded as not only a cheaper tool to identify the major hotspots of the food supply chain, but also a proper method to start improving the sustainability of the 99% of the food and beverage small- and medium-sized enterprises [33]. Moreover, since the collection of primary data involves time, cost, and resource efforts, especially for SMEs, the use of secondary data might be more than sufficient for a preliminary identification of the major hotspots of a process or product, as shown in the production of bread [34] and lager beer [35].

The bibliometric map also showed a scientific interest towards other environmental impacts not exclusively related to fossil energy use or climate impacts, such as those linked to the agricultural production affecting the biodiversity, soil quality, water release of nutrients (mainly nitrogen and phosphorus), and pesticides. For instance, the GHG emissions associated to such a phase are in descending order due to: (1) crop residue decomposition;

(2) inorganic fertilizers applied; (3) manufacture, storage, and transportation of inorganic N and P fertilizers, and pesticides to the farm gate; (4) main farming operations, such as tillage, seeding, pesticide spraying, and crop harvesting; (5) soil carbon gains or losses from various cropping systems; and (6) emissions of N_2O from fallow areas destined to the growth of the next year crop [36]. Several crop rotation systems, lower N fertilizer rates and reduced tillage seemed to be effective to mitigate the carbon footprint of several crops such as durum wheat [37–39]. Even if organic farming is regarded as low-carbon agriculture [40], its lower productivity with respect to conventional one requires more cultivated land, this greatly enhancing the damage to the ecosystem quality, as observed in the case of organic durum wheat [41].

Figure 3 shows an important correlation between nutrition (diet) and meat and food security. Since the world population is expected to grow from about 7 billion to 9.6 billion people in 2050, as well as the global meat and milk consumption, especially in China and India, the promotion of healthy diets can reduce the environmental footprint of food consumption [42,43], as in the case of the Mediterranean diet, which is not only protective against lifestyle diseases, such as cardio-vascular disease, obesity, type 2 diabetes mellitus and certain cancers, but also responsible for a more favorable impact on the environment [44].

As shown in Figure 3, the utmost studied foods coincided with the most impacting ones, such as meat and milk. In this group also rice and wheat were included, even if they are staple food with a relatively low environmental impact but a high worldwide production and consumption.

Over the latest five years (Figure 5), other terms, such as *water, energy demand* and *food waste*, underlines the contribution of the consumer and post-consumer steps of the life cycle of different foods and beverages. As concerning the consumer phase, any mitigation action of its environmental impact would ask for the diffusion of more appropriate cooking systems. In the case of dry pasta, its cooking energy consumption might be significantly reduced by using quite smaller water-to-pasta ratios than the conventional one of 10 L per kg of dry pasta [45–47] or adopting novel home eco-sustainable pasta cookers [48,49]. On the contrary, in the case of coffee the use of ground and roasted coffee instead of coffee pods or capsules would drastically cut the GHGs emitted to produce their packaging materials and dispose of post-consumer packaging wastes [50].

Finally, an increasing number of studies focused not only on the analysis of food waste to measure their environmental impacts and suggested some mitigation options, but also on *water* and *land use*, the latter being especially related to *meat* and *milk* production. Food is lost and wasted along the whole supply chain from farms to processing, retailing and consumption at home and restaurants. Food waste not only involves the loss of valuable and often scarce resources, such as water, soil, and energy, but also contributes to climate change. In the European Union, about 88 million metric tons of food are wasted every year, equivalent to 173 kg per person, 53% and 19% of which being wasted by households and processing, respectively [51]. Per capita food waste generation by households was found to be practically independent of the country income, thus any action on food waste would be equally relevant in high, upper-middle and lower-middle income countries [52]. Food waste should be handled to avoid a negative impact on the environment or human health. According to the waste hierarchy set out at Article 4 of the revised Waste Framework [53], food waste formation should be limited as much as possible using for example less material in design and manufacture. Once formed, its entire apparatus or replacement parts should be refurnished to be re-used, recycled, submitted to other recovery options, and finally disposed of via landfilling or incineration with no energy recovery, as the least preferred option.

From such a map analysis, the food and beverage industry is expected in the near future to bear an ever-increasing responsibility towards consumer and environment and to invest more to prevent the development of more serious and costly adverse effects on food security.

5. Conclusions

The bibliometric data analysis, articulated through various indicators and combined to literature mapping clustering tools, allowed a graphical and numerical assessment of the research panorama on the environmental assessment of food production, as well as the geographic origin of this kind of research and growth or erosion of the scientific impact of specific countries. It represented an objective evaluation of this particular research topic, researchers or research Institutions, and could help researchers to address their research work and select the appropriate journals for their papers. The data highlighted the presence of few specialized journals with relevant citations per paper and H-index, as well as many other multidisciplinary journals with H-Indexes like those of the specialized ones. Over the latest five years this research field was mainly aimed at managing the agricultural practices, mitigating global warming and water use, assuring food security and sustainable food consumption, while minimizing food waste formation.

Supplementary Materials: The following are available online at https://www.mdpi.com/article/10.3390/su132111633/s1, Table S1: List of terms used for the query string: Category 1—Selected Terms of two thematic groups, A—Standard methods and B—Generic Terms, Table S2: List of terms used for the query string: Category 2—Food products, Crop production Terms, Table S3: List of terms used for the query string: Category 2—Food products, category Crop processed, Live Animals, Livestock Primary, Livestock Processed, Table S4: List of terms used for the query string: Category 2—Food products, category Generic, Food and beverage, Table S5: List of terms used for textual analysis of abstract, Table S6: Scientific production in EAFP for the top 5 nations ranked by the number of documents published related to their population and governmental investment in R&D.

Funding: This research received no external funding.

Institutional Review Board Statement: Not applicable.

Informed Consent Statement: Not applicable.

Data Availability Statement: Data available on request due to restrictions e.g., privacy or ethical.

Conflicts of Interest: The author declares that the research was conducted in the absence of any commercial or financial relationships that could be construed as a potential conflict of interest.

References

1. Crist, E.; Mora, C.; Engelman, R. The interaction of human population, food production, and biodiversity protection. *Science* **2017**, *356*, 260–264. [CrossRef]
2. IPCC (Intergovernmental Panel on Climate Change). Sixth Assessment Report (AR6): Climate Change 2022. 2017. Available online: https://www.ipcc.ch/site/assets/uploads/2018/11/AR6_WGII_outlines_P46.pdf (accessed on 2 October 2021).
3. Miedzinski, M.; Allinson, R.; Arnold, E.; Harper, J.; Doranova, A.; Giljum, S.; Griniece, E.; Kubeczko, K.; Mahieu, B.; Markandya, A. *Assessing Environmental Impacts of Research and Innovation Policy, Study for the European Commission*; Directorate-General for Research and Innovation: Brussels, Belguim, 2013. [CrossRef]
4. EEA (European Environment Agency). The European Environment—State and Outlook 2010. Available online: https://www.eea.europa.eu/soer/2010/europe/land-use/download (accessed on 3 October 2021).
5. Rogissart, L.; Foucherot, C.; Bellassen, V. *Estimating Greenhouse Gas Emissions from Food Consumption: Methods and Results*; I4CE (Institute for Climate Economics): Paris, France, 2019; Available online: https://www.i4ce.org/wp-core/wp-content/uploads/2019/03/0318-I4CE2984-EmissionsGES-et-conso-alimentaire-Note-20p-VA_V2.pdf (accessed on 1 October 2021).
6. Cowell, S.J.; Clift, R. A methodology for assessing soil quantity and quality in life cycle assessment. *J. Clean. Prod.* **2000**, *8*, 321–331. [CrossRef]
7. BIER. Research on the Carbon Footprint of Beer. Beverage Industry Environmental Roundtable. 2012. Available online: https://www.bieroundtable.com/publication/beer/ (accessed on 30 September 2021).
8. Notarnicola, B.; Tassielli, G.; Renzulli, P.A. Modeling the agri-food industry with life cycle assessment. In *Life Cycle Assessment Handbook*; Curran, M.A., Ed.; Wiley: Hoboken, NY, USA, 2012; pp. 159–184.
9. Di Matteo, G.; Nardi, P.; Grego, S.; Guidi, C. Bibliometric analysis of climate change vulnerability assessment research. *Environ. Syst. Decis.* **2018**, *38*, 508–516. [CrossRef]
10. Haunschild, R.; Bornmann, L.; Marx, W. Climate change research in view of bibliometrics. *PLoS ONE* **2016**, *11*, e0160393. [CrossRef] [PubMed]
11. Wang, B.; Pan, S.Y.; Ke, R.Y.; Wang, K.; Wei, Y.M. An overview of climate change vulnerability: A bibliometric analysis based on Web of Science database. *Nat. Hazards* **2014**, *74*, 1649–1666. [CrossRef]

12. Verma, S.; Singh, K. Food security in India: A bibliometrics study. *Lib. Herald.* **2019**, *57*, 379–392. [CrossRef]
13. Sweileh, W.M. Bibliometric analysis of peer-reviewed literature on food security in the context of climate change from 1980 to 2019. *Agric. Food Secur.* **2020**, *9*, 1–15. [CrossRef]
14. Zhang, M.; Gao, M.; Yue, S.; Zheng, T.; Gao, Z.; Ma, X.; Wang, Q. Global trends and future prospects of food waste research: A bibliometric analysis. *Environ. Sci. Pollut. Res.* **2018**, *25*, 24600–24610. [CrossRef]
15. Huang, L.; Chen, K.; Zhou, M. Climate change and carbon sink: A bibliometric analysis. *Environ. Sci. Pollut. Res.* **2020**, *2*, 8740–8758. [CrossRef]
16. Li, X.; Li, Y.; Li, G. A scientometric review of the research on the impacts of climate change on water quality during 1998–201 *Environ. Sci. Pollut. Res.* **2020**, *27*, 14322–14341. [CrossRef]
17. Durieux, V.; Gevenois, P.A. Bibliometric indicators: Quality measurements of scientific publication. *Radiology* **2010**, *255*, 342–35 [CrossRef]
18. Noyons, C.M. Science maps within a science policy context. In *Handbook of Quantitative Science and Technology Research*; Springer Dordrecht, The Netherlands, 2004; pp. 237–255. [CrossRef]
19. van Eck, N.J.; Waltman, L. Text mining and visualization using VOSviewer. *ISSI Newsl.* **2011**, *7*, 50–54.
20. Pallottino, F.; Cimini, A.; Costa, C.; Antonucci, F.; Menesatti, P.; Moresi, M. Bibliometric analysis and mapping c publications on brewing science from 1940 to 2018. *J. Inst. Brew.* **2020**, *126*, 394–405.
21. De Robbio, A. Analisi citazionale e indicatori bibliometrici nel modello Open Access. *Boll. Aib* **2007**, 257–288.
22. Van Eck, N.J.; Waltman, L. Software survey: VOSviewer, a computer program for bibliometric mapping. *Scientometrics* **2010**, *8*, 523–538. [CrossRef] [PubMed]
23. Waltman, L.; Van Eck, N.J.; Noyons, E.C. A unified approach to mapping and clustering of bibliometric networks. *J. Informet* **2010**, *4*, 629–635. [CrossRef]
24. Mattsson, B.; Olsson, P. Environmental audits and life cycle assessment. In *Auditing in the Food Industry*; Dillon, M., Griffith, C Eds.; Woodhead Publishing: Cambridge, UK, 2001; Chapter 10; pp. 174–194.
25. Nemecek, T.; Jungbluth, N.; Canals, L.M.; Schenck, R. Environmental impacts of food consumption and nutrition: Where are w and what is next? *Int. J. Life Cycle Assess.* **2016**, *21*, 607–620. [CrossRef]
26. Burck, J.; Hagen, U.; Höhne, N.; Nascimento, L.; Bals, C. *Climate Change Performance Index. Results 2020*; Germanwatch Berlin, Germany, 2019; Available online: https://www.germanwatch.org/en/17281 (accessed on 3 April 2021).
27. Crippa, M.; Guizzardi, D.; Muntean, M.; Schaaf, E.; Solazzo, E.; Monforti-Ferrario, F.; Olivier, J.G.J.; Vignati, E. *Fossil CO_2 Emission of All World Countries—2020 Report*; EUR 30358 EN; Publications Office of the European Union: Luxembourg, 2020; Available online: https://edgar.jrc.ec.europa.eu/overview.php?v=booklet2020 (accessed on 3 October 2021).
28. Moresi, M.; Cibelli, M.; Cimini, A. Standard methods effectively useful to mitigate the environmental impact of food industry In *Environmental Impact of Agro-Food Industry and Food Consumption*; Galanakis, C., Ed.; Academic Press: San Diego, CA USA, 2021; Chapter 1; pp. 1–30.
29. FAO. Water for Sustainable Food and Agriculture A Report Produced for the G20 Presidency of Germany. 2017. Available online http://www.fao.org/3/i7959e/i7959e.pdf (accessed on 3 October 2021).
30. BMUB/UBA/TUB. BMUB/UBA/TUB Position Paper on EU Product and Organisation Environmental Footprint Proposa as Part of the Communication Building the Single Market for Green Products (COM/2013/0196 Final). 2014. Available online: https://webgate.ec.europa.eu/fpfis/wikis/display/EUENVFP/Steering+Committee+workspace?preview=%2F63542 841%2F66782536%2FPosition+paper+on+PEF_TUB_BMUB_UBA.pdf (accessed on 3 October 2021).
31. Galatola, M.; Pant, R. Product environmental footprint—Breakthrough or breakdown for policy implementation of life cycle assessment? *Int. J. Life Cycle Assess.* **2014**, *19*, 1356–1360. [CrossRef]
32. Sala, S.; Cerutti, A.K.; Pant, R. *Development of a Weighting Approach for the Environmental Footprint*; Publications Office of the European Union: Luxembourg, 2018; Available online: https://ec.europa.eu/environment/eussd/smgp/documents/2018 _JRC_Weighting_EF.pdf (accessed on 3 October 2021).
33. Cimini, A.; Moresi, M. Are the present standard methods effectively useful to mitigate the environmental impact of the 99% EU food and drink enterprises? *Trends Food Sci. Technol.* **2018**, *77*, 42–53. [CrossRef]
34. Espinoza-Orias, N.; Stichnothe, H.; Azapagic, A. The carbon footprint of bread. *Int. J. Life Cycle Assess.* **2011**, *16*, 351–365 [CrossRef]
35. Cimini, A.; Moresi, M. Effect of brewery size on the main process parameters and cradle-to-grave carbon footprint of lager beer *J. Ind. Ecol.* **2017**, *22*, 1139–1155. [CrossRef]
36. Liu, T.; Wang, Q.; Su, B. A review of carbon labeling: Standards, implementation, and impact. *Renew. Sustain. Energy Rev.* **2016**, *53*, 68–79. [CrossRef]
37. Alhajj Ali, S.; Tedone, L.; De Mastro, G. Optimization of the environmental performance of rainfed durum wheat by adjusting the management practices. *J. Clean. Prod.* **2015**, *87*, 105–118. [CrossRef]
38. Failla, S.; Ingrao, C.; Arcidiacono, C. Energy consumption of rainfed durum wheat cultivation in a Mediterranean area using three different soil management systems. *Energy* **2020**, *195*, 116960. [CrossRef]
39. Gan, Y.; Liang, C.; Hamel, C.; Cutforth, H.; Wang, H. Strategies for reducing the carbon footprint of field crops for semiarid areas A review. *Agron. Sustain. Dev.* **2011**, *31*, 643–656. [CrossRef]

0. Chiriacò, M.V.; Grossi, G.; Castaldi, S.; Valentini, R. The contribution to climate change of the organic versus conventional wheat farming: A case study on the carbon footprint of wholemeal bread production in Italy. *J. Clean. Prod.* **2017**, *153*, 309–319. [CrossRef]

1. Cibelli, M.; Cimini, A.; Moresi, M. Environmental profile of organic dry pasta. *Chem. Eng. Trans.* **2021**, *87*, 397–402. [CrossRef]

2. FAO. *Building Climate Resilience for Food Security and Nutrition*; Food and Agriculture Organization of the United Nations: Rome, Italy, 2018; Available online: http://www.fao.org/3/I9553EN/i9553en.pdf (accessed on 3 October 2021).

3. WRI (World Resources Institute). *Creating a Sustainable Food Future. A Menu of Solutions to Sustainably Feed More than 9 Billion People by 2050*; World Resources Report 2013–14: Interim Findings; World Resources Institute: Washington, DC, USA, 2013.

4. Moresi, M. Assessment of the life cycle greenhouse gas emissions in the food industry. *Agro Food Ind. Hi-Tech* **2014**, *25*, 53–62.

5. Cimini, A.; Cibelli, M.; Messia, M.C.; Marconi, E.; Moresi, M. Cooking quality of commercial spaghetti: Effect of the water-to-dried pasta ratio. *Eur. Food Res. Technol.* **2018**, *245*, 1037–1045. [CrossRef]

6. Cimini, A.; Cibelli, M.; Messia, M.C.; Moresi, M. Commercial short-cut extruded pasta: Cooking quality and carbon footprint vs. water-to-pasta ratio. *Food Bioprod Process* **2019**, *116*, 150–159. [CrossRef]

7. Cimini, A.; Cibelli, M.; Moresi, M. 2019b. Reducing the cooking water-to-dried pasta ratio and environmental impact of pasta cooking. *J. Sci. Food Agric.* **2019**, *99*, 1258–1266. [CrossRef]

8. Cimini, A.; Cibelli, M.; Moresi, M. Development and assessment of a home eco-sustainable pasta cooker. *Food Bioprod. Process.* **2020**, *122*, 291–302. [CrossRef]

19. Cimini, A.; Cibelli, M.; Taddei, A.R.; Moresi, M. Effect of cooking temperature on cooked pasta quality and sustainability. *J. Sci. Food Agric.* **2021**, *101*, 4946–4958. [CrossRef] [PubMed]

50. Cibelli, M.; Cimini, A.; Cerchiara, G.; Moresi, M. Carbon Footprint of different methods of coffee preparation. *Sustain. Prod. Consum.* **2021**, *27*, 1614–1625. [CrossRef]

51. EU Parliament News. Food Waste: The Problem in the EU in Numbers. 2017. Available online: https://www.europarl.europa.eu/news/en/headlines/society/20170505STO73528/food-waste-the-problem-in-the-eu-in-numbers-infographic (accessed on 3 October 2021).

52. United Nations Environment Programme. Food Waste Index Report 2021. *Nairobi.* 2021. Available online: https://www.unep.org/resources/report/unep-food-waste-index-report-2021 (accessed on 3 October 2021).

53. Directive 2008/98/EC of the European Parliament and of the Council of 19 November 2008 on Waste and Repealing Certain Directives. *Official Journal of the European Union.* L 312/3-30. Available online: https://eur-lex.europa.eu/legal-content/EN/TXT/PDF/?uri=CELEX:32008L0098&from=EN (accessed on 3 October 2021).

*sustainability* 

Article

A Quantitative Modeling and Prediction Method for Sustained Rainfall-PM$_{2.5}$ Removal Modes on a Micro-Temporal Scale

Tingchen Wu [1,2,3,4], Xiao Xie [5,*], Bing Xue [5] and Tao Liu [1,2,3,4]

1 Faculty of Geomatics, Lanzhou Jiaotong University, Lanzhou 730070, China; giswtchen@163.com (T.W.); ltaochina@foxmail.com (T.L.)
2 National-Local Joint Engineering Research Center of Technologies and Applications for National Geographic State Monitoring, Lanzhou 730070, China
3 Gansu Provincial Engineering Laboratory for National Geographic State Monitoring, Lanzhou 730070, China
4 Faculty of Geosciences and Environmental Engineering, Southwest Jiaotong University, Chengdu 611756, China
5 Key Lab for Environmental Computation and Sustainability of Liaoning Province, Institute of Applied Ecology, Chinese Academy of Sciences, Shenyang 110016, China; xuebing@iae.ac.cn
* Correspondence: xiexiao@iae.ac.cn

Abstract: PM$_{2.5}$ is unanimously considered to be an important indicator of air quality. Sustained rainfall is a kind of typical but complex rainfall process in southern China with an uncertain duration and intervals. During sustained rainfall, the variation of PM$_{2.5}$ concentrations in hour-level time series is diverse and complex. However, existing analytical methods mainly examine overall removals at the annual/monthly time scale, missing a quantitative analysis mode that applies micro-scale time data to describe the removal phenomenon. In order to further achieve air quality prediction and prevention in the short term, it is necessary to analyze its micro-temporal removal effect for atmospheric environment quality forecasting. This paper proposed a quantitative modeling and prediction method for sustained rainfall-PM$_{2.5}$ removal modes on a micro-temporal scale. Firstly, a set of quantitative modes for sustained rainfall-PM$_{2.5}$ removal mode in a micro-temporal scale were constructed. Then, a mode-constrained prediction of the sustained rainfall-PM$_{2.5}$ removal effect using the factorization machines (FM) was proposed to predict the future sustained rainfall removal effect. Moreover, the historical observation data of Nanjing city at an hourly scale from 2016 to January 2020 were used for mode modeling. Meanwhile, the whole 2020 year observation data were used for the sustained rainfall-PM$_{2.5}$ removal phenomenon prediction. The experiment shows the reasonableness and effectiveness of the proposed method.

Keywords: sustained rainfall; PM$_{2.5}$ removal mode; micro-temporal scale; quantitative modeling; mode prediction

Citation: Wu, T.; Xie, X.; Xue, B.; Liu, T. A Quantitative Modeling and Prediction Method for Sustained Rainfall-PM$_{2.5}$ Removal Modes on a Micro-Temporal Scale. *Sustainability* **2021**, *13*, 11022. https://doi.org/10.3390/su131911022

Academic Editor: Constantinos Cartalis

Received: 27 August 2021
Accepted: 1 October 2021
Published: 5 October 2021

Publisher's Note: MDPI stays neutral with regard to jurisdictional claims in published maps and institutional affiliations.

1. Introduction

Air pollution is currently a major environmental challenge for both developed and developing countries worldwide, with increasing industrialization, growing urbanization and energy consumption posing a serious threat to public health [1,2]. According to a report from the World Health Organization, PM$_{2.5}$ is unanimously considered to be an important indicator of air quality [3,4]. The rainfall processes are typically regarded as strong drivers to remove PM$_{2.5}$ so as to improve the air quality.

However, in the existing available research on an observation data analysis, there are obvious inconsistency between rainfall and the PM$_{2.5}$ removal effect, especially for sustained rainfall, a kind of typical but complex rainfall process in southern China with an uncertain duration and intervals. For example, Kao et al. [5] found that in a rainforest environment, summers with high precipitation are negatively correlated with PM$_{2.5}$ levels; Preethi et al. [6] proposed that the effect of simulated Indian monsoon rainfall removal depends to a large extent on climatic wind speeds; Neal et al. [7] pointed out that there has been no systematic improvement in air quality in mid Wales for 17 years in the face of increased rainfall. In addition to the complexity of the sustained rainfall itself, the variations in regional environmental conditions are possibly another important reason. Therefore, studying the relationship between sustained rainfall and the $PM_{2.5}$ removal effect becomes a multi-factor related scientific issue.

When expressing the effect of rainfall on the removal of $PM_{2.5}$ concentrations in the air, the observed time series is usually the most intuitive expression [8,9]. The observed value will be considered as a sign of the presence of the rainfall process [10]. However, sustained rainfall, which is characterized by two or more sustained rainfall events within a given period of time, is intermittent, slow to change and uncertain in its length of formation, and its complexity needs to be fully considered [11,12]. During sustained rainfall, the variation of $PM_{2.5}$ concentrations in hour-level time series is diverse and complex. However, existing analytical methods mainly examine overall removals at the annual/monthly time scale and mainly use the correlation analysis (CA) between rainfall amount and $PM_{2.5}$ concentration based on macroscopic monitoring data, which refers to the rainfall time series and $PM_{2.5}$ time series sampled on an annual or monthly temporal scale [13,14]. For example, using the data of a macro-temporal scale, Shaibu et al. [15] confirmed that monthly $PM_{2.5}$ concentrations in the Niger Delta region of Nigeria show a significant positive correlation with monthly rainfall. Similarly, through a study of annual datasets from five air quality monitoring stations in Bahrain from 2006 to 2012, Jassim et al. [16] indicated little correlation between rainfall and $PM_{2.5}$, leading to a year-on-year increase in $PM_{2.5}$ concentrations. They are all missing a quantitative analysis mode that applies micro-scale time data to describe the removal phenomenon. In order to further achieve air quality prediction and prevention in the short term, it is necessary to analyze its micro-temporal removal effect for atmospheric environment quality forecasting.

In addition, due to the sustained rainfall and the atmospheric pollution particulate matter itself being complex, the large-scale temporal analysis lacks guidance for specific sustained rainfall-$PM_{2.5}$ removal processes [17]. On this basis, a part of the study proposes to extract the historical single rainfall process using hourly observation data and adopt a predetermined calculation model to quantify the removal effect of the rainfall process. Chhavi et al. [18] analyzed the wet removal effect by calculating the $PM_{2.5}$ concentration difference before and after rainfall, including the positive and negative removal; Kapwata et al. [19], based on the intensity of rainfall, delineate the rainfall classes, counting the percentage of positive removal to summarize the removal effect according to the influence factors, such as rainfall duration and rainfall volume, which have certain guiding significance. The above methods do not take into account the changes in the effects produced by complex sustained rainfall processes at different stages, including effects such as hygroscopic growth and the secondary transformation of gaseous pollutants which cause $PM_{2.5}$ concentrations to rise or rebound [20–22]; at the same time, they lack universal law exploration, having difficulties in serving the scientific prediction and early warning of air quality systems and active prevention [23,24].

In this paper, we propose a quantitative modeling and prediction method for sustained rainfall-$PM_{2.5}$ removal modes on a micro-temporal scale. The detailed contributions are as follows:

- A novel micro-scale analytical framework for quantitatively elucidating the mechanism of $PM_{2.5}$ removal by sustained rainfall was proposed. Compared with the yearly, monthly and daily time scales, the hourly scale is a more suitable form of information for decision making; therefore, the framework would more clearly express the complex characteristics of sustained rainfall than the analysis methods of large-scale data. The innovative hourly scale data analysis in this paper is more useful for practical applications in predicting and assessing air quality.
- A set of quantitative $PM_{2.5}$ removal modes based on a micro-analysis are proposed. The modes would highlight the specific and high-level patterns of the removal effect

of sustained rainfall at the micro-scale than the traditional micro-scale data analysis methods. During sustained rainfall, the variation of $PM_{2.5}$ concentrations in an hourly time series is diverse and complex. The analysis of hourly scales reveals new characteristic modes that are different from the traditional large scale. These "declining, rebounding, or rising" modes not only allow the analysis of historical data from different regions, but also allow the prediction of $PM_{2.5}$ removal at hourly intervals using future hourly rainfall, which can help the relevant systems and departments to make timely decisions on air pollution control.

This paper is organized as follows. The study area and data on analytical framework are viewed in Section 2. Section 3 introduces the quantitative definition of the sustained rainfall-$PM_{2.5}$ removal mode on a micro-temporal scale, then presents the rainfall-$PM_{2.5}$ removal phenomenon predicting algorithm based on the quantitative model and Section 4 discusses the experimental results. Finally, the discussions are presented in Section 5.

2. Study Area and Data

2.1. Study Area

The experimental area of this paper is located in Nanjing, which is in the middle of the lower reaches of the Yangtze River and has a subtropical monsoon climate with cold and dry winters, high temperatures and rainy summers and a relatively large rainfall variability, making it one of the regions in China with more droughts and floods [25,26]. Some studies have shown that (i) due to the influence of strong convective weather and global warming, there is a significant increase in the amount and duration of rainfall in Nanjing, which has a different degree of the purifying effect on air pollutants [27]; (ii) at the same time, according to the results of a pollutant source analysis in Nanjing, $PM_{2.5}$ water-soluble ion concentrations have diurnal and seasonal differences, making the average mass concentration of water-soluble ions higher during the day than at night, and the flushing effect of spring and summer rainfall is generally lower than that of autumn and winter [28,29]; (iii) according to the statistics, the annual average humidity in Nanjing is nearly 80%, and some rainfall in this environment may lead to hygroscopic growth of particulate matter, resulting in higher pollutant concentrations [30].

2.2. Dataset

The experimental data in this paper were obtained from hourly $PM_{2.5}$ concentrations and meteorological observations released by the China General Environmental Monitoring Station and the National Meteorological Information Centre for the period from January 2016 to December 2020. Based on the geographical location of the monitoring stations in Nanjing and the completeness of the data, the $PM_{2.5}$ data observed at the CCM (Caochangmen) site (32.0572 °N, 118.7486 °E) and the rainfall data observed at the PK (Pukou) site (32.177 °N, 118.706 °E) were used for the experiment.

Figure 1 demonstrates the phenomenon of $PM_{2.5}$ concentration removal by rainfall observed on a macro-temporal scale (yearly and monthly) to a micro-temporal scale (hourly): from the annual scale, the average annual $PM_{2.5}$ concentration in Nanjing decreases year by year from 2016 to 2020, while the pollutant concentration is relatively lower in years with a higher average annual rainfall; from the monthly scale, the monthly rainfall shows an overall stable trend of first increasing and then decreasing, and the average monthly $PM_{2.5}$ concentration in the period of concentrated rainfall are significantly lower than those during periods of low rainfall, indicating that rainfall has a significant effect on the removal of air pollutants; on an hourly scale, the hour-by-hour $PM_{2.5}$ concentration changes during a single rainfall process, thus rainfall has an obvious effect on pollutant removal while there are situations that cause concentrations to increase, and it is not possible to obtain the mode of the effect of regional rainfall on $PM_{2.5}$ from a particular rainfall process. In summary, the mechanism of $PM_{2.5}$ removal by sustained rainfall in Nanjing is complex, and it is difficult to satisfy the study of the mechanism of $PM_{2.5}$ removal by rainfall with the macro time statistics and the time series of a single rainfall process, so the micro time-series effect model proposed in this paper is applied for a deeper analysis.

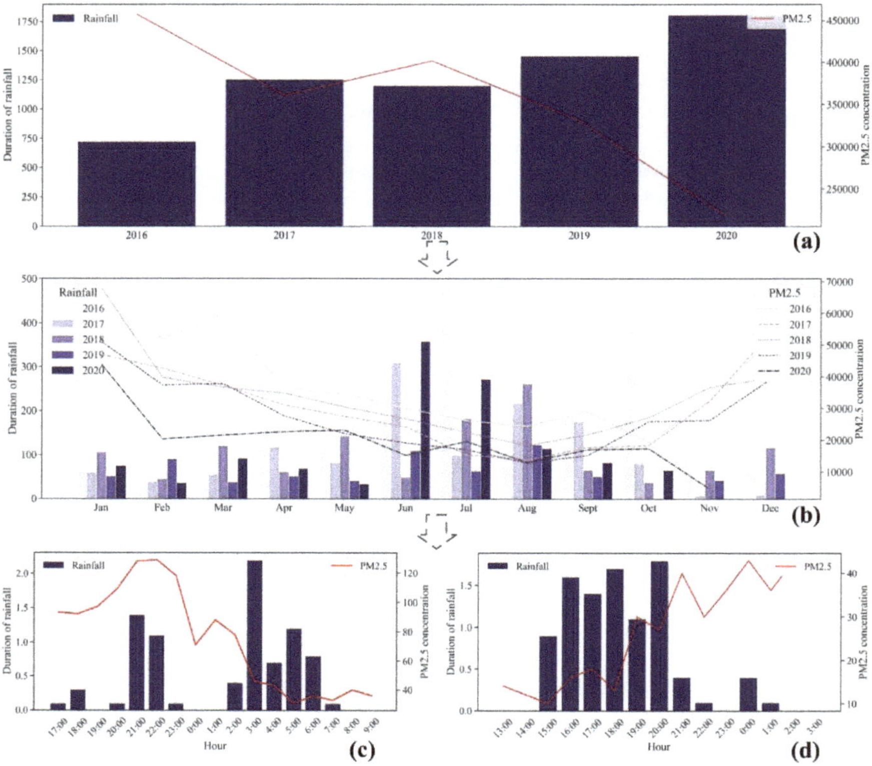

Figure 1. Macro/micro-temporal scale of sustained rainfall-PM$_{2.5}$ observation data series and removal phenomenon in Nanjing from 2016 to 2020: (**a**) Yearly temporal-level; (**b**) Monthly temporal-level; (**c**,**d**) Hourly temporal-level.

3. Methods

3.1. A Quantitative Modeling for Sustained Rainfall-PM$_{2.5}$ Removal Mode in Micro-Temporal Scale

3.1.1. Overview

In order to accurately describe the study object and the study boundary from the observation dataset, this section first defines a microscopic time-series fragment model of the sustained rainfall process, on the basis of which the time-series process of the removal effect is structured, further modeling the concomitant factors affecting the role of rainfall in PM$_{2.5}$ removal, and finally establishes an evaluation mode to quantitatively describe the removal effect.

3.1.2. Micro-Temporal Modeling of Sustained Rainfall Process

The model needs to accurately identify a complete sustained rainfall process in micro-scale observations due to the large variation in time duration and the existence of uncertain intervals. The sustained rainfall process is characterized by intermittent multi-fragmentation on hourly observation data series.

Time-series segments (TS) are mathematical frameworks for describing and modeling event sequences in the time domain, and are commonly used to construct interpretable time-series analysis models, such as trend prediction and anomaly detection [31,32]. In this paper, the sustained rainfall time-series segment (SRS) is defined at the micro-temporal scale rainfall: $\{R_t : t = hour\}$. SRS visualizes the variation of rainfall in the time dimension of a sustained rainfall process, as shown in Figure 2, and is defined as follows:

$$SRS = \left\{ [R_{t'}, \ldots, R_{t''}] \, | \, t', \Delta t, t'' \right\} \tag{1}$$

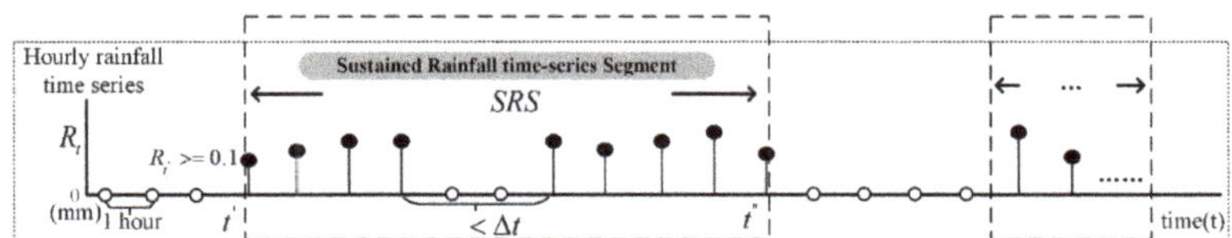

Figure 2. Schematic diagram of a micro-temporal model of the sustained rainfall process.

The quantitative measurement steps are:

Step 1: Using the hourly rainfall values as a benchmark, the moment of the first occurrence of 0.1 mm and above rainfall is taken as the starting point t' of the time-series fragment.

Step 2: Since rainfall and its resulting effects will remain in space for a certain period of time, a threshold value Δt is set to indicate the intermittent duration of the sustained rainfall process, i.e., the sequence before and after when rainfall is zero does not exceed Δt is regarded as the same time-series fragment.

Step 3: Satisfy the above conditions and the last occurrence of rainfall greater than zero is the end point t'', and obtain a complete SRS.

The micro-temporal model of the sustained rainfall process provides a quantitative basis for determining the baseline interval for the clearance effect analysis.

3.1.3. Sustained Rainfall Removal Concomitant Factor Modeling

Sustained rainfall processes are often concomitant by changes in the accompanying meteorological factors such as temperature, humidity, wind speed and direction, which will affect the removal effect to varying degrees [33]. Therefore, in this paper, we consider the effect and intensity of the existing influence factors [34–37], and balance the availability of their own observational data and the predictability of future trends to establish a set of accompanying influence factors F, including the (i) direct factor (F_D) to describe rainfall characteristics, and the (ii) indirect factor (F_I) to describe environmental characteristics. The factors and impact effects are shown in Table 1.

On this basis, the removal effects time-series segment (RES) is defined in conjunction with the PM$_{2.5}$ time series. The RES refers to the process and subsequent effects of rainfall on PM$_{2.5}$ at the scale of the mechanism of removal, while avoiding the chance of PM$_{2.5}$ concentration values before rainfall anomalies by expanding the series values for t and t' forward and backward for n hours, respectively (Figure 3):

Table 1. Table of qualitative descriptions of concomitant factor.

	Class	Factor	Label	Impact Effects
F	F_D	Rainfall Total	P	High correlation with air pollutant concentrations, which can directly influence the removal effect
		Rainfall Duration	D	
	F_I	Temperature	$\overline{T}$	When the temperature near the ground is high, atmospheric convection is intensified, which tends to reduce $PM_{2.5}$ concentrations, and conversely $PM_{2.5}$ is not easily dispersed
		Humidity	$\overline{H}$	Changes in $PM_{2.5}$ are closely related to the moisture content of the air, with "hygroscopic increase" occurring due to the adsorption of particulate matter concentrations
		Wind Power	W	Stronger winds also facilitate the dilution and uplift of pollutants
		Initial $PM_{2.5}$	C_1	The effect of removal is influenced by the magnitude of $PM_{2.5}$ concentrations before rainfall, and has little effect on particulate concentrations when air quality is good
		Seasonal	S	The removal effect is mostly higher at night than during the day; the total positive removal of sustained rainfall will be slightly higher in autumn than in other seasons
		Day and night	K	

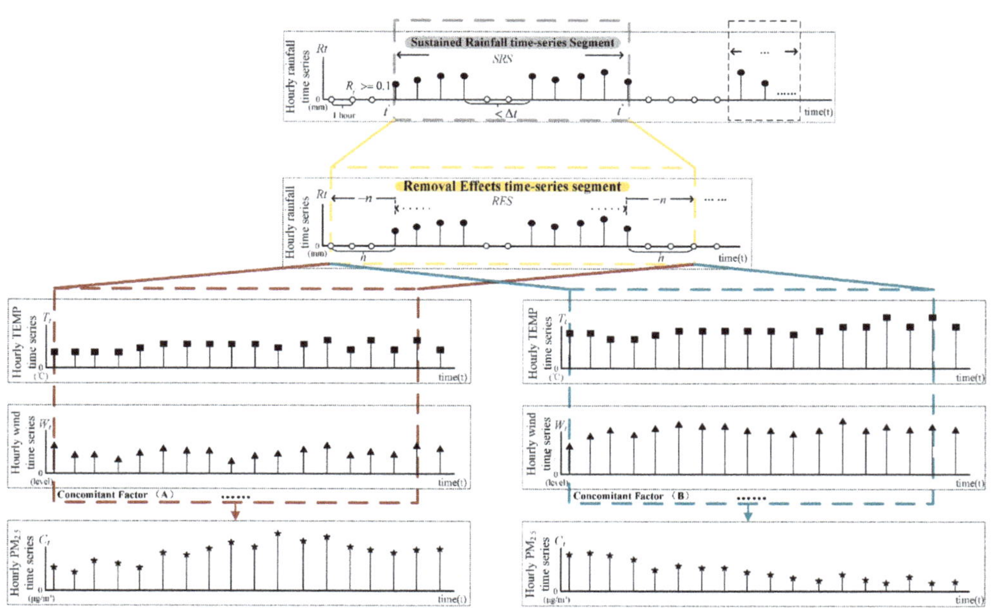

Figure 3. Schematic diagram of a micro-temporal model of the removal effects process and concomitant factor.

The concomitant factor model $\{F, RES\}$ for sustained rainfall removal identifies the information dimensions and effect intervals for the quantitative modeling.

3.1.4. Quantitative Evaluation Modeling of Removal Effects

The effect evaluation refers to the evaluation of the change in the effect produced by a geographical process considering the influence of relevant factors, and usually uses quantitative indicators to measure and analyze its dynamic change characteristics in spatial and temporal modes [38,39]. The quantitative evaluation model of removal effects is proposed based on the above-mentioned models (SRS, RES):

$$M = \left\{ M_{i \in (T,P,D,U,R,L)} \middle| \langle RP, RF, RS \rangle, (TSR, TSE) \right\} \tag{2}$$

(1) Quantitative Measurement of Removal Effect Indicators

Based on the model (SRS, RES), the following measurable evaluation indicators (RP, RF, RS) that quantitatively characterize the temporal variability in the removal processes are defined below:

(i)   Rate of process RP, the ratio of the difference between the very small value of $PM_{2.5}$ concentration change when it exists and the pre-start $PM_{2.5}$ concentration.

(ii) Rate of final RF, the ratio of the difference between the initial $PM_{2.5}$ concentration before the start and the concentration after the end, the magnitude of which allows a quantitative evaluation of the intensity of removal.

(iii) Rate of rebound RR, the ratio of the difference between the minimum value and the ending concentration for the entire rainfall process.

(2) Modeling of Removal Mode based on Effect Indicators

The (SRS, RES) and (RP, RF, RR) were used to modeling the removal process of sustained rainfall and the indicators were combined with zero boundaries, corresponding to the rising and falling changes in $PM_{2.5}$ concentrations in Figure 4, and combined with the ensemble theory to classify the effect patterns in Table 2.

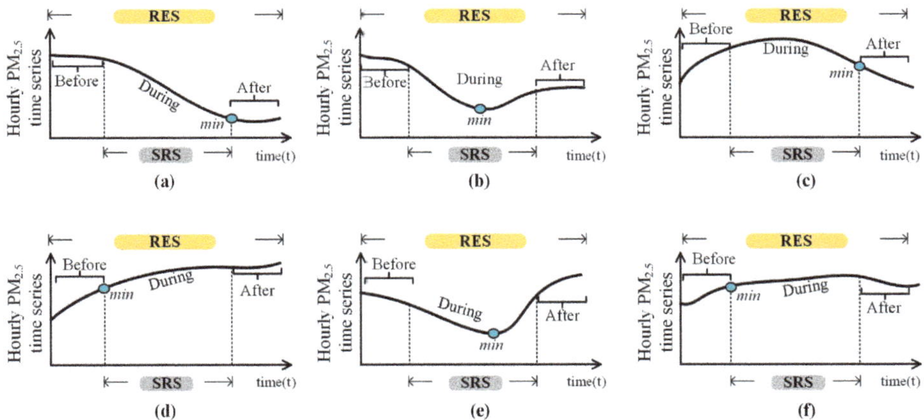

Figure 4. Schematic diagram of the sustained rainfall-$PM_{2.5}$ removal modes. (**a**) Totally Removal Mode; (**b**) Partly Removal Mode; (**c**) Delayed Removal Mode; (**d**) Totally Ascent Mode; (**e**) Rebounding Ascend Mode; (**f**) Lasting Ascent Mode.

Table 2. Table of sustained rainfall-$PM_{2.5}$ removal modes, quantitative classification and qualitative description.

Removal Mode M_i	Effect Indicators			Trends in $PM_{2.5}$ Concentrations		
	RP	RS	RR	During	After	Min
M_T	>0	<0	>0	Continued decline	Decline	Non-existent
M_P	>0	>0	>0	Decline, rebound	Decline	Existent
M_D	<0	<0	>0	Continued rise	Decline	Existent
M_A	<0	>0	<0	Continued rise	Rise	Non-existent
M_R	>0	>0	<0	Decline, rebound	Rise	Existent
M_L	<0	<0	<0	Continued rise	Rise	Existent

The removal mode, classified according to the above rules, covers all of the combinations of indicators and corresponds to realistic removal phenomena, providing a category a priori information for predicting future sustained rainfall.

3.2. The Mode Predicting of Sustained Rainfall-PM$_{2.5}$ Removal Effect Using the Quantitative Model

3.2.1. Overview

In this section, based on the quantitative model of sustained rainfall-PM$_{2.5}$ removal mode, the future rainfall is further fitted using a multiple regression to fit the removal modes, the core steps are as follows (Figure 5): Step 1, using the time-series window according to the constitute time-series sample of removal modes with high-dimensional characteristics; Step 2, extracting the principal components and component features of the pattern classification features through the principal component analysis method for the concomitant factor intensity; Step 3, establishing factorization classification regression machines for removal modes according to the factor intensity, which can predict the removal phenomenon based on the future meteorological and current air quality information.

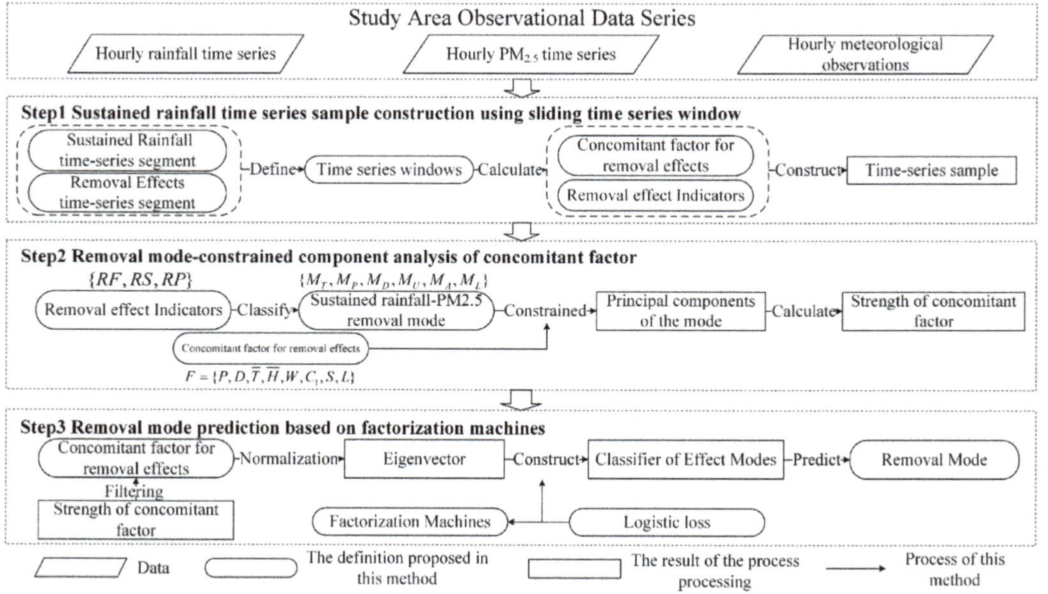

Figure 5. Flowchart for the prediction of sustained rainfall-PM$_{2.5}$ removal effect using the quantitative model.

3.2.2. Sustained Rainfall Time-Series Sample Construction Using Sliding Time-Series Window

Firstly, according to the definitions of the SRS and RES proposed in Section 3.1, a small-scale sliding window is applied to process the hourly rainfall time series R_t, hourly PM$_{2.5}$ time series C_t and hourly meteorological observations, setting up to capture all of the sustained rainfall processes in the historical data [40].

Secondly, the statistics of the corresponding direct and indirect factors are calculated within the extracted time-series range (SRS, RES) according to the concomitant factor model F proposed in this paper, as Table 3.

Table 3. Table of quantitative calculation of concomitant factor.

Class		Label	Time-Series Range	Quantitative Calculation	
F	F_D	P		$P = \sum_{t=t'}^{t''} R_t$	(3)
		D		$D = \sum_{t=t'}^{t''} t$	(4)
	F_I	$\overline{T}$	SRS	$\overline{T} = \frac{\sum_{t=t'}^{t''} T_t}{t''-t'}$	(5)
		$\overline{H}$		$\overline{H} = \frac{\sum_{t=t'}^{t''} H_t}{t''-t'}$	(6)
		W		$W \in \{0,1,\ldots,17\}$	(7)
		C_1	RES	$C_1 = \frac{\sum_{t=(t'-n)}^{t'} C_t}{n}$	(8)
		S	SRS	$S \in \{Spr, Sum, Fal, Win\}$	(9)
		K		$K \in \{Am, Pm\}$	(10)

The quantitative evaluation indicators (RP, RF, RR) of the extracted sustained rainfall processes were calculated based on the $PM_{2.5}$ concentration series C_t according to the evaluation model of removal effects described in this paper.

$$\begin{cases} RP = \frac{(C_1 - C_{min})}{C_1} \\ RR = \frac{(C_2 - C_{min})}{C_2} \\ RF = \frac{(C_1 - C_2)}{C_1} \end{cases} \qquad (11)$$

$$C_2 = \frac{\sum_{t=t'}^{t''+n} C_t}{n} \qquad (12)$$

$$C_{min} = min(C_{t'}, \ldots, C_{t''}) \qquad (13)$$

where C_1 is the initial $PM_{2.5}$ concentration during the sustained rainfall process, C_2 is the $PM_{2.5}$ concentration at the end of the sustained rainfall process and C_{min} is the minimum value for the whole sustained rainfall process.

Finally, sustained rainfall time-series samples $\vec{V}$ were constructed in the form of multi-dimensional feature vectors, as follows:

$$\vec{V} = \left(P, D, \overline{T,H}, W, C_1, S, K, [RF, RS, RP]\right) \qquad (14)$$

It provides a computable basis for the accurate extraction of analytical reference intervals from hourly observations.

3.2.3. Removal Mode-Constrained Component Analysis of Concomitant Factor

According to the removal mode division rules in Table 2, the $\vec{V}$ were classified to obtain six types of labeled samples corresponding to M_i, as shown in the following equation. Due to the different degrees of influence of the removal factors on effect indicators in different effect modes, such as the change of effective removal rate magnitude with increasing rainfall duration in complete removal mode, a multi-factor analysis is needed to explore its change pattern.

$$\vec{V} = \left(P, D, \overline{T,H}, W, C_1, S, K \mid M_i\right) \qquad (15)$$

The principal component analysis is a statistical method that reduces the information of multi-dimensional variables to a few characteristic components by a linear transformation and reflects as much information of the original variables as possible [41]. In this section, the main components $\{F_1, F_2, \ldots, F_n\}$ of the sample features under each mode will be extracted separately, and the weights of each effect factor will be calculated by the eigenvalues λ of each principal component and the variance contribution ratio $E(\lambda)$, and the relative strength S of the clearance factors under different modes will be evaluated quantitatively and used to screen the effect factors involved in the construction of the mode classifier.

$$S_{F_i} = \frac{A_{F_i}}{\sqrt{\lambda_{F_i}}} \times \frac{E(\lambda_{F_i})}{\sum_k^n E(\lambda_{F_k})} \tag{16}$$

A_{F_i} in the above equation is the loading of the initial factor i on the principal component.

The removal mode-constrained component analysis of concomitant factor enables the extraction of effective and stable effect factors that map mode characteristics, reducing the redundancy and sparsity of the factor information and improving the accuracy of the clearance effect prediction.

3.2.4. Removal Mode Prediction Based on Factorization Machines

Factorization machines (FM) are a classification method with a good learning ability for sparse data, solving the sparse information generated by the one-hot coding of categorical features, while taking into account the two-two correlation features between the features to meet the correlation and between the environmental factors in the rainfall process [42].

We use the factors filtered by the factor strengths calculated in Section 3.2.3, normalized by features (scaling all factors to between -1 and 1) with label coding to form the feature vector $\vec{X}_i$, which is used to build the factor decomposer $\hat{y}$ of the time-series effects model.

$$\begin{cases} \hat{y} = w_0 + \sum_{i=1}^{n} w_i + \vec{X}_i + \sum_{i=1}^{n-1} \sum_{j=i+1}^{n} v_i, v_j \vec{X}_i \vec{X}_j \\ w_{i,j} = v_i v_j \end{cases} \tag{17}$$

Assuming that the sustained rainfall time-series sample has n features, v_i and v_j in the above equation are the implicit vectors of the feature matrix decomposition, and $w_{i,j}$ is the interrelationship between the two features i, j.

Further construction of a classifier oriented to the removal mode: the loss function loss Equation (18) is designed using the logistic regression theory, the $\hat{y}$ is mapped into different classes by the step function sigmoid and the logistic loss is used as the criterion for optimization.

$$\begin{cases} loss(\hat{y}, y) = \sum_{i=1}^{m} -ln\sigma(\hat{y}^{(i)}, y^{(i)}) \\ \sigma(x) = \frac{1}{1-e^{-x}} \end{cases} \tag{18}$$

We can use the factor classifier constructed from the mode information to achieve a prediction of the removal effect that will result from a sustained rainfall process based on the rainfall forecast information.

4. Experimental and Analysis Section

This paper uses the historical observation data of Nanjing from January 2016 to January 2020 as $Data_{train}$ for the model analysis and uses the data from January 2020 to December of the same year $Data_{test}$ to verify the feasibility and effectiveness of the method of this paper.

4.1. Construct the Sustained Rainfall Time-Series Sample

Firstly, all of the datasets were processed in a uniform manner.

A sliding window was designed according to Section 3.2.2 to extract the sustained rainfall process, with all of the rainfall durations greater than 1 h, where the general residence time of particulate matter in the air and a window threshold of Δt was set to 3 h, rainfall with an interruption of no more than 3 h was considered as a complete sustained process (starting moment t' and ending moment t''), with a total of 427 sustained rainfalls collected.

While calculating the removal in each concomitant factors for each process, including the rainfall total, the rainfall duration and initial $PM_{2.5}$ concentration are shown according to Table 3, with seasonal and diurnal values used for labeling; further combining Equation (3) to calculate the removal effect indicators, and using these to classify the above rainfall samples into six modes according to the removal mode classification rules in Table 2, the result is shown in Table 4.

Table 4. A sample table of the sustained rainfall time series in Nanjing.

$\vec{V}$	$[t'-n, t''+n]$	P	D	$\overline{T}$	$\overline{H}$	W	C_1	S	K	$[RF, RS, RP]$	$\mathbf{M}$
$\vec{V_1}$	2016/01/04/18:00- 2016/01/05/07:00	18.0	12	8.21	4.91	4	276.33	4	2	$[0.96, 0.31, 0.95]$	M_P
$\vec{V_2}$	2016/01/10/21:00- 2016/01/11/08:00	7.0	12	6.04	3.67	3	150.67	4	2	$[0.19, -0.03, 0.22]$	M_T
... ...	... ...						... ...				... ...
$\vec{V_{426}}$	2020/10/15/16:00- 2016/10/16/17:00	17.6	24	14.02	2.31	2	15.33	3	2	$[0.93, 0.83, 0.61]$	M_P
$\vec{V_{427}}$	2020/10/21/05:00- 2020/10/21/10:00	3.8	6	17.38	4.75	2	38.33	3	0	$[0.08, -0.02, 0.11]$	M_T

The findings show 85 times of M_T, 191 times of M_P, M_D 80 times, M_A 85 times, M_R 12 times and M_L 14 times. By plotting the numerical distribution of concomitant factors for different modes (Figure 6), the removal mechanism of sustained rainfall in Nanjing was explored.

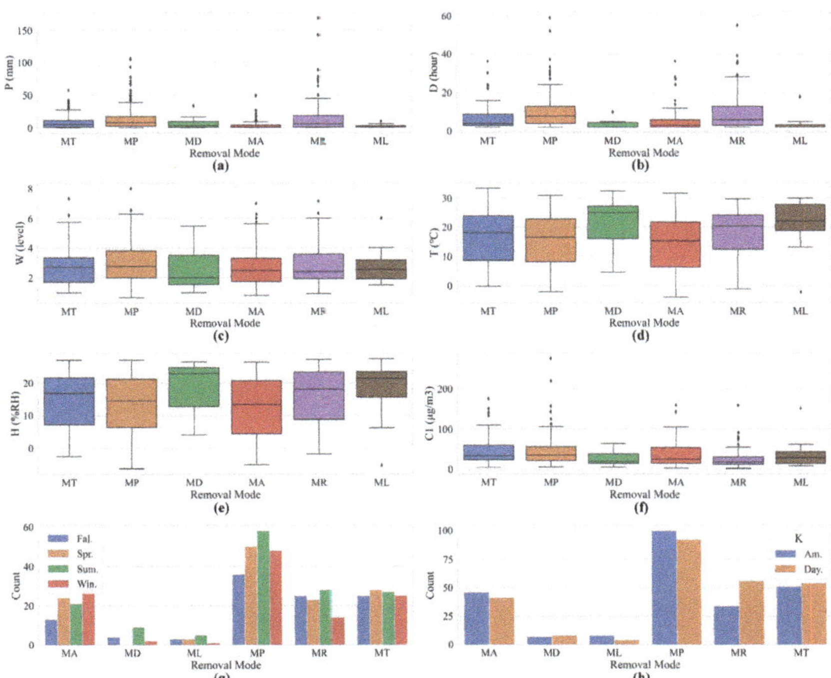

Figure 6. Distribution of the numerical distribution of concomitant factors for different removal modes. (**a**) Rainfall Total P and removal modes; (**b**) Rainfall Duration D and removal modes; (**c**) Wind Power W and removal modes; (**d**) Temperature $\overline{T}$ and removal modes; (**e**) Humidity $\overline{H}$ and removal modes; (**f**) Initial $PM_{2.5}$ C_1 and removal modes; (**g**) Seasonal factor S and removal modes; (**h**) Day and night factor K and removal modes.

4.2. Principal Component Analysis of the Removal Mode

Due to the large differences in the values of the effectors, the rainfall samples were first normalized by the eigenvector $\vec{V}$, and the high-dimensional feature samples of each type of mode were further subjected to the principal component analysis. According to the criteria for selecting the principal components, in this case, when the eigenvalues of the components were all greater than 1 and the cumulative contribution rate was 85%, there were the four types of extracted components. The results are shown in Table 5.

Table 5. Explanatory table for the principal component analysis of removal mode.

Component	Eigenvalue (λ)	Contribution of Variance $E(\lambda)$	Cumulative Contribution (%)
Principal component 1	2.774	35.67	35.67
Principal component 2	2.101	27.56	63.26
Principal component 3	1.662	16.28	79.54
Principal component 4	1.079	11.46	91.00
Principal component 5	0.662	6.061	97.06
Principal component 6	0.156	2.94	100.00

Based on the principal component eigenvalues and variance contribution rates, the factor strengths of different modes were calculated Equation (16) and the strengths of the effectors were ranked, and the total rainfall P, rainfall duration D, initial $PM_{2.5}$ concentration C_1 and wind speed scale W were selected to participate in the construction of the mode classifier (Table 6).

Table 6. Concomitant factor intensity of the sustained rainfall time series sample with removal mode.

	M_T	M_P	M_D	M_A	M_R	M_L
P	0.223	0.339	0.232	0.250	0.306	0.421
D	0.214	0.294	0.453	0.321	0.276	0.559
$\overline{T}$	0.127	0.168	−0.045	−0.174	−0.164	−0.103
$\overline{H}$	0.079	0.176	−0.137	−0.076	−0.037	0.218
W	0.263	0.132	0.232	0.271	0.386	0.372
C_1	0.193	0.266	0.411	0.263	0.442	0.421
S	0.074	0.135	0.033	0.173	−0.154	0.032
K	0.031	0.095	0.127	0.041	−0.076	0.093

4.3. Predict the Removal Mode and Phenomenon

We combine the above removal concomitant factor strengths with the factorization machines in Section 3.2.4 to build a classifier for predicting the removal mode and phenomenon: the total rainfall P, rainfall duration D, initial $PM_{2.5}$ concentration C_1 and wind speed scale W in $Data_{train}$ form the feature vector $\vec{X}$ and the classifier $\hat{y}$ in Equation (17), and the loss function in Equation (18) is combined to train the classifier according to the stochastic gradient descent (SGD) method. Figure 7 shows the effect of the removal modes classification, where $\langle C_1, W \rangle$ and $\langle P, D \rangle$ are the factors that form the modal length of the vector, which is used to present the modes distribution based on the effector features; the classifier designed in this paper can classify the sustained precipitation process into the six types described according to the effectors.

To validate the removal effect classifier, all of the sustained rainfall extracted from January 2020 to December of the same year in Nanjing were classified using the samples in $Data_{test}$, and the accuracy of the classification results was evaluated by plotting the receiver operating characteristic (ROC) curve and the curve area, area under the curve (AUC), to evaluate the accuracy of the classification results (Figure 8), where the true positive rate (*TPR*) and false positive rate (*FPR*) of the tested samples were calculated as follows:

$$\begin{cases} TPR = \frac{TP}{TP+FN} \\ FPR = \frac{TP}{FP+TN} \end{cases} \tag{19}$$

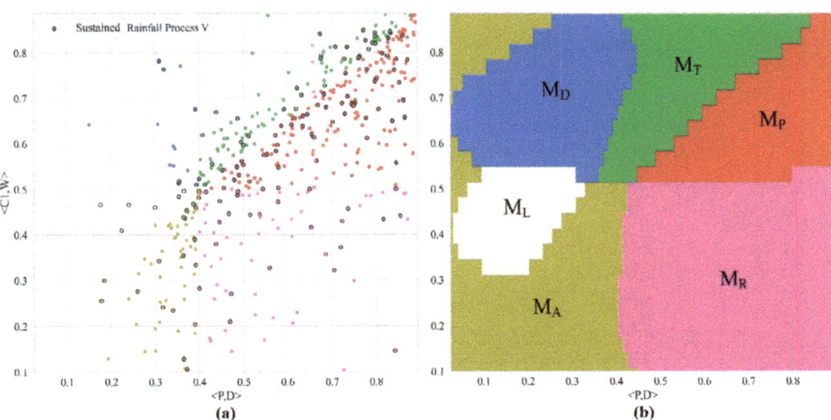

Figure 7. FM classifier result graph for removal modes, (**a**) Scatter plot of modes classification results; (**b**) Interpolated rendering of modes classification results.

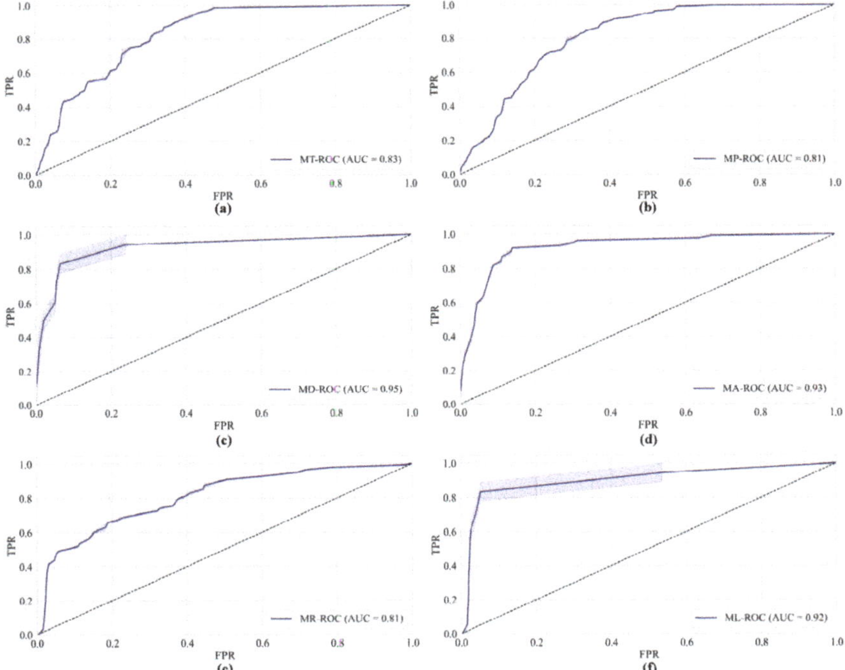

Figure 8. The ROC and AUC of removal modes classifier, (**a**) ROC of Totally Removal Mode; (**b**) ROC of Removal Mode; (**c**) ROC of Delayed Removal Mode; (**d**) ROC of Totally Ascend Mode; (**e**) ROC of Rebounding Ascend Mode; (**f**) ROC of Lasting Ascend Mode.

5. Discussion

In this paper, we mainly analyze the effect of sustained rainfall on $PM_{2.5}$ removal from the perspective of a microscopic temporal scale with historical observation data at the hourly scale. By combining previous research results, we propose a model analysis framework to quantitatively describe the removal effect from a more refined perspective for combining the mechanism of sustained rainfall effect on $PM_{2.5}$. The primary conclusions are summarized as follows.

In this paper, we use hourly scale observations for proposing models to quantitatively express sustained rainfall processes with intermittent duration and relative complexity. It is able to provide data boundaries for studying the role of rainfall removal. Moreover, we consider a large number of environmental influences and construct a concomitant factor model F, which can improve the accuracy and information dimension of the analysis. Based on the above considerations, we conclusively propose the removal modes for a quantitative description of the removal phenomenon.

- M_T, the $PM_{2.5}$ concentration change has a continuous decreasing trend during the rainfall process, which has good improvement of the air quality for a period of time after the precipitation.
- M_P, the $PM_{2.5}$ concentration change is due to the fact that when the removal of particulate pollutants by prolonged precipitation reaches its limit [29], a small portion of the particulate matter does not completely settle to the ground and floats into the air again, thus showing a slight rebound of the concentration values.
- M_D, $PM_{2.5}$ concentrations continue to rise during rainfall, but drop sharply after the end and are lower than the average concentration values before it.
- M_A, $PM_{2.5}$ concentration changes in a continuous upward trend when the rainfall duration is too short or small; the humid air will make the suspended pollutants expand, which is more likely to cause the accumulation of pollutants and make the $PM_{2.5}$ concentration rise.
- M_R, due to the longer duration of the process, there is often a short gap or the secondary precipitation is weak precipitation and other phenomena, which will cause a serious concentration rebound, making the concentration of particulate matter higher than before the precipitation.
- M_L, $PM_{2.5}$ concentrations continue to rise without rebound during rainfall, and the rise tends to scale off after the end, eventually making the $PM_{2.5}$ concentrations rise.

The method in this paper is able to classify the proposed model by historical observation data. The results show that of the sustained rainfall processes occurring in Nanjing from 2016 to 2020, only 85 were able to provide complete removal of $PM_{2.5}$, 63.4% of the precipitation processes resulted in $PM_{2.5}$ rebound and up to 177 sustained rainfall processes ultimately led to elevated $PM_{2.5}$ concentrations.

Based on the above quantitative modeling framework, we construct a classifier in combination with the model identification method, considering it for the accurate forecasting of future air quality in short periods. The accuracy evaluation results of the model show that the ROC of our constructed classifier performs well, and the AUC refers to more than 0.85, showing the reasonableness and effectiveness of the method in this paper.

Due to the limited data acquisition, more years of hourly and environmental data for $PM_{2.5}$ are lacking in this paper. Therefore, it lacks the samples of the lasting ascent mode and the delayed removal mode. Future studies are expected to obtain more hourly temporal observations for the purpose of removal modes construction and acquisition, ultimately to improve the accuracy of the prediction classifiers.

6. Conclusions

Rainfall is an effective way to remove major air pollutants such as $PM_{2.5}$. However, most studies on the relationship between rainfall data and $PM_{2.5}$ concentrations have only focused on the changes in the air quality under the influence of long-time span rainfall.

ignoring the effects of single rainfall processes that lead to increases or rebound changes in the $PM_{2.5}$ concentrations, in addition to the effect of wet deposition.

Therefore, based on the definition and generalization of the sustained rainfall process and its effects, this paper uses a time-series statistical method to extract and calculate the single sustained rainfall process and its removal effect factors based on microscopic time-scale observation data; in this process, the effect evaluation index is specifically proposed to quantitatively describe the degree of the removal effect, so as to establish a time-series effect model of $PM_{2.5}$ concentration removal by rainfall. The potential, deep-seated effect of rainfall processes on $PM_{2.5}$ concentrations is explored. The model is further combined with pattern recognition theory to design an effect pattern classifier for the sample characteristics of the rainfall process, and finally realize the micro-temporal prediction of air quality after a single rainfall. Using the hourly observation data of Nanjing from 2016 to 2020, a total of 427 sustained rainfall processes were collected using this micro-temporal time-series effects model, and the rate of process, rate of rebound and rate of final were calculated and classified into six types of modes: 85 totally removal mode, 191 partly removal mode, 14 delayed removal mode, 85 rebounding sscend mode and 12 lasting ascent mode. The classifier was constructed based on the factors, indicators and model categories, and the ROC evaluation index showed that the classifier has good performance and is capable of quantitatively predicting future $PM_{2.5}$ concentration decreases, increases and rebound effects using easily accessible rainfall and $PM_{2.5}$ concentration forecast information, with a view to providing decision-making information for future regional ambient air quality forecasting and refined control. For the acquisition of hour-by-hour $PM_{2.5}$ concentration forecasts, further investigation of the finer variation characteristics within rainfall periods is required on the basis of this study.

Author Contributions: Conceptualization, X.X.; methodology, X.X. and T.W.; software, T.W.; validation, T.W. and X.X.; formal analysis, X.X. and T.W.; investigation, T.W.; resources, X.X.; data curation, B.X.; writing—original draft preparation, T.W. and X.X.; writing—review and editing, X.X. and T.W.; visualization, T.W.; supervision, X.X. and T.L.; project administration, B.X.; funding acquisition, B.X. All authors have read and agreed to the published version of the manuscript.

Funding: This research was funded by National Natural Science Foundation of China (41971166) and LZJTU EP (201806).

Institutional Review Board Statement: Ethical review and approval were waived because necessary permissions were obtained from the local governments to which the schools are affiliated for this study. Informed Consent Statement: Informed consent was obtained from all subjects involved in the study.

Informed Consent Statement: Studies not involving human (Not applicable).

Data Availability Statement: Data available on request due to ethical restrictions. The data presented in this study are available on request from the corresponding author. The data are not publicly available due to not having participants consent to share their anonymized data with a third party.

Acknowledgments: We express our sincere appreciation to Qing Zhu from the Faculty of Geosciences and Environmental Engineering, Southwest Jiaotong University and the anonymous reviewer for their constructive comments.

Conflicts of Interest: The authors declare no conflict of interest. The funders had no role in the design of the study; in the collection, analyses, or interpretation of data; in the writing of the manuscript, or in the decision to publish the results.

References

1. Wu, H.; Gao, X. Multimodal Data Based Regression to Monitor Air Pollutant Emission in Factories. *Sustainability* **2021**, *13*, 2663. [CrossRef]
2. Wang, Z.; Liang, L.; Wang, X. Spatiotemporal evolution of $PM_{2.5}$ concentrations in urban agglomerations of China. *J. Geogr. Sci.* **2021**, *31*, 878–898. [CrossRef]
3. Pui, D.; Chen, S.C.; Zuo, Z. $PM_{2.5}$ in China: Measurements, sources, visibility and health effects, and mitigation. *Particuology* **2014**, *13*, 1–26. [CrossRef]

4.   Amrita, T. Study of Ambient Air Quality Trends and Analysis of Contributing Factors in Bengaluru, India. *Orient. J. Chem.* **2017**, *33*, 1051–1056.

5.   Kao, Y.H.; Lin, C.W.; Chiang, J.K. Predictive Meteorological Factors for Elevated PM$_{2.5}$ Levels at an Air Monitoring Station Near a Petrochemical Complex in Yunlin County, Taiwan. *Open J. Air Pollut.* **2019**, *8*, 1–17. [CrossRef]

6.   Preethi, B.; Kripalani, R.H.; Kumar, K.K. Indian summer monsoon rainfall variability in global coupled ocean-atmospheric models. *Clim. Dyn.* **2010**, *35*, 1521–1539. [CrossRef]

7.   Neal, C.; Reynolds, B.; Neal, M.; Pugh, B.; Hill, L.; Wickham, H. Long-term changes in the water quality of rainfall, cloud water and stream water for moorland, forested and clear-felled catchments at Plynlimon, mid-Wales. *Hydrol. Earth Syst. Sci.* **1999**, *5*, 459–476. [CrossRef]

8.   Liu, Z.; Shen, L.; Yan, C.; Du, J.; Li, Y.; Zhao, H. Analysis of the Influence of Precipitation and Wind on PM$_{2.5}$ and PM10 in the Atmosphere. *Adv. Meteorol.* **2020**, *2020*, 1–13. [CrossRef]

9.   Yamamoto, T.; Hori, K.; Tatebayashi, J. An experimental investigation of the PM adhesion characteristics in a fluidized bed type PM removal device. *Powder Technol.* **2016**, *289*, 31–36. [CrossRef]

10.   Qin, S.; Liu, F.; Wang, J.; Sun, B. Analysis and forecasting of the particulate matter (PM) concentration levels over four major cities of China using hybrid models. *Atmos. Environ.* **2014**, *98*, 665–675. [CrossRef]

11.   Fu, S.M.; Sun, J.H.; Ling, J.; Wang, H.J.; Zhang, Y.C. Scale interactions in sustaining persistent torrential rainfall events during the Mei-yu season. *J. Geophys. Res. Atmos.* **2016**, *121*, 12856–12876. [CrossRef]

12.   Liu, Z.Z.; Yan, Z.X.; Duan, J.; Qiu, Z.H. Infiltration regulation and stability analysis of soil slope under sustained and small intensity rainfall. *J. Cent. South Univ.* **2013**, *20*, 2519–2527. [CrossRef]

13.   Das, A.V.; Basu, S. Temporal trend of microsporidial keratoconjunctivitis and correlation with environmental and air pollution factors in India. *Indian J. Ophthalmol.* **2021**, *69*, 1089.

14.   Sharma, R.C.; Sharma, N. Statistical Investigation of Effect of Rainfall on Air Pollutants in the Atmosphere, Haryana State, Northern India. *Am. J. Environ. Prot.* **2018**, *6*, 14–21.

15.   Shaibu, V.O.; Weli, V.E. Relationship between PM$_{2.5}$ and Climate Variability in Niger Delta, Nigeria. *Am. J. Environ. Prot.* **2017**, *5*, 20–24. [CrossRef]

16.   Jassim, M.S.; Coskuner, G.; Munir, S. Temporal analysis of air pollution and its relationship with meteorological parameters in Bahrain. *Arab. J. Geosci.* **2018**, *11*, 1–15. [CrossRef]

17.   Witkowska, A.; Lewandowska, A.U. Water soluble organic carbon in aerosols (PM1, PM2.5, PM10) and various precipitation forms (rain, snow, mixed) over the southern Baltic Sea station. *Sci. Total Environ.* **2016**, *573*, 337–346. [CrossRef]

18.   Wei, L.A.; Yza, B.; Ping, L.C. Numerical simulation of the influence of major meteorological elements on the concentration of air pollutants during rainfall over Sichuan Basin of China. *Atmos. Pollut. Res.* **2020**, *11*, 2036–2048.

19.   Kapwata, T.; Wright, C.Y.; du Preez, D.J.; Kunene, Z.; Mathee, A.; Ikeda, T.; Landmandi, W.; Maharajj, R.; Sweijdk, N.; Minakawa, N.; et al. Exploring rural hospital admissions for diarrhoeal disease, malaria, pneumonia, and asthma in relation to temperature, rainfall and air pollution using wavelet transform analysis. *Sci. Total Environ.* **2021**, *2021*, 148207.

20.   Zhu, T.; Shang, J.; Zhao, D.F. The roles of heterogeneous chemical processes in the formation of an air pollution complex and gray haze. *Sci. China-Chem.* **2011**, *54*, 145–153. [CrossRef]

21.   Liu, F.; Tan, Q.; Jiang, X.; Yang, F.; Jiang, W. Effects of relative humidity and PM$_{2.5}$ chemical compositions on visibility impairment in Chengdu, China. *J. Environ. Sci.* **2019**, *86*, 15–23. [CrossRef]

22.   Guan, P.; Zhang, H.; Zhang, Z.; Chen, H.; Li, Y. Assessment of Emission Reduction and Meteorological Change in PM$_{2.5}$ and Transport Flux in Typical Cities Cluster during 2013–2017. *Sustainability* **2021**, *13*, 5685. [CrossRef]

23.   Zhang, C.; Pei, H.; Jia, Y.; Bi, Y.; Lei, G. Effects of air quality and vegetation on algal bloom early warning systems in large lakes in the middle–lower Yangtze River basin. *Environ. Pollut.* **2021**, *2012*, 117455. [CrossRef]

24.   Amelija, D.; Nenad, Z.; Emina, M.; Jasmina, R.; Miomir, R.; Ljiljana, Z. The effect of pollutant emission from district heating systems on the correlation between air quality and health risk. *Therm. Sci.* **2011**, *15*, 293–310. [CrossRef]

25.   Li, Y.; Zhu, K. Spatial dependence and heterogeneity in the location processes of new high-tech firms in Nanjing, China. *Pap. Reg. Sci.* **2017**, *96*, 519–536. [CrossRef]

26.   Zhai, K.; Gao, X.; Zhang, Y.; Wu, M. Perceived Sustainable Urbanization Based on Geographically Hierarchical Data Structures in Nanjing, China. *Sustainability* **2019**, *11*, 2289. [CrossRef]

27.   Chen, K.; Yan, Y.; Hu, Z. Influence of Air Pollutants on Fog Formation in Urban Environment of Nanjing, China. *Procedia Eng.* **2011**, *24*, 654–657. [CrossRef]

28.   Wang, G.; Huang, L.; Gao, S.; Song, T. Characterization of water-soluble species of PM10 and PM2.5 aerosols in urban area in nanjing, china. *Atmos. Environ.* **2002**, *36*, 1299–1307. [CrossRef]

29.   Chen, T.; He, J.; Lu, X.; She, J.; Guan, Z. Spatial and temporal variations of PM$_{2.5}$ and its relation to meteorological factors in the urban area of nanjing, china. *Int. J. Environ. Res. Public Health* **2016**, *13*, 921. [CrossRef]

30.   Zou, X.; Qian, Y.; Zhang, S. Spatiotemporal Variations of PM$_{2.5}$ Concentration and Relationship with Other Criteria Pollutants in Nanjing, China. *Nat. Environ. Pollut. Technol.* **2018**, *17*, 499–505.

31.   Guijo-Rubio, D.; Duran-Rosal, A.M.; Gutierrez, P.A.; Troncoso, A.; Hervas, M.C. Time-Series Clustering Based on the Characterization of Segment Typologies. *IEEE Trans. Cybern.* **2020**, *99*, 1–14. [CrossRef] [PubMed]

2.   Leles, M.C.; Sansão, J.P.H.; Mozelli, L.A.; Guimarães, H.N. Improving reconstruction of time-series based in singular spectrum analysis: A segmentation approach. *Digit. Signal Process.* **2017**, *77*, 63–76. [CrossRef]

3.   Onuorah, C.U.; Leton, T.G.; Momoh, Y. Influence of meteorological parameters on particle pollution (PM2.5 and PM10) in the tropical climate of port harcourt, nigeria. *Arch. Curr. Res. Int.* **2019**, 1–12. [CrossRef]

4.   Kayes, I.; Shahriar, S.A.; Hasan, K.; Akhter, M.; Salam, M.A. The relationships between meteorological parameters and air pollutants in an urban environment. *Glob. J. Environ. Sci. Manag.* **2019**, *5*, 265–278.

5.   Uchiyama, R.; Okochi, H.; Ogata, H.; Katsumi, N.; Nakano, T. Characteristics of trace metal concentration and stable isotopic composition of hydrogen and oxygen in "urban-induced heavy rainfall" in downtown Tokyo, Japan; the implication of mineral/dust particles on the formation of summer heavy rainfall. *Atmos. Res.* **2018**, *217*, 73–80. [CrossRef]

6.   Gautam, S.; Yadav, A.; Pillarisetti, A.; Smith, K.; Arora, N. Short-term introduction of air pollutants from fireworks during diwali in Rural Palwal, Haryana, India: A case study. *IOP Conf. Ser.: Earth Environ. Sci.* **2018**, *120*, 012009. [CrossRef]

7.   Yang, W.; He, Z.; Huang, H.; Huang, J. A clustering framework to reveal the structural effect mechanisms of natural and social factors on $PM_{2.5}$ concentrations in China. *Sustainability* **2021**, *13*, 1428. [CrossRef]

8.   Kashki, A.; Karami, M.; Zandi, R.; Roki, Z. Evaluation of the effect of geographical parameters on the formation of the land surface temperature by applying ols and gwr, a case study shiraz city, Iran. *Urban Clim.* **2021**, *37*, 100832. [CrossRef]

9.   Zhang, Y.; Zhang, F.; Zhu, H.; Guo, P. An optimization-evaluation agricultural water planning approach based on interval linear fractional bi-level programming and iahp-topsis. *Water* **2019**, *11*, 1094. [CrossRef]

10. Kulanuwat, L.; Chantrapornchai, C.; Maleewong, M.; Wongchaisuwat, P.; Boonya-Aroonnet, S. Anomaly detection using a sliding window technique and data imputation with machine learning for hydrological time series. *Water* **2021**, *13*, 1862. [CrossRef]

11. Cai, T.T.; Ma, Z.; Wu, Y. Sparse pca: Optimal rates and adaptive estimation. *Ann. Stat.* **2012**, *41*, 3074–3110. [CrossRef]

12. Wang, S.; Li, C.; Zhao, K.; Chen, H. Learning to context-aware recommend with hierarchical factorization machines. *Inf. Sci.* **2017**, *409–410*, 121–138. [CrossRef]

 *sustainability*

Article

PM$_{2.5}$ Concentration Prediction Based on Spatiotemporal Feature Selection Using XGBoost-MSCNN-GA-LSTM

Hongbin Dai , Guangqiu Huang *, Huibin Zeng and Fan Yang

School of Management, Xi'an University of Architecture and Technology, Xi'an 710055, China; daihongbin@xauat.edu.cn (H.D.); zenghuibin@xauat.edu.cn (H.Z.); yangfan@xauat.edu.cn (F.Y.)
* Correspondence: gqhuang@xauat.edu.cn; Tel.: +86-152-7710-7077

Abstract: With the rapid development of China's industrialization, air pollution is becoming more and more serious. Predicting air quality is essential for identifying further preventive measures to avoid negative impacts. The existing prediction of atmospheric pollutant concentration ignores the problem of feature redundancy and spatio-temporal characteristics; the accuracy of the model is not high, the mobility of it is not strong. Therefore, firstly, extreme gradient lifting (XGBoost) is applied to extract features from PM$_{2.5}$, then one-dimensional multi-scale convolution kernel (MSCNN) is used to extract local temporal and spatial feature relations from air quality data, and linear splicing and fusion is carried out to obtain the spatio-temporal feature relationship of multi-features. Finally, XGBoost and MSCNN combine the advantages of LSTM in dealing with time series. Genetic algorithm (GA) is applied to optimize the parameter set of long-term and short-term memory network (LSTM) network. The spatio-temporal relationship of multi-features is input into LSTM network, and then the long-term feature dependence of multi-feature selection is output to predict PM$_{2.5}$ concentration. A XGBoost-MSCGL of PM$_{2.5}$ concentration prediction model based on spatio-temporal feature selection is established. The data set comes from the hourly concentration data of six kinds of atmospheric pollutants and meteorological data in Fen-Wei Plain in 2020. To verify the effectiveness of the model, the XGBoost-MSCGL model is compared with the benchmark models such as multilayer perceptron (MLP), CNN, LSTM, XGBoost, CNN-LSTM with before and after using XGBoost feature selection. According to the forecast results of 12 cities, compared with the single model, the root mean square error (RMSE) decreased by about 39.07%, the average MAE decreased by about 42.18%, the average MAE decreased by about 49.33%, but R^2 increased by 23.7%. Compared with the model after feature selection, the root mean square error (RMSE) decreased by an average of about 15% On average, the MAPE decreased by 16%, the MAE decreased by 21%, and R^2 increased by 2.6%. The experimental results show that the XGBoost-MSCGL prediction model offer a more comprehensive understanding, runs deeper levels, guarantees a higher prediction accuracy, and ensures a better generalization ability in the prediction of PM$_{2.5}$ concentration.

Keywords: XGBoost; MSCNN; genetic algorithm; LSTM; feature selection; spatiotemporal feature extraction

Citation: Dai, H.; Huang, G.; Zeng, H.; Yang, F. PM$_{2.5}$ Concentration Prediction Based on Spatiotemporal Feature Selection Using XGBoost-MSCNN-GA-LSTM. *Sustainability* **2021**, *13*, 12071. https://doi.org/10.3390/su132112071

Academic Editors: Jun Yang, Ayyoob Sharifi, Baojie He and Chi Feng

Received: 18 September 2021
Accepted: 28 October 2021
Published: 1 November 2021

Publisher's Note: MDPI stays neutral with regard to jurisdictional claims in published maps and institutional affiliations.

1. Introduction

With the increasing of environmental pollution, the weather issue of haze is spreading in China's major cities. PM$_{2.5}$ has become a major problem of air pollution. Recent studies have shown that PM$_{2.5}$ leads to the occurrence of respiratory diseases, immune diseases, cardiovascular and cerebrovascular diseases and tumors [1,2]. Accurate prediction and early warnings of the concentration of PM$_{2.5}$ are of great significance. Many scholars have begun to integrate multiple data features, but too many data and factor features will affect the prediction effect, and redundant features will affect the performance of model prediction. Therefore, many scholars have begun to use feature selection to make predictions. For example: In power system, cooperative search algorithm is used to select

103

power load features [3], and minimum redundancy and maximum correlation are used to obtain the best feature set of power load [4]; In wind energy, the multi-agent feature selection method is used to establish the wind speed prediction model [5]; In the stock market, using random forest combined with depth generation model is used to predict the daily stock trend [6]; In tourism, random forest is used for feature selection to predict the number of visitors [7]; In agriculture, model feature (MF) and principal component analysis (PCA) is combined with regression algorithm to predict the water content of rice canopy [8]; In the economy, the genetic algorithm-based feature selection (GAFS) method combined with random forest is used to estimate the per capita medical expenses [9]; In the aspect of transportation, XGBoost (extreme gradient enhancement) screening feature combined with long-term and short-term memory network is used to predict airport passenger flow [10].

Aiming at air quality prediction, the main models used in the existing research include linear regression model [11], grey model [12], geographical weighted regression model [13], mixed effect model [14] and generalized weighted mixed model [15]. In essence, these statistical models are still linear, although the complex relationship between $PM_{2.5}$ and other factors is simplified in the model, the prediction result of $PM_{2.5}$ concentration still remain uncertain. With the development of computer technology, machine learning (including deep learning) methods are increasingly used in $PM_{2.5}$ concentration estimation due to their strong nonlinear modeling ability, such as support vector regression (SVR) [16], k-nearest neighbor (KNN) [17], random forest (RF) [18], multilayer neural network (MLP) [19], artificial neural network (ANN) [20], long-term memory network (LSTM) [21], convolution neural network (CNN) [22], and chemical transport model (CTM) [23]. These models all show better performance than traditional statistical models in predicting $PM_{2.5}$ concentration, and have stronger nonlinear expression capabilities.

In order to better predict air quality, many scholars have also begun to apply feature selection to air pollutants. Jin et al. [24] proposed a hybrid deep learning prediction that decomposes $PM_{2.5}$ data through empirical mode decomposition (EMD) and Convolutional Neural Network (CNN) so that an air pollution prediction model can be established. Masmoudi et al. [25] combined the multi-objective regression method with random forest to perform feature selection and predict the concentration of multiple air pollutants. Mehdi et al. [26] studied the impact of feature importance on $PM_{2.5}$ prediction in Teheran urban area, and used random forest, XGBoost and deep learning technology, of which XGBoost was used to obtain the best model. Zhang et al. [27] used XGBoost model to screen out the most critical characteristics and predict the $PM_{2.5}$ pollutant concentration in Beijing in the next 24 h. Ma et al. [28] used XGBoost and grid importance to predict $PM_{2.5}$ in the Northwestern United States. Zhai et al. [29] used XGBoost for feature screening and predicted the daily average concentration of $PM_{2.5}$ in Beijing area of GA-MLP. Gui et al. [30] used XGBoost model to build a virtual ground-based $PM_{2.5}$ observation network at 1180 meteorological stations in China, as a result, he found that XGBoost model has strong robustness and accuracy for $PM_{2.5}$ prediction.

At present, some researchers use deep learning method to estimate the spatial and temporal distribution of $PM_{2.5}$ concentration. Although the traditional prediction model adds multivariate features, it ignores the impact of redundant features on the prediction results, resulting in the impact of features with little correlation and importance on the prediction results. The scale of relevant models is still relatively small, and it still relies on artificial feature selection to a large extent, and does not make full use of deep learning method to give full play to the advantages of deep learning through deeper and wider network structure. In the related research on the prediction of $PM_{2.5}$ using feature selection, the prediction is mainly based on a single model, not the perspective of spatio-temporal features, and the importance of feature selection is too emphasized in the related research on the application of feature selection for prediction. The problem of insufficient precision still remains unsolved. The single or combined $PM_{2.5}$ concentration prediction model does not show strong robustness, and the degree of model optimization is not high. Existing researches are limited to cities in specific regions, ignoring the predictive performance of the model itself, resulting in poor applicability and migration of the model used.

The main contributions of this paper are as follows:

(1) In terms of the research object, the air quality of Fenwei plain is worse than that of other regions in China. Therefore, it is typical to predict and analyze the $PM_{2.5}$ concentration of the cities in this region. In this paper, the $PM_{2.5}$ concentration of 12 cities in this region is predicted. Through the simulation and comparison in 12 cities, the portability and applicability of this study are verified.

(2) In terms of prediction model, firstly, Pearson correlation analysis and XGBoost are used to select the features of $PM_{2.5}$ to solve the problem of feature redundancy, and the optimal features are extracted through one-dimensional multi-scale convolution kernel to solve the local time relationship and spatial feature relationship in air quality data. Then the parameters of LSTM are optimized by genetic algorithm to solve the accuracy problem of the model. Finally, the extracted features are input into LSTM for prediction. An XGBoost MSCGL (XGBoost-MSCNN-GA-LSTM) model is proposed to improve the $PM_{2.5}$ prediction of Fenwei plain. The combined model constructed in this paper not only conforms to the temporal characteristics of prediction data, solves the problem of feature redundancy and insufficient accuracy of the traditional machine model, but also follows the optimal and simplest principle in the nesting of the model.

(3) In terms of prediction results, the experiment also discusses the $PM_{2.5}$ h concentration prediction under the influence of different characteristics. The prediction results show that appropriate input characteristics will help to improve the prediction accuracy of the model, and the model has been proved for many times that the prediction accuracy of the combined prediction model proposed in this paper is higher than that of a single deep learning model. After many experiments, it is found that the prediction results of XGBoost mscgl are better than XGBoost CNN, XGBoost LSTM, XGBoost MLP and XGBoost CNN LSTM models. The advantages of the proposed model are verified from multiple angles and multiple evaluation indexes, and the experimental results show that the proposed model has good robustness.

2. Study Area and Data

2.1. Study Area

Fenwei plain is the general name of Fenhe plain, Weihe plain and its surrounding terraces in the Yellow River Basin. It ranges from the north, Yangqu County in Shanxi Province to the south, Qinlin Moutains in Shaanxi Province, and to the west, Baoji City in Shaanxi Province. It is distributed in Northeast southwest direction, about 760 km long and 40–100 km wide. It has a population of 55,5445., including Xi'an, Baoji, Xianyang, Weinan and Tongchuan in Shaanxi Province, Taiyuan, Jinzhong, Lvliang, Linfen and Yuncheng in Shanxi Province, and Luoyang and Sanmenxia in Henan Province. Since 2019, Fenwei plain is still the area with the highest $PM_{2.5}$ concentration in China. The average $PM_{2.5}$ concentration in autumn and winter is about twice as much as other seasons, and the days of heavy pollution account for more than 95% of the whole year [31]. In 2020, the average concentration of $PM_{2.5}$ in Fenwei plain was 70 $\mu g/m^3$, and serious pollution occurred in 152 days. Table 1 shows the factors of air pollutants [32].

Table 1. Air Pollutant Factors of $PM_{2.5}$ Concentration Prediction Model.

Variable	Unit	Variable	Unit
$PM_{2.5}$	$\mu g/m^3$	CO	mg/m^3
PM_{10}	$\mu g/m^3$	NO_2	$\mu g/m^3$
SO_2	$\mu g/m^3$	O3_8h	$\mu g/m^3$

2.2. Study Data

2.2.1. Air Quality Data

Since December 2013, the China Environmental Protection Agency (EPA) has published open air quality observation data from China's ground monitoring stations. The study data in this article comes from the atmospheric pollutants of 12 cities in Xi'an, Baoji, Xianyang, Weinan, Tongchuan, Taiyuan, Jinzhong, Luliang, Linfen, Yuncheng, Luoyang, and Sanmenxia from 1 January 2020 to 31 December 2020 ($PM_{2.5}$, PM_{10}, NO_2, SO_2, O_3, CO hourly concentration data set, Table 1 is the atmospheric pollutant factors of $PM_{2.5}$ concentration prediction model. There are 2,838,240 pieces of air quality data and meteorological data in 12 cities.

2.2.2. Meteorological Data

The meteorological data of this paper come from the Chinese weather website platform. As shown in Table 2, through data preprocessing, 21 types of meteorological factors are selected in this paper, and they are average surface temperature, maximum surface temperature, minimum surface temperature, daily average wind speed, daily maximum wind speed, daily maximum wind direction, maximum wind speed, maximum wind direction, daily precipitation of maximum wind speed, 20–8 h (mm) precipitation, 8–20 h (mm) precipitation, 20–20 h (mm) precipitation, average temperature, maximum temperature, minimum temperature, daily average pressure, daily maximum pressure, daily minimum pressure, sunshine hours, daily average relative humidity, daily minimum relative humidity, and season.

Table 2. Meteorological Factors of $PM_{2.5}$ Concentration Prediction Model.

Variable	Unit	Abbreviation	Variable	Unit	Abbreviation
Average surface temperature	°C	avg (ST)	Average temperature	°C	avg (T)
Maximum surface temperature	°C	high (ST)	Maximum temperature	°C	high (T)
Lowest surface temperature	°C	low (ST)	Minimum temperature	°C	low (T)
Average wind speed	m/s	avg (m/s)	Sunshine duration	h	sunshine (h)
Maximum wind speed	m/s	high (m/s)	Average humidity	%	avg (%)
Daily maximum wind speed and direction	-	highdirection	Lowest humidity	%	low (%)
Extreme wind speed	m/s	extrem (m/s)	Average air pressure	hPa	avg (hPa)
Extreme wind direction	-	extremdirection	Maximum daily pressure	hPa	high (hPa)
20–8 h (mm) precipitation	mm	20–8 (mm)	Lowest daily pressure	hPa	low (hPa)
8–20 h (mm) precipitation	mm	8–20 (mm)	Season	-	season
20–20 h (mm) precipitation	mm	20–20 (mm)			

2.3. Data Processing

2.3.1. Division of Data Set

The data set needs to be divided before it can be input to the model for training. Otherwise, the prediction model will have no additional data for effect evaluation, and the training results may be overfitted due to training on all data. In the experiment, each data set is divided into training set and test set, after that, the training set is divided into training set and verification set. The data ratio of training set, test set, and verification set is 6:2:2. The training set mainly learns the sample data set and establishes a classifier by matching some parameters. A classification method is established, which is mainly used to train the model. The verification set is used to determine the network structure or the parameters controlling the complexity of the model, and select the number of hidden units in the neural network. The test set is used to test the performance of the finally selected optimal model. It mainly tests the resolution of the trained model (recognition rate, etc.).

2.3.2. Raw Data Processing

Identification and Processing of Abnormal Data

Abnormal data may be caused by errors in the process of collecting and recording data. Abnormal data will affect the prediction accuracy of the model, so it is necessary to identify and process the abnormal data. Outlier detection is used to find outliers. Here, quartile analysis is used to identify outliers. First, the first quartile and the third quartile of variables are solved. If there is a value less than the first quartile or greater than the third quartile, the value is determined as an outlier. The horizontal processing method is used to correct the abnormal data.

The calculation formula of horizontal treatment method is shown in Equations (1) and (2) If,

$$\begin{cases} |y_i - y_{i-1}| < \varepsilon_a \\ |y_i - y_{i+1}| > \varepsilon_a \end{cases} \tag{1}$$

Then,

$$y_t = \frac{y_{t+1} + y_{t-1}}{2} \tag{2}$$

Among them, y_i represents the concentration of air pollutants in a certain day or hour, y_{i-1} represents the concentration of air pollutants in the previous day or hour, and y_{i+1} represents the concentration of air pollutants in the next day or hour, ε_a represents the threshold.

Data Normalization

Due to the different meanings and dimensions of physical quantities such as air pressure and evaporation, the input to the prediction model will have an impact on results., so it is necessary to normalize such data. The input of normalized data into the prediction model can effectively reduce the training time of the model, accelerate the convergence speed of the model, and further improve the prediction accuracy of the model. The normalized calculation formula of the data is shown in Equation (3). This method realizes the equal scaling of the original data [33]:

$$x_{norm} = \frac{x - x_{\min}}{x_{\max} - x_{\min}} \tag{3}$$

Among them, x_{norm} is the normalized value, x is the original data, $x_{\min}$ is the minimum value in the original data, $x_{\max}$ is the maximum value in the original data, and the size of the normalized data is constrained between 0 to 1 interval.

3. Method

3.1. XGBoost

XGBoost is an extreme gradient boosting decision tree, which belongs to a machine learning algorithm. The algorithm introduces regular items during the generation period and prunes at the same time, making the algorithm more efficient and more accurate. [34]

XGBoost (eXtreme Gradient Boosting) can be expressed in a form of addition, as shown in Equation (4):

$$\hat{y}_i = \sum_{k=1}^{K} f_k(x_i), f_k \in F \tag{4}$$

Among them, $\hat{y}_i$ represents the predicted value of the model; K represents the number of decision trees, f_k represents the k sub-models, x_i represents the i-th input sample, F represents the set of all decision trees. The objective function of XGBoost consists of two parts: a loss function and a regular term, as shown in Equations (5) and (6):

$$L(\varphi)^t = \sum_{i=1}^{n} l(y_i, \hat{y}_i^{(t-1)} + f_t(x_i)) + \Omega(f_k) \tag{5}$$

$$\Omega(f_k) = \gamma T + \frac{1}{2}\lambda\|\omega\|^2 \tag{6}$$

Among them, $L(\varphi)^t$ represents the objective function of the tth iteration, $\hat{y}_i^{(t-1)}$ represents the predicted value of the $(t-1)$ iteration; $\Omega(f_k)$ represents the regular term of the model of the tth iteration, which plays a role in reducing overfitting; γ and λ represent the regular term Coefficient to prevent the decision tree from being too complicated, T represents the number of leaf nodes of the model.

Using Taylor's formula to expand the objective function shown in Equation (7), we can get:

$$\begin{aligned} L(\phi) &\cong \sum_{i=1}^{n} [g_i f_t(x_i) + \tfrac{1}{2} h_i f_t^2(x_i)] + \gamma T + \tfrac{1}{2}\lambda \sum_{j=1}^{T} \omega_j^2 \\ &\cong \sum_{j=1}^{T} [(\sum_{i \in I_j} g_i)\omega_j + \tfrac{1}{2}(\sum_{i \in I_j} h_i + \lambda)\omega_j^2] + \gamma T \end{aligned} \tag{7}$$

Among them, g_i represents the first derivative of sample x_i; h_i represents the second derivative of sample x_i; ω_j represents the output value of the j-th leaf node, and I_j represents the sample subset of the value of the j-th leaf node.

It can be seen from Equation (7) that the objective function is a convex function. Taking the derivative of ω_j and making the derivative function equal to zero, the objective function can reach the minimum value of ω_j, as shown in Equation (8):

$$\omega_j^* = -\frac{\sum\limits_{i \in I_j} g_i}{\sum\limits_{i \in I_j} h_i + \lambda} \tag{8}$$

Equation (9) can be used to evaluate the quality of a tree model. The smaller the value, the better the tree model. It can be easily concluded that we can obtain the scoring formula for the tree to split the node:

$$\hat{L}(\phi)_{\min} = -\frac{1}{2}\sum_{j=1}^{T} \frac{\left(\sum\limits_{i \in I_j} g_i\right)^2}{\sum\limits_{i \in I_j} h_i + \lambda} + \gamma T \tag{9}$$

Equation (10) is used to calculate the split node of the tree model.

$$Gain = -\frac{1}{2}\left[\frac{(\sum_{i\in I_j} g_i)^2}{\sum_{i\in I_j} h_i + \lambda} + \frac{(\sum_{i\in I_R} g_i)^2}{\sum_{i\in I_R} h_i + \lambda} + \frac{(\sum_{i\in I} g_i)^2}{\sum_{i\in I} h_i + \lambda}\right] - \gamma \tag{10}$$

3.2. One-Dimensional Multi-Scale Convolution Kernel (MSCNN)

Convolutional neural network has been successfully applied to image recognition direction, which verifies that the network has a strong extraction of feature map. Based on the analysis of the data set, it is found that the characteristics of the data are multi features, shown in the form of numerical value, rather than in the form of feature map. Therefore, this study preprocesses the data, combines the characteristics of the data into a feature map, and inputs it to the convolution neural network to complete the extraction of the spatial and temporal characteristics of the air pollutant concentration data and meteorological factors [35]. The spatiotemporal feature extraction of single factor $PM_{2.5}$ is shown in Figure 1. Among them, the feature map is traversed from left to right on the data feature axis through a one-dimensional multi-scale convolution kernel to complete the convolution operation, the number of steps is 1, and the feature vectors output by different convolution kernels are spliced and fused to obtain a single factor. The spatial characteristics of the relationship. On the time axis, as the convolution kernel traverses from top to bottom to complete the convolution operation, the number of steps is 1, and the local trend of the single factor changing over time can be obtained. Finally, the spliced and fused feature vectors are merged in the data feature direction, and the spatio-temporal features of multi-site $PM_{2.5}$ are output.

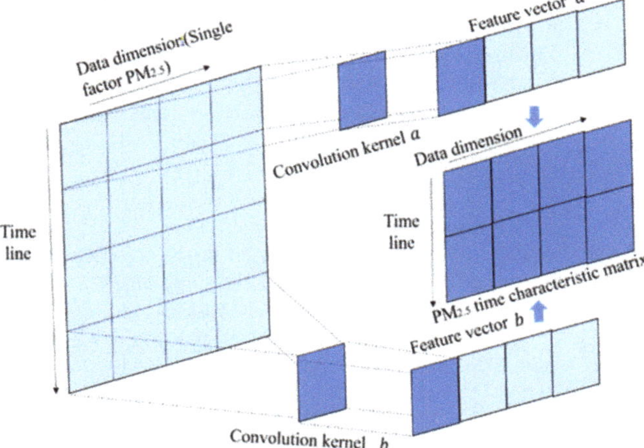

Figure 1. One-dimensional convolution feature extraction process diagram.

The following is the formula derivation of MSCNN's convolution operation on the special whole. The feature map contains N sample data and M air pollutant factors. Then the feature map formula of single factor i is as shown in Equations (11) and (12):

$$X_i = [x_i^1, x_i^2, x_i^3, \ldots, x_i^N]^T \tag{11}$$

$$X_i^{t:t+T-1} = [x_i^t, x_i^{t+1}, x_i^{t+2}, \ldots, x_i^{t+T-1}]^T \tag{12}$$

In the formula, $X_i^t = [x_i^t, x_i^{t+1}, x_i^{t+2}, \ldots, x_i^{t+T-1}] \in R$ represents the vector of the single factor i at time t, $X_i^{t:t+T-1}$ represents the T group vector of X_i in the $[t, t+T-1]$ time zone, and T represents the matrix transpose.

The convolution operation multiplies the weight matrix W_j by $X_i^{t:t+T-1}$.

(1) Single-factor spatial feature relationship: multiply W_j by $X_i^{t:t+T-1}$ on the data feature axis.

(2) Single factor time change feature: multiply x by y on the time feature axis.

When the first convolution kernel traverses the entire feature map on the time axis and the number of steps is 1, the feature vector a_i^j is obtained, and its size is $N - T + 1$, and the eigenvectors obtained by multiple convolution kernels Z merge $[N - T + 1] \times Z$ size A_i in the data feature direction, and A_i represents the single-factor spatiotemporal feature matrix, as shown in Equations (13) and (14).

$$a_i^j = \left[a_{t+T-1,j}^j, a_{t+T,j}^j, a_{t+T+1,j}^j, \ldots, a_N^j \right] \tag{13}$$

$$A_i = \left[a_n^1, a_n^2, a_n^3, \ldots, a_n^Z \right] \tag{14}$$

So far, the single-factor spatiotemporal feature extraction has been completed, but the data set also contains other features, such as NO2, SO2, CO, etc. A total of M factors, so we can extract the M factors through the same operation as above, and then they can be extracted. Single-feature spatio-temporal feature matrix, and then linearly splicing and fusion them to form a multi-factor fusion spatio-temporal feature matrix A, as shown in Equation (15):

$$A = [A_1, A_2, A_3, \ldots, A_M] \tag{15}$$

Based on MSCNN convolution neural network, the space-time characteristics of air quality data are extracted. This method makes a simple transformation of the two dimensional feature map to form a side-by-side one-dimensional feature map, which makes the network training show better generalization ability. Meanwhile, the convolution neural network automatic feature extraction method replaces the traditional artificial feature selection method, which makes the feature extraction more comprehensive and deeper.

3.3. Genetic Algorithm

The genetic algorithm is a method to perform crossover and mutation operations on feasible solutions in the population, so the objective function of the genetic algorithm does not require derivable or continuous conditions. The genetic algorithm applies a probabilistic optimization method to automatically obtain and guide the optimized search space and adaptively adjust the search direction. The genetic algorithm is simple, universal, and suitable for parallel processing. The specific steps of the algorithm are shown in Figure 2.

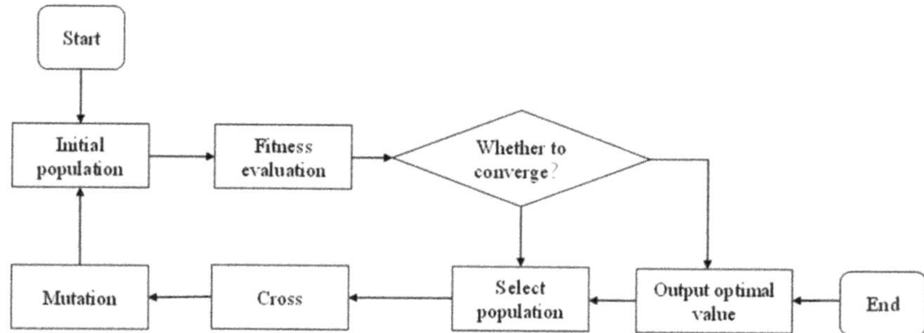

Figure 2. GA algorithm flow.

The GA process can be divided into six stages: initialization, fitness calculation, checking termination conditions, crossover, selection, and mutation. In the initialization phase, a chromosome is selected arbitrarily in the search space, and then the fitness of the chromosome is determined according to the preset fitness function. For optimization algorithms such as GA, the fitness function is a key factor that affects the performance of the model. Chromosomes are randomly selected based on the fitness of the fitness function. Dominant chromosomes have a higher chance of being inherited to the next generation. The selected dominant chromosomes can produce offspring through the exchange of similar segments and changes in gene combinations.

3.4. LSTM

Long Short-Term Memory (LSTM) is an improvement of Recurrent Neural Network (RNN) [36]. RNN has a higher probability of gradient disappearance and gradient explosion during training, and there is a long-term dependence problem. LSTM can effectively solve this problem. LSTM introduces a gate mechanism, which makes LSTM have a longer-term memory than RNN and can be more effective in learning. In LSTM, each neuron is equivalent to a memory cell (cell, c_t). LSTM controls the state of the memory cell through a "gate" mechanism, thereby increasing or deleting the information in it. The structure of LSTM is shown in Figure 3.

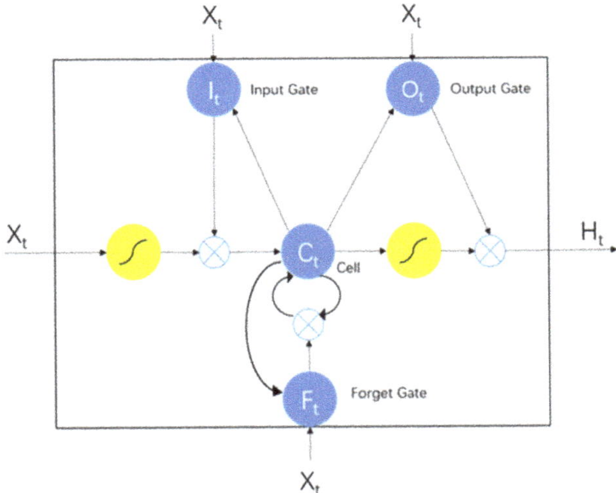

Figure 3. LSTM Unit Structure.

In the LSTM cell structure, the Input Gate (i_t) is used to determine what information is added to the cell, and the Forget Gate (f_t) is used to determine what information is deleted from the cell. The Output Gate (o_t) is used to determine what information is output from the cell. The complete training process of LSTM is that at each time t, the three gates receive the input vector x_t at time t and the hidden state h_{t-1} of the LSTM at time $t-1$ and the information of the memory unit c_t and then perform the received information Logical operation, the logical activation function σ decides whether to activate i_t, and then synthesize the processing result of the input gate and the processing result of the forgetting gate to generate a new memory unit c_t, and finally obtain the final output result h_t through the nonlinear operation of the output gate. The calculation formula for each process as shown in Equations (16)–(20).

Input Gate calculation formula:

$$i_t = \sigma(W_{xi}^T x_t + W_{hi}^T h_{t-1} + b_i) \tag{16}$$

Forget Gate calculation formula:

$$f_t = \sigma(W_{xf}^T x_t + W_{hf}^T h_{t-1} + b_f)$$ (17)

output gate calculation formula:

$$o_t = \sigma(W_{xo}^T x_t + W_{ho}^T h_{t-1} + b_o)$$ (18)

Memory unit calculation formula, the internal hidden state:

$$c_t = f_t \times c_{t-1} + i_t \times \tanh(W_{xc}^T x_t + W_{hc}^T h_{t-1} + b_c)$$ (19)

Hidden state calculation formula:

$$h_t = o_t \tanh(c_t)$$ (20)

Among them, σ represents generally a nonlinear activation function, such as a sigmoid or tanh function. W_{xi}, W_{xf}, W_{xo}, W_{xc} represents the weight matrices of nodes connected to the input vector W_t for each layer, W_{hi}, W_{hf}, W_{ho}, W_{hc} represents the weight matrices connected to the previous short-term state h_{t-1} for each layer, b_i, b_f, b_o, b_c represents the offset terms of each layer node. In short, the input gate in LSTM can identify important inputs, and the forget gate can reasonably retain important information and extract it when needed. Therefore, this feature of LSTM can effectively identify long-term patterns such as time series, making training convergence faster.

3.5. XGBoost-MSCGL Model

Figure 4 shows the XGBoost-MSCGL process. First, the atmospheric pollutant data and meteorological data are normalized and processed with missing values. Secondly, Pearson analyzes the correlation of the original data and uses XGBoost to select the importance of features. Furthermore, input the data after feature selection into MSCNN, and use the MSCNN algorithm to extract the temporal and spatial features of the data. At the same time, GA is used to optimize the parameters of the LSTM, the best fitness output of the chromosome is used as the global optimal parameter combination of the LSTM network, and then the data extracted from the spatiotemporal features are input into the optimized LSTM for prediction. In order to better verify the effect of the model, finally combined models such as XGBoost-MLP, XGBoost-LSTM, XGBoost-CNN are used for comparison, and then RMSE, MAE, MAPE and other indicators are used for evaluation.

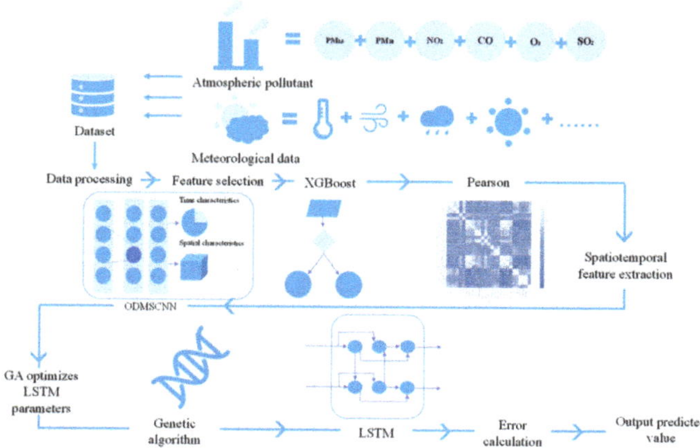

Figure 4. XGBoost-MSCGL Model Process.

3.6. Evaluation Index

In order to measure the accuracy of the prediction model, this paper uses Root Mean Square Error (RMSE), Mean Absolute Error (MAE) and Mean Absolute Percentage Error (MAPE) as evaluation indicators. The formulas are shown in Equations (21)–(23).

$$\text{RMSE} = \sqrt{\frac{1}{N}\sum_{i=1}^{N}\left(X_i - \overline{X}\right)^2} \tag{21}$$

$$MAE = \frac{1}{N}\sum_{i=1}^{N}\left|X_i - \overline{X}\right|^2 \tag{22}$$

$$MAPE = \frac{100}{N}\sum_{i=1}^{N}\left|\frac{X_i - \overline{X}_i}{X_i}\right| \tag{23}$$

$$R_2 = 1 - \frac{\sum_i \left(\hat{y}_i - y_i\right)^2}{\sum_i \left(\overline{y} - y_i\right)^2} \tag{24}$$

Where $\hat{y}$ represents the predicted value, y_i is the true value, and N is the number of test samples. The ranges of RMSE, MAE, and MAPE are all $[0, +\infty)$. Generally, the larger the value of RMSE and MAE, the greater the error and the lower the prediction accuracy of the model. MAPE is the most intuitive prediction accuracy criterion. When MAPE tends to 0%, it means the model is perfect, when MAPE tends to 100%, it means that the model is inferior. Generally, it can be considered that the prediction accuracy is higher when the MAPE is less than 10% [37]. R^2 measures the applicability of the model to sample values and can test the prediction ability of the model. The closer to 1, the higher the fitness of the model, and the closer to 0, the lower the fitness of the model.

4. Results

4.1. Analysis of Factor Characteristics

In order to better analyze the characteristics of the model input factors, the Pearson correlation method is used for analysis. As shown in Figure 5, the factors for the correlation coefficient of $PM_{2.5}$ in Yuncheng are PM_{10} (0.9) and CO (0.8), which are highly positively correlated. Further, SO_2 (0.5), average humidity (0.5), and seasons (0.5) are moderately positively correlated, and the average surface temperature (-0.5), the highest surface temperature (-0.5), the duration of sunshine (-0.5), the average temperature (-0.5), the lowest temperature (-0.5), and the highest temperature (-0.5) have a moderately negative correlation. The factors of the correlation coefficient of Xianyang $PM_{2.5}$ are that CO (0.9) is highly positively correlated. Further, PM_{10} (0.6), the lowest humidity (0.5), season (0.6), etc. are moderately correlated, the average surface temperature (-0.5), the surface, the lowest temperature (-0.5), the highest surface temperature (-0.5), the average temperature (-0.5), the lowest temperature (-0.5), and the highest temperature (-0.5) have a moderately negative correlation. The correlation coefficients of $PM_{2.5}$ in Xi'an are PM_{10} (0.8), CO (0.9) and SO_2 (0.7), which are highly positively correlated, and season (0.5) is moderately correlated. The average surface temperature (-0.6), minimum surface temperature (-0.6), extremely high wind speed (-0.5), average temperature (-0.6), minimum temperature (-0.5) and maximum temperature (-0.6) are moderately negatively correlated. The correlation coefficient of $PM_{2.5}$ in Weinan is that PM_{10} (0.8) and CO (0.8) are highly positively correlated. average humidity (0.5), minimum humidity (0.5) and season (0.5) are moderately correlated, sunshine duration (-0.6) is highly negatively correlated, and maximum surface temperature (-0.5), maximum wind speed (-0.5), average temperature (-0.5) and maximum temperature (-0.5) are moderately negatively correlated. The factors of the correlation coefficient of Taiyuan $PM_{2.5}$ are PM_{10} (0.9) and CO (0.9), which are highly positively correlated. NO_2 (0.5), SO_2 (0.6), average humidity (0.6), minimum humidity (0.6), season (0.5) It is moderately correlated. The highest surface temperature (-0.5), highest wind speed (-0.5), extremely high wind speed (-0.5), and sunshine duration (-0.5) are moderately negatively correlated The correlation coefficient of $PM_{2.5}$ in Tongchuan had a high positive correlation with CO (0.9), moderate correlation with PM_{10} (0.6), average humidity (0.5), minimum humidity (0.5) and season (0.5), and moderate negative correlation with maximum surface temperature (-0.5), maximum wind speed (-0.5), extremely high wind speed (-0.5) and sunshine duration (-0.5). The factors of the correlation coefficient of $PM_{2.5}$ in Sanmenxia are PM_{10} (0.7) and CO (0.8), which are highly positively correlated, the average humidity (0.5), the lowest humidity (0.5), and the season (0.5) are moderately positively correlated and the average surface temperature (-0.5), the highest surface temperature (-0.5), extremely high wind speed (-0.5), average temperature (-0.5), and highest temperature (-0.5) have a moderately negative correlation. The factors of the correlation coefficient of Lvliang $PM_{2.5}$ are PM_{10} (0.6), CO (0.7), average humidity (0.6), minimum humidity (0.6) and season (0.6), which are highly positively correlated. The average surface temperature (-0.5) and the surface average temperature (-0.5), extremely high wind speed (-0.5), average temperature (-0.5), maximum temperature (-0.5), sunshine duration (-0.5) are moderately negatively correlated, average humidity (0.5), minimum humidity (0.5), season (0.5), etc. have a moderate correlation. Luoyang $PM_{2.5}$ correlation coefficient factors are PM_{10} (0.8) and CO (0.9) are highly positively correlated, the average surface temperature (-0.5), the highest surface temperature (-0.5), the average temperature (-0.5), the highest temperature (-0.5), sunshine duration (-0.5) are moderately negatively correlated, and average humidity (0.5), minimum humidity (0.5), season (0.5), etc. are moderately correlated. Linfen $PM_{2.5}$ correlation coefficient factors PM_{10} (0.9) and CO (0.9) are extremely highly positively correlated. Further, SO_2 (0.7), average humidity (0.6), minimum humidity (0.7), season (0.6) are highly positively correlated. The average surface temperature (-0.6), the highest surface temperature (-0.7), the average temperature (-0.6), the lowest temperature (-0.5), the highest temperature (-0.6), and the duration of sunshine (-0.5) have a moderately negative correlation. The correlation coefficients of $PM_{2.5}$, PM_{10} (0.9) and CO (0.9), $SO2$ (0.7), average humidity (0.6) and minimum humidity (0.7) in Jinzhong were highly positively correlated. The average surface temperature (-0.5), maximum surface temperature (-0.6), average temperature (-0.5), minimum temperature (-0.4), maximum temperature (-0.5), and sunshine duration (-0.5) were moderately negatively correlated The correlation coefficient of $PM_{2.5}$ was PM_{10} (0.7) and CO (0.8), and the average humidity (0.5), minimum humidity (0.5) and season (0.5) were moderately correlated. The average temperature (-0.5), the maximum temperature (-0.5), the average temperature (-0.5), the minimum temperature (- 0.4), and the maximum temperature (-0.5) were moderately negatively correlated.

Meteorological elements affect air quality by affecting the accumulation, diffusion and elimination of pollutants. In the studies of $PM_{2.5}$ and PM_{10} concentration, they are found closely related to meteorological elements (such as temperature, precipitation, wind speed, etc.). According to existing studies, relative humidity has an important a key factor to fine particle concentration [38]. At higher relative humidity, pollutants are attached to the surface of water vapor easier. Water solution is a good place for chemical reaction [39]. Wind direction and speed affect the dispersion of particulate matter in the air [40]. Chen et al. made predictions on $PM_{2.5}$ concentration in Zhejiang Province, finding that meteorological factors such as air temperature, air pressure, evaporation, humidity are remarkably correlated with $PM_{2.5}$ concentration [41]. Zhang Zhifei et al. found that O_3 h mass concentration has positive correlation with air temperature, solar radiation, visibility and wind speed, whereas NO_2 concentration is positively correlated with relative humidity and atmospheric pressure [42]. Precipitation [43], season [44], precipitation [45], sunshine duration [46], and other factors have remarkable impacts on the concentrations of air pollutants. Different city characteristics will also have different impacts on PM2.5, the correlation coefficient of PM_{10} and CO in Jinzhong and Linfen is 0.9. Further, the two numbers in Lv Liang are 0.6 and 0.7. The correlation coefficients of 12 cities show that temperature, surface temperature, atmospheric pressure, air humidity, and sunshine duration all affect PM$_{2.5}$. Further analysis is needed in selecting appropriate features for the model.

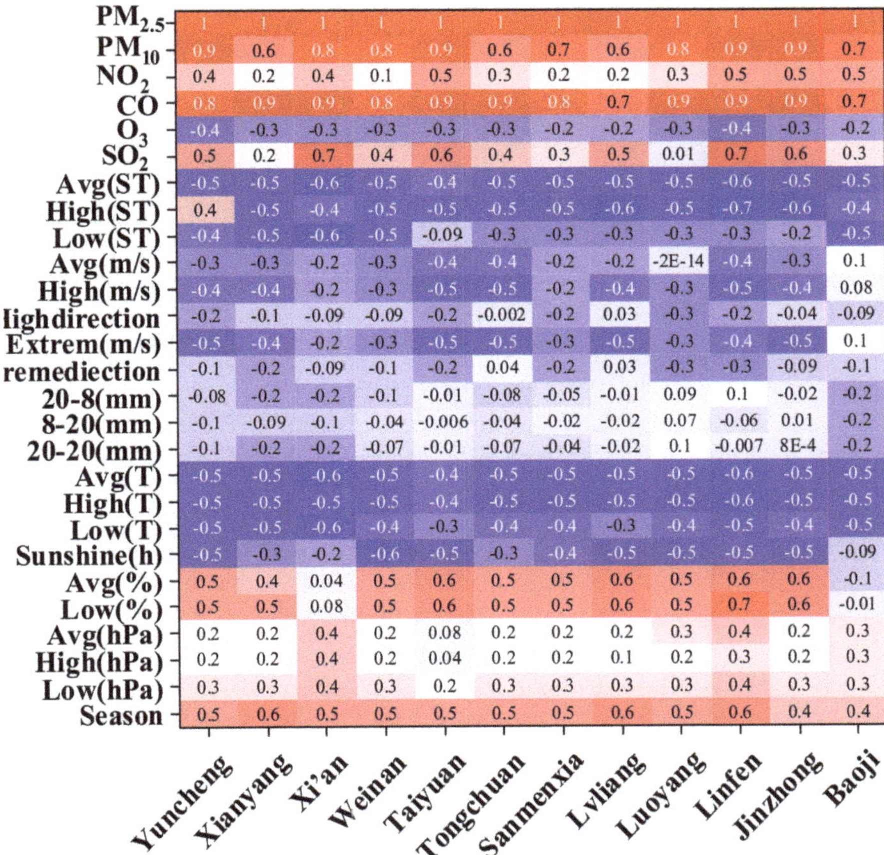

Figure 5. Pearson Analysis of Atmospheric Pollutants and Meteorological Factors in 12 Cities of Fenwei Plain.

4.2. Feature Selection

Through Pearson analysis, it is found that addition to the traditional six atmospheric pollutants, meteorological factors are also main factors to PM$_{2.5}$ concentration, such as surface temperature, temperature, sunshine duration, humidity, and so on. Consider unrelated and redundant factors, which may obscure the role of important factors and require the mining and refinement of raw data.

4.2.1. Feature Importance Sorting Principle

The traditional GBDT algorithm uses first derivative, while XGBoost expands the error function with second-order Taylor, using both first-order and second-order derivatives. XGBoost uses a second-order Taylor expansion of the error function, and XGBoost uses column sampling of features to select the proportion of features used in training and to prevent over-fitting effectively. The parallel approximate histogram algorithm for XGBoost's feature split gain calculation can make full use of multicore CPUs for parallel computation. Traditional feature selection models iterate continuously during operation, and new trees will be generated after each iteration. When dealing with complex datasets, they may iterate over hundreds of thousands of times, so they are not efficient. To overcome this disadvantage, the XGBoost algorithm uses a regression tree to build models. This system is based on the Boosting algorithm, which has made great breakthroughs in prediction accuracy and training speed. In fact, XGBoost calculates which feature to select as the split point based on the gain of the structure fraction. The importance of a feature is the sum of times it occurs in all trees. The more an attribute is used to build a decision tree in a model, the more important it is. Using gradient enhancement makes it relatively easy to retrieve the importance for each attribute after building an enhanced tree. Generally importance represents a score, indicating the usefulness or value of a feature in the process of building an enhanced tree in a model. The more attributes used for key decisions in a decision tree, the higher its relative importance is. Generally speaking, importance provides a score indicating how useful or valuable each attribute is in building an enhanced decision tree in a model. The more times attributes are used to make key decisions using a decision tree the higher the relative importance is. This importance is explicitly calculated for each attribute in the dataset so attributes can be ranked and compared with each other. The importance of a single decision tree is calculated by increasing the number of performance indicators per attribute split point, weighted by the number of observations the node is responsible for.

4.2.2. Experimental Process and Analysis of Feature Selection

We conduct a feature filtration on some parts of training set, and divided data sets into training sets and validation sets. First, we make XGBoost model which contains that contains all the feature training sets, use the five-fold cross validation to find the optimal parameters, and sort the features based on Fscore. Then we filter the sorted feature sets, evaluate whether a feature can be preserved under Fscore value, and delete the feature set which is scored lowest one by one. The AUC value of the validation set under the new feature subset is used to determine whether the predicted results of the remaining features are better or not. Both the number of features and the model improvement effect should be taken into consideration when selecting features. As some features have limited improvement effect on models, this experiment should use features that have greater impact on prediction of $PM_{2.5}$ concentration. The threshold h is set (the exact value of H is set according to the experimental results) to select the features. If the AUC value of the validation set increases more than *h*, the recently deleted features are saved. If the AUC value increases less than h or decreases, the deleted features are still removed. The algorithm can filter out the features that have a greater impact on the target variable and reduce the redundancy between the features.

As shown in the Figure 6, features are filtered by XGBoost. we use the "*importance_type = gain*" method to calculate the importance of features. We use five-fold cross validation meothod and grid search to find the optimal parameters of XGBoost model. The parameters of XGBoost algorithm are according to the weight of features. The importance of a feature can be used as a model explanatory value. This method represents the average gain from the presence of a feature as a split point in all trees.

In all trees, the number of times a feature is used to split nodes is *Weight*, and the total gain that a feature brings each time it splits a node is *Total_gain*. F Score formula is shown in Equation (25):

$$\text{F Score} = Total\_gain / weight \tag{25}$$

Average gain is calculated as Equation (26):

$$\text{AverageGain} = \frac{Total\_gain}{Fscore} \tag{26}$$

XGBoost calculates which feature to select as the split point based on the increment of the structure fraction. The importance of a feature is the sum of the number of times it occurs in all trees. The more an attribute is used to construct a decision tree in a model, the more important it is. Using XGBoost to rank the feature importance, as shown in

Figure 6, the top 10 cities are the 12 cities with different feature importance of $PM_{2.5}$. We input the filtered features into MSCNN-GA-LSTM.

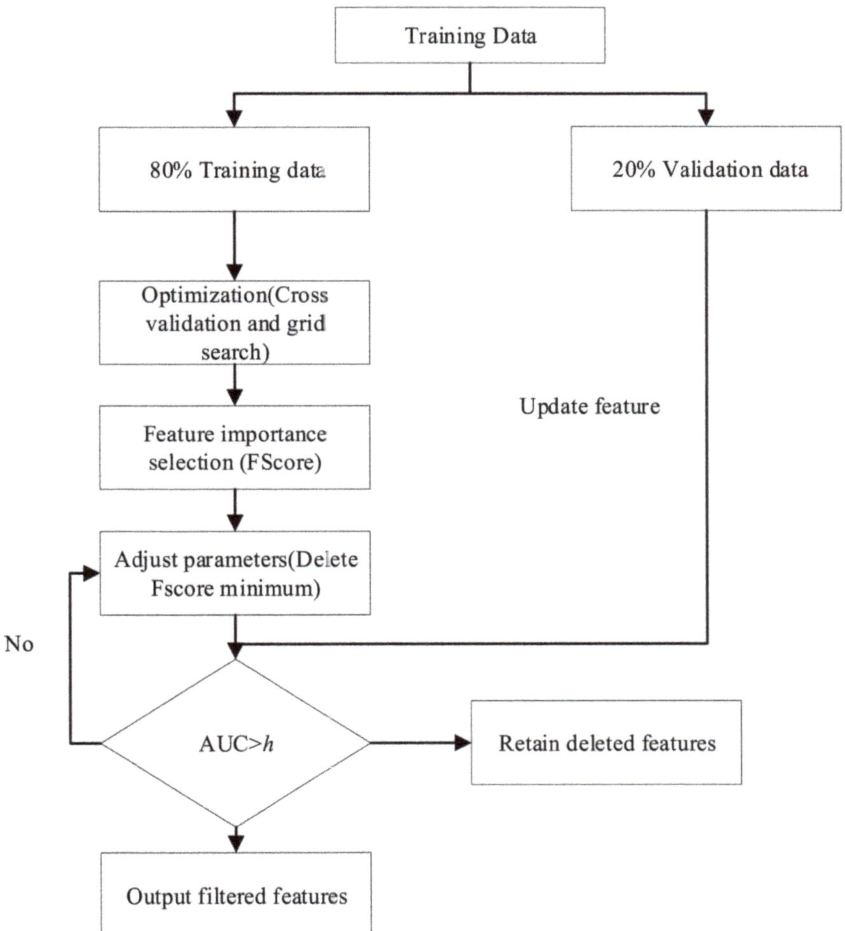

Figure 6. XGBoost model flow.

The importance of features is sorted by XGBoost, and the threshold h is set to 0.002. As shown in Figure 7 the y-axis represents each city, and the x-axis coordinates represent each feature. The numbers in the box represent the value of features importance in different cities. The color depth of the box represents the size of Fscore. The darker the color, the more important the feature. The lighter the color, the less important the feature. The top 10 feature importance of 12 cities are listed in the chart. Consistent with the previous Pearson correlation analysis, we found that the air pollutant characteristics with strong correlation, such as PM_{10} and CO, ranked as first and second in 12 cities in the feature importance ranking, while the factors with strong negative correlation, such as maximum temperature and average wind speed, are also of high importance. The feature importance of PM2.5 varied in different cities. We input the filtered features into MSCNN-GA-LSTM.

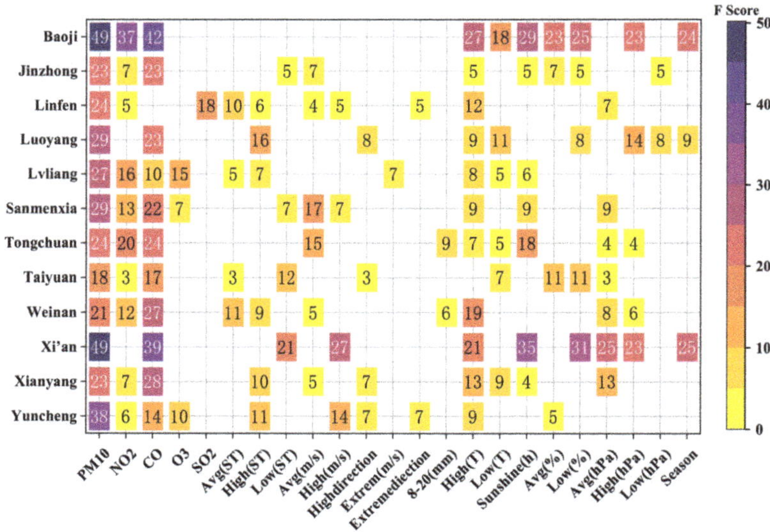

Figure 7. Atmospheric Pollutant Factors and Meteorological Factors in 12 cities of Fenwei Plain.

4.3. GA Optimize LSTM Optimal Parameters

In the prediction model, genetic algorithm is introduced to globally optimize the initial parameters of the LSTM network. Using traditional experience to set parameter value will make algorithm convergence easily fall into local optimum in the late period of algorithm iteration. To overcome this problem, we dynamically set the initialization parameters of the genetic algorithm. Further, we use a larger probability of perturbation, and avoid local optimum as the number of times of iteration gradually increases. Our repeated experiments and the final optimization parameters are listed in the Table 3—a good convergence effect has been achieved.

Table 3. GA Optimized LSTM Optimal Parameters.

City	Generations	Chromosome	Adaptability	First Layer	Second Layer	Third Layer	Dense Layer
Baoji	3	2	1464	223	215	172	225
Jinzhong	6	5	842	147	92	237	141
Linfen	20	15	1529	93	241	-	168
Luoyang	13	5	2679	159	77	73	196
Lvliang	18	11	1197	120	18	-	226
Sanmenxia	10	5	1195	181	226	-	140
Taiyuan	6	16	1653	59	-	-	187
Tongchuan	13	17	986	97	-	-	220
Weinan	12	4	2605	224	39	122	91
Xi'an	14	14	2179	65	238	-	255
Xianyang	14	18	1360	194	89	-	122
Yuncheng	1	6	1779	146	53	-	139

4.4. Forecast Results

4.4.1. Model Comparison before and after Feature Selection

At the beginning of this section, we evaluate the performance of different models by using the predictions from the test set. Figures 8 and 9 show the simulated prediction results of PM$_{2.5}$ in 12 cities using nine models. First, PM$_{2.5}$ test set data are input into four single trained models for calculation, and the PM$_{2.5}$ h predictions are compared with the measured results. The predicted PM$_{2.5}$ h concentration is close to the measured value when the measured value of PM2.5 h concentration increases rapidly, the predicted values deviate from the measured values significantly. This may be due to the redundancy of features and the influence of space-time characteristics. It is difficult to accurately predict if the model is not trained to filter feature values. The MLP model is similar to the LSTM model in that the predicted values deviate greatly from the measured values when the measured values increase or decrease sharply. The main reason why XGBoost model is not efficient is that it cannot achieve accurate prediction over time series data. When the measured values are small, the predicted values of $PM_{2.5}$ concentration are consistent with the measured values, and when the measured values are large, the predicted values are larger than the measured values. Comparing the predicted values of PM2.5 concentration of four single models in 12 cities, the LSTM model has the best predicted results.

$PM_{2.5}$ test set data are input into five trained combination models to calculate. The predicted $PM_{2.5}$ h concentration values of 12 cities are compared with the measured values which are shown in Figures 8 and 9. In the figure, the predicted values of the XGBoost-MSCGL model $PM_{2.5}$ are consistent with the measured values, even when some individual $PM_{2.5}$ h concentration values increase or decrease sharply, the predicted values are close to the measured values. XGBoost-LSTM prediction is similar to XGBoost-MSCGL model in that when the measured value increases or decreases sharply, the predicted value has a smaller deviation from the measured value, but the predicted result is slightly worse than that of XGBoost-MSCGL model. When the measured value of XGBoost-MLP model is higher or lower, the predicted value has a larger deviation from the measured value and the predicted value is smaller than that of the measured value. CNN-LSTM model performs better when the measured value increases or decreases sharply. However, compared with the other eight models, its prediction effect is the worst. For PM2.5 average concentration prediction, the predicted value is larger than the measured value. Comparing XGBoost-MLP, XGBoost-LSTM, XGBoost-CNN, XGBoost-MSCGL with CNN, LSTM, MLP, and CNN-LSTM, we found that the predicted value of the model after feature selection is closer to the measured value than that before feature selection, with a greater increase in accuracy, and a marked decrease in derivation value. Comparing the predicted values of PM2.5 h concentration of the nine models with their corresponding measured values, the XGBoost-MSCGL model had the best prediction effect.

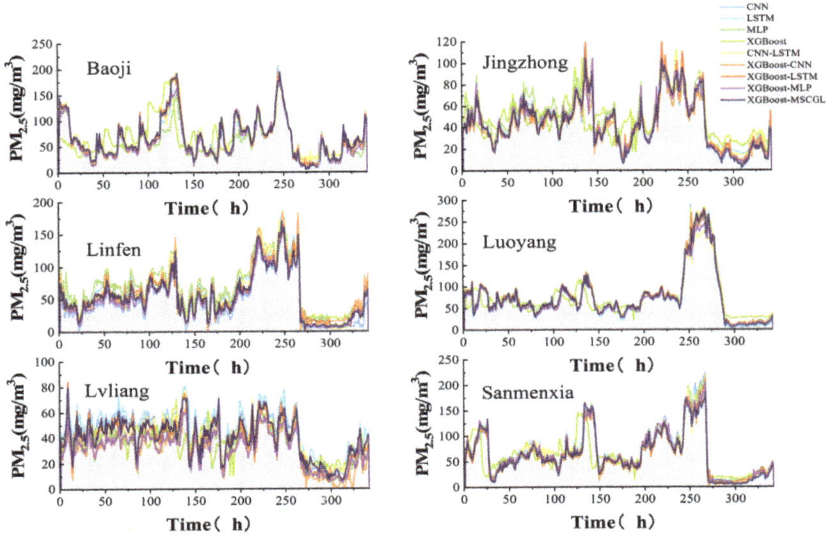

Figure 8. Predicted and Measured $PM_{2.5}$ h Concentration Values of Nine Models in Six Cities: Baoji, Jinzhong, Linfen, Luoyang, Lvliang, and Sanmenxia.

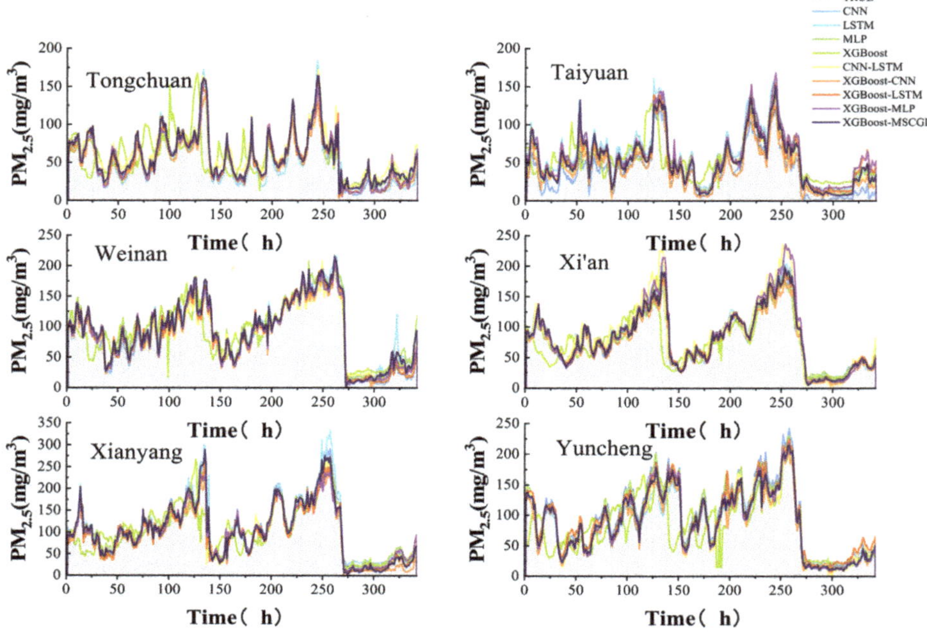

Figure 9. Predicted and Measured PM$_{2.5}$ h Concentration Values of Nine Models in Six Cities: Tongchuan, Taiyuan, Weinan, Xi'an, Xianyang, and Yuncheng.

4.4.2. Model Accuracy Evaluation

The accuracy of the four models was evaluated by RMSE, MAPE, MAE, and R^2. The smaller the RMSE, MAPE, and MAE, the higher the accuracy of the model, and the larger the R^2, the higher the accuracy of the model. In order to better evaluate the error, prediction effect, and prediction accuracy of the nine models, we selected four evaluation indexes to evaluate the performance results of each model in each city, as shown in Table 4.

Table 4. Accuracy comparison of nine models in 12 cities.

City	Evaluation Index	CNN	LSTM	MLP	XGBoost	CNN-LSTM	XGBoost-CNN	XGBoost-LSTM	XGBoost-MLP	XGBoost-MSCGL
BAOJI	RMSE	11.45	10.43	18.28	12.82	11.10	9.38	9.23	8.50	8.15
	MAE	7.87	7.06	11.34	8.92	8.89	5.94	5.96	5.55	5.19
	MAPE	13.41	11.82	15.89	25.45	23.23	11.87	12.36	10.68	10.14
	R2	82.15%	83.48%	79.98%	90.09%	92.62%	94.73%	94.90%	95.67%	96.02%
JINZHONG	RMSE	9.23	11.45	18.28	10.01	11.10	10.75	7.55	11.74	7.14
	MAE	5.96	7.87	11.34	7.64	8.89	7.36	4.94	8.78	4.80
	MAPE	12.36	13.41	15.89	12.68	23.23	10.64	7.94	20.37	7.39
	R2	94.90%	92.15%	79.98%	78.06%	92.62%	96.01%	98.03%	95.23%	98.24%
LINFEN	RMSE	12.70	12.70	19.12	11.34	10.92	11.85	10.15	10.73	10.09
	MAE	10.07	10.07	15.88	8.47	7.62	8.20	7.21	7.54	6.74
	MAPE	24.26	24.12	44.35	40.36	19.65	29.1	23.2	17.6	17.1
	R2	88.70%	88.70%	74.38%	90.98%	91.64%	90.16%	92.78%	91.92%	92.86%

Table 4. *Cont.*

City	Evaluation Index	CNN	LSTM	MLP	XGBoost	CNN-LSTM	XGBoost-CNN	XGBoost-LSTM	XGBoost-MLP	XGBoost-MSCGL
LUOYANG	RMSE	12.36	15.47	10.04	19.50	9.36	9.38	7.64	11.82	7.64
	MAE	7.10	6.98	6.95	14.63	6.00	5.10	5.03	4.90	4.90
	MAPE	9.91	12.84	14.52	38.39	8.27	9.79	9.78	7.05	7.75
	R2	85.63%	83.16%	87.12%	89.13%	94.50%	97.48%	96.33%	96.01%	98.33%
LVLIANG	RMSE	9.19	10.62	6.18	10.39	6.35	8.47	7.08	5.89	5.31
	MAE	7.87	8.92	5.17	7.63	5.23	6.99	5.91	4.75	4.26
	MAPE	22.87	27.78	16.03	26.16	17.71	22.65	17.22	15.08	13.43
	R2	57.46%	43.19%	80.76%	45.58%	79.68%	63.88%	74.75%	82.52%	85.78%
SANMENXIA	RMSE	13.71	11.39	11.71	26.03	12.01	11.88	10.48	11.00	11.61
	MAE	8.79	7.79	7.50	16.65	7.66	8.37	7.07	6.99	7.24
	MAPE	25.51	21.55	26.21	51.50	20.47	15.98	21.79	17.22	12.71
	R2	81.35%	84.03%	83.69%	68.83%	93.37%	93.50%	94.94%	94.43%	93.80%
TAIYUAN	RMSE	16.49	11.76	10.61	20.97	7.39	12.43	8.55	7.66	7.08
	MAE	14.43	9.85	8.27	15.02	5.35	10.16	6.25	5.46	5.07
	MAPE	36.22	23.93	21.66	44.90	12.47	22.30	14.64	12.11	11.54
	R2	74.30%	86.93%	89.35%	58.42%	94.84%	85.39%	93.08%	94.44%	95.25%
TONGCHUAN	RMSE	16.49	11.76	8.55	20.97	10.61	12.43	7.66	7.39	7.08
	MAE	14.43	9.85	6.25	15.02	8.27	10.16	5.46	5.35	5.07
	MAPE	36.22	23.93	14.64	44.90	21.66	22.30	12.11	12.47	11.54
	R2	74.30%	86.93%	93.08%	58.42%	89.35%	85.39%	94.44%	94.84%	95.25%
WEINAN	RMSE	11.25	14.23	14.47	23.47	11.56	10.05	9.50	9.17	8.90
	MAE	8.25	10.27	10.59	16.85	9.31	7.62	6.77	6.79	6.66
	MAPE	15.86	24.35	22.49	35.85	20.81	13.79	10.99	11.03	11.27
	R2	84.96%	81.95%	81.67%	78.09%	84.68%	95.98%	96.41%	96.66%	96.85%
XI'AN	RMSE	10.80	12.37	11.19	26.78	16.51	9.33	6.56	12.01	6.07
	MAE	7.55	9.06	8.39	16.59	11.21	6.85	4.65	7.30	3.94
	MAPE	12.50	16.30	16.68	25.98	16.62	10.38	8.16	8.97	5.95
	R2	84.99%	83.43%	84.63%	69.21%	88.29%	96.27%	98.15%	93.81%	98.42%
XIANYANG	RMSE	16.55	22.58	15.78	41.72	12.79	16.12	13.04	13.80	12.65
	MAE	11.34	14.85	12.05	26.81	8.56	11.68	8.88	9.41	8.25
	MAPE	17.91	27.03	22.50	34.86	12.32	16.87	11.36	13.27	11.31
	R2	83.02%	87.02%	83.66%	55.69%	85.84%	93.38%	95.67%	95.15%	95.92%
YUNCHENG	RMSE	15.21	11.90	11.19	38.39	11.08	10.75	11.74	7.55	7.14
	MAE	12.58	9.08	7.97	28.79	8.47	7.36	8.78	4.94	4.80
	MAPE	20.26	13.29	11.75	42.15	13.15	10.64	20.37	7.94	7.39
	R2	82.01%	85.11%	85.67%	49.05%	85.76%	96.01%	95.23%	98.03%	98.24%

Among the nine models which predicted PM2.5 h concentration value, XGBoost-MSCGL had the best prediction effect. The average MAE (8.26), RMSE (5.6), MAPE (9.9), R^2 (0.95) in 12 models were the highest, while the XGBoost model had the worst predictive effect in nine models, with the average MAE (21.67), RMSE (15.25), MAPE (31.94%), R^2

(0.69) in 12 cities. R^2 was the smallest of the nine models. The correlation coefficient R2 of the four single models was 79.07%, which may be related to the unstable time series of PM_2 concentration and no screening of features during the model building process, resulting in no further improvement of model accuracy. From the prediction effect after feature selection, the overall prediction effect of the combination of feature selection based on XGboost with a single model has been remarkably improved. XGBoost-CNN and XGBoost in 12 cities prediction compared with CNN, LSTM, MLP, XGBoost-MSCGL, the values of -LSTM, XGBoost-MSCGL, and CNN-LSTMRMSE decreased by 13.25%, 28.63%, 20.16%, 21.64%, the values of MAPE decreased by 14.86%, 29.96%, 27.31%, 26.25%, respectively, and the values of MAE decreased by 17.02%, 24.90%, 32.26%, 33.68%. R^2 Values increased by 11.98%, 16.62%, 12.70%, and 6.80%. The results show that feature selection based on XGBoost can effectively improve the accuracy of prediction model and reduce the error. For $PM_{2.5}$ concentration prediction, feature selection can effectively improve the accuracy and reduce the overestimation or underestimation caused by redundant features.

Among the four combination models which predicted PM2.5 h concentration value after feature selection, XGBoost-MSCGL has the best prediction effect. Compared with XGBoost-CNN model and XGBoost-LSTM model, XGBoost-MLP model has slightly higher prediction accuracy with correlation coefficient R^2(0.83). The prediction results of XGBoost-CNN model are the worst among the four models, MAE (7.98), RMSE (11.07), MAPE (13.96), R^2(0.9). XGBoost-MSCGL is better than XGBoost-MLP, XGBoost-LSTM, and XGBoost-CNN in predicting performance, with RMSE, MAE, and MAPE decreasing 11.11%, 15.97%, 15.36%, and R^2 increasing 3%, respectively, in 12 cities. Overall, XGBoost-MSCGL is better than XGBoost-MLP, XGBoost-LSTM, and XGBoost-CNN in predicting performance. As for cities, XGBoost-MSCGL performed best in Xi'an with MAE (3.94), MAPE (5.59), R^2 (0.98). The worst in Xianyang was RMSE (12.65), MAE (8.25), and Lv Liang's R^2 (85.78).

By analyzing the predicted data of 12 cities in the Fenwei Plain, it is noted that different prediction models have different performances in reducing errors and improving consistency of changes in different cities. The prediction error may be related to the different city characteristics that we choose, and to a different dispersion of air pollutant concentration values in each season. Using four deep learning combination models for training and validating the prediction accuracy, the results show that XGBoost-MSCGL has the highest prediction accuracy for most city training sets, and its prediction performance is better than other models. Through the three indicators of RMSE, MAE, and MAPE, we can see that XGBoost-MSCGL has better prediction performance than XGBoost-MLP, XGBoost-LSTM, XGBoost-CNN. In 12 cities, RMSE, MAE, and MAPE decreased by 11.11%, 15.97%, and 15.36%, respectively. However, XGBoost-LSTM in Xianyang, XGBoost-MLP in Weinan, and XGBoost-CNN in Jin are slightly higher than XGBoost-MSCGL in MAPE. XGBoost-MSCGL in Xi'an, Taiyuan, Sanmenxia, and other cities declined significantly. Overall, the error value of XGBoost-MSCGL in the four combined prediction models is small, the performance is outstanding, and the prediction effect is better.

Through the analysis of the prediction data of the 12 cities in the Fenwei plain, we noticed that the performances of different prediction models were different in reducing errors and improving the consistency of changes of changes in different cities. The prediction errors might have something to do with the different types of city characteristics and the degree of dispersion of the concentration of air pollution in different seasons. We used four models of deep learning to train and test. The results show that XGBoost-MSCGL has the highest prediction accuracy for most of the city's test sets, and it is better than other models in terms of prediction performance.

5. Discussion

In this study, the $PM_{2.5}$ feature selection based on XGBoost, combined with MSCNN to extract temporal and spatial features, and GA optimized LSTM, were used to establish the XGBoost-MSCGL air pollutant concentration prediction model. Compared with other machine learning, feature selection combined with feature extraction and combined with deep learning is an effective method for processing big data (especially spatio-temporal feature data). Combining spatio-temporal feature and models can improve the performance of spatio-temporal data prediction to a certain extent. In different cities, the importance of PM 2.5 influencing factors are different. It is necessary to select $PM_{2.5}$ influencing factors in different cities and propose redundant features and delete redundant features in order to avoid influencing the accuracy of the prediction model. The prediction method proposed in this paper is feasible for the $PM_{2.5}$ h concentration data prediction in multiple cities, and the method can be used in multiple regions and predictions on different atmospheric pollutant concentration. In terms of input variables, regular monitoring data from the National Environmental Monitoring Station, and China Meteorological Administration are used. In terms of modeling methods, machine learning and deep learning algorithms are combined. On the premise of eliminating redundant features, space and time features are considered, and a genetic algorithm is used to optimize the parameters of the LSTM network, enabling it to capture optimal parameters better. With stronger capturing ability, the long-term dependence relationship hidden by air quality data is more accurate, and the prediction accuracy is further improved.

The shortcoming of this research is that in different cities, the performances of XGBoost-MSCGL model may be different due to driving factors, spatio-temporal characteristics, model types, model structure, and model development methods. We find that in cities such as Xi'an, the model performs well. However, in some cities, their performances cannot achieve the same accuracy and prediction effect. The dispersion of PM 2.5 concentration data in different cities and other city air pollutants may also affect the prediction performance of the model. So, it was necessary to further analyze the reason for the difference. At the same time, the data volume, the dispersion between air pollutant concentration values and space features might also affect the performance of model prediction. So, it was necessary to further analyze the reason for the difference. In this study, the range and interval prediction of air pollutants concentrations are not taken into consideration. In following researches, it needs to be discussed in detail. Only in this way could the relevant government and enterprises better monitor and manage the release of air pollution.

6. Conclusions

In this study, based on the hourly concentration data and meteorological data of six air pollutants in 12 cities in the Fenwei Plain in 2020, a PM 2.5 concentration prediction model, based on XGBoost-MSCGL, was established, and the performance of the model was compared with XGBoost-MLP, XGBoost-LSTM, and XGBoost-CNN. The main research results are as follows: In the $PM_{2.5}$ concentration prediction, the XGBoost-MSCGL model performs better in 12 cities in the Fenwei Plain, with smaller error values and better prediction results. As for feature selection, compared with the prediction of all influencing factors, the prediction effect of the former is significantly improved for the factors of feature selection. From the perspective of spatio-temporal characteristics, the hourly concentration prediction performance of the 12 cities considering spatio-temporal characteristics is better than the prediction model that does not consider spatio-temporal characteristics. From the perspective of the optimized model, the accuracy of the optimized model is significantly improved compared to the unoptimized model. In general, based on feature selection, screening the influencing factors of $PM_{2.5}$ according to their importance helps to reduce the feature redundancy of the data set. In terms of overall performance, the prediction performance of the XGBoost-MSCGL model is generally better than that of the XGBoost-MLP, XGBoost-LSTM, and XGBoost-CNN models. Compared with other prediction methods, the PM2.5 concentration prediction, based on the XGBoost-MSCGL model, has better performance in accurately predicting the actual data in different cities. Compared with other models, it has a higher accuracy improvement and achieves better prediction especially when the data are at extremely high and low points in the sharp fluctuations. The migration of the model is verified by the prediction results of 12 cities in Fenwei plain. The concentration change direction and volatility of PM2.5 need to be further considered in future research.

Author Contributions: The article was written through the contributions of all authors. H.D.: conceptualization, methodology, modelling, analysis, writing—original draft preparation. G.H.: conceptualization, modelling, writing—reviewing and editing, revision. H.Z.: modelling, analysis, writing—original draft preparation, writing—reviewing and editing. F.Y.: analysis writing—reviewing and editing, revision. All authors have read and agreed to the published version of the manuscript.

Funding: This paper is funded by the National Natural Science Foundation of China (71874134). Key Project of Basic Natural Science Research Plan of Shaanxi Province (2019JZ-30).

Institutional Review Board Statement: Not applicable.

Informed Consent Statement: Not applicable.

Data Availability Statement: Both air pollutant data and meteorological data come from public data provided by the Ministry of Ecology and Environment of the People's Republic of China (http://www.mee.gov.cn/, accessed on 18 September 2021) and China Meteorological Administration (http://www.cma.gov.cn/, accessed on 18 September 2021).

Conflicts of Interest: The authors declare no conflict of interest.

References

1. Kim, Y.; Manley, J.; Radoias, V. Medium-and long-term consequences of pollution on labor supply: Evidence from Indonesia. *J. Labor. Econ.* **2017**, *6*, 1–15. [CrossRef]
2. Braithwaite, I.; Zhang, S.; Kirkbride, J.B.; Osborn, D.P.; Hayes, J.F. Air pollution (particulate matter) exposure and associations with depression, anxiety, bipolar, psychosis and suicide risk: A systematic review and meta-analysis. *Environ. Health. Persp.* **2019**, *127*, 1–23. [CrossRef] [PubMed]
3. Niu, W.-J.; Feng, Z.-K.; Li, S.-S.; Wu, H.-J.; Wang, J.-Y. Short-term electricity load time series prediction by machine learning model via feature selection and parameter optimization using hybrid cooperation search algorithm. *Environ. Res. Lett.* **2021**, *16*, 055032. [CrossRef]
4. Dai, Y.; Zhao, P. A hybrid load forecasting model based on support vector machine with intelligent methods for feature selection and parameter optimization. *Appl. Energy* **2020**, *279*, 115332. [CrossRef]
5. Liu, X.; Zhang, H.; Kong, X.; Lee, K.Y. Wind speed forecasting using deep neural network with feature selection. *Neurocomputing* **2020**, *397*, 393–403. [CrossRef]
6. Haq, A.U.; Zeb, A.; Lei, Z.; Zhang, D. Forecasting daily stock trend using multi-filter feature selection and deep learning. *Expert Syst. Appl.* **2021**, *168*, 114444. [CrossRef]
7. Peng, L.; Wang, L.; Ai, X.-Y.; Zeng, Y.-R. Forecasting Tourist Arrivals via Random Forest and Long Short-term Memory. *Cogn. Comput.* **2021**, *13*, 125–138. [CrossRef]
8. Elsherbiny, O.; Fan, Y.; Zhou, L.; Qiu, Z. Fusion of Feature Selection Methods and Regression Algorithms for Predicting the Canopy Water Content of Rice Based on Hyperspectral Data. *Agriculture* **2021**, *11*, 51. [CrossRef]
9. Ceylan, Z.; Atalan, A. Estimation of healthcare expenditure per capita of Turkey using artificial intelligence techniques with genetic algorithm-based feature selection. *J. Forecast.* **2021**, *40*, 279–290. [CrossRef]
10. Yang, Z.; Wang, Y.; Li, J.; Liu, L.; Ma, J.; Zhong, Y. Airport Arrival Flow Prediction considering Meteorological Factors Based on Deep-Learning Methods. *Complexity* **2020**, *2020*, 6309272. [CrossRef]
11. Baker, K.R.; Foley, K.M. A nonlinear regression model estimating single source concentrations of primary and secondarily formed PM2.5. *Atmos. Environ.* **2011**, *45*, 3758–3767. [CrossRef]
12. Zhou, W.; Wu, X.; Ding, S.; Ji, X.; Pan, W. Predictions and mitigation strategies of PM2.5 concentration in the Yangtze River Delta of China based on a novel nonlinear seasonal grey model. *Environ. Pollut.* **2021**, *276*, 116614. [CrossRef] [PubMed]
13. Wu, J.; Yao, F.; Li, W.; Si, M. VIIRS-based remote sensing estimation of ground-level PM2.5 concentrations in Beijing–Tianjin–Hebei: A spatiotemporal statistical model. *Remote Sens. Environ.* **2016**, *184*, 316–328. [CrossRef]
14. Ma, Z.; Liu, Y.; Zhao, Q.; Liu, M.; Zhou, Y.; Bi, J. Satellite-derived high resolution PM2.5 concentrations in Yangtze River Delta Region of China using improved linear mixed effects model. *Atmos. Environ.* **2016**, *133*, 156–164. [CrossRef]
15. Kloog, I.; Koutrakis, P.; Coull, B.A.; Lee, H.J.; Schwartz, J. Assessing temporally and spatially resolved PM2.5 exposures for epidemiological studies using satellite aerosol optical depth measurements. *Atmos. Environ.* **2011**, *45*, 6267–6275. [CrossRef]
16. Lai, X.; Li, H.; Pan, Y. A combined model based on feature selection and support vector machine for PM2.5 prediction. *J. Intell. Fuzzy Syst.* **2021**, *40*, 10099–10113. [CrossRef]

7. Yazdi, M.D.; Kuang, Z.; Dimakopoulou, K.; Barratt, B.; Suel, E.; Amini, H.; Lyapustin, A.; Katsouyanni, K.; Schwartz, J. Predicting Fine Particulate Matter (PM2.5) in the Greater London Area: An Ensemble Approach using Machine Learning Methods. *Remote Sens.* **2020**, *12*, 914. [CrossRef]

8. Bi, J.; Stowell, J.; Seto, E.Y.W.; English, P.B.; Al-Hamdan, M.Z.; Kinney, P.L.; Freedman, F.R.; Liu, Y. Contribution of low-cost sensor measurements to the prediction of PM2.5 levels: A case study in Imperial County, California, USA. *Environ. Res.* **2020**, *180*, 108810. [CrossRef]

9. Mao, X.; Shen, T.; Feng, X. Prediction of hourly ground-level PM 2.5 concentrations 3 days in advance using neural networks with satellite data in eastern China. *Atmos. Pollut. Res.* **2017**, *8*, 1005–1015. [CrossRef]

10. Zhou, Y.; Chang, F.-J.; Chen, H.; Li, H. Exploring Copula-based Bayesian Model Averaging with multiple ANNs for PM2.5 ensemble forecasts. *J. Clean. Prod.* **2020**, *263*, 121528. [CrossRef]

11. Dhakal, S.; Gautam, Y.; Bhattarai, A. Exploring a deep LSTM neural network to forecast daily PM2.5 concentration using meteorological parameters in Kathmandu Valley, Nepal. *Air Qual. Atmos. Health* **2021**, *14*, 83–96. [CrossRef]

12. Park, Y.; Kwon, B.; Heo, J.; Hu, X.; Liu, Y.; Moon, T. Estimating PM2.5 concentration of the conterminous United States via interpretable convolutional neural networks. *Environ. Pollut.* **2020**, *256*, 113395. [CrossRef]

13. Lv, B.; Cobourn, W.G.; Bai, Y. Development of nonlinear empirical models to forecast daily PM2.5 and ozone levels in three large Chinese cities. *Atmos. Environ.* **2016**, *147*, 209–223. [CrossRef]

14. Jin, X.-B.; Yang, N.-X.; Wang, X.-Y.; Bai, Y.-T.; Su, T.-L.; Kong, J.-L. Deep Hybrid Model Based on EMD with Classification by Frequency Characteristics for Long-Term Air Quality Prediction. *Mathematics* **2020**, *8*, 214. [CrossRef]

15. Masmoudi, S.; Elghazel, H.; Taieb, D.; Yazar, O.; Kallel, A. A machine-learning framework for predicting multiple air pollutants' concentrations via multi-target regression and feature selection. *Sci. Total. Environ.* **2020**, *715*, 136991. [CrossRef] [PubMed]

16. Joharestani, M.Z.; Cao, C.; Ni, X.; Bashir, B. Talebiesfandarani S. PM2.5 Prediction Based on Random Forest, XGBoost, and Deep Learning Using Multisource Remote Sensing Data. *Atmosphere* **2019**, *10*, 373. [CrossRef]

17. Zhang, Y.; Zhang, R.; Ma, Q.; Wang, Y.; Wang, Q.; Huang, Z.; Huang, L. A feature selection and multi-model fusion-based approach of predicting air quality. *ISA Trans.* **2020**, *100*, 210–220. [CrossRef] [PubMed]

18. Ma, J.; Cheng, J.C.; Xu, Z.; Chen, K.; Lin, C.; Jiang, F. Identification of the most influential areas for air pollution control using XGBoost and Grid Importance Rank. *J. Clean. Prod.* **2020**, *274*, 122835. [CrossRef]

19. Zhai, B.; Chen, J. Development of a stacked ensemble model for forecasting and analyzing daily average PM2.5 concentrations in Beijing, China. *Sci. Total. Environ.* **2018**, *635*, 644–658. [CrossRef]

20. Gui, K.; Che, H.; Zeng, Z.; Wang, Y.; Zhai, S.; Wang, Z.; Luo, M.; Zhang, L.; Liao, T.; Zhao, H.; et al. Construction of a virtual PM2. 5 observation network in China based on high-density surface meteorological observations using the Extreme Gradient Boosting model. *Environ. Int.* **2020**, *141*, 105801. [CrossRef]

21. Ministry of Ecology and Environment of the People's Republic of China. Notice on the Issuance of the "Beijing-Tianjin-Hebei and Surrounding Areas, and the Fenwei Plain, 2020–2021 Autumn and Winter Comprehensive Management of Air Pollution Action Plan". Available online: http://www.mee.gov.cn/xxgk2018/xxgk/xxgk03/202011/t20201103_806152.html (accessed on 18 September 2021).

22. Ministry of Ecology and Environment of the People's Republic of China. The Air Quality Objectives of the Three Key Regions in Autumn and Winter of 2019–2020 Are All over Fulfilled. Available online: http://www.mee.gov.cn/ywdt/hjywnews/202004/t20200427_776493.shtml. (accessed on 18 September 2021).

23. Kong, Z.; Zhang, C.; Lv, H.; Xiong, F.; Fu, Z. Multimodal Feature Extraction and Fusion Deep Neural Networks for Short-Term Load Forecasting. *IEEE Access* **2020**, *8*, 185373–185383. [CrossRef]

24. Sheridan, R.P.; Wang, W.M.; Liaw, A.; Ma, J.; Gifford, E.M. Extreme Gradient Boosting as a Method for Quantitative Structure–Activity Relationships. *J. Chem. Inf. Model.* **2016**, *56*, 2353–2360. [CrossRef] [PubMed]

25. Zhang, L.; Zhang, J.; Niu, J.; Wu, Q.; Li, G. Track Prediction for HF Radar Vessels Submerged in Strong Clutter Based on MSCNN Fusion with GRU-AM and AR Model. *Remote Sens.* **2021**, *13*, 2164. [CrossRef]

26. Kong, W.; Dong, Z.Y.; Jia, Y.; Hill, D.J.; Xu, Y.; Zhang, Y. Short-Term Residential Load Forecasting Based on LSTM Recurrent Neural Network. *IEEE Trans. Smart Grid* **2019**, *10*, 841–851. [CrossRef]

27. Lu, H.; Azimi, M.; Iseley, T. Short-term load forecasting of urban gas using a hybrid model based on improved fruit fly optimization algorithm and support vector machine. *Energy Rep.* **2019**, *5*, 666–677. [CrossRef]

28. Cheng, Y.; He, K.; Du, Z.; Zheng, M.; Duan, F.; Ma, Y. Humidity plays an important role inthe PM2.5 pollution in Beijing. *Environ. Pollut.* **2015**, *197*, 68–75. [CrossRef]

29. Brown, S.G.; Hyslop, N.P.; Roberts, P.T.; McCarthy, M.C.; Lurmann, F.W. Wintertime vertical vari-ations in particulate matter (PM) and precursor concentrations inthe San Joaquin Valley during the California regional coarse PM /Fine PM air quality study. *J. Air Waste Manag.* **2006**, *56*, 1267–1277. [CrossRef]

30. Li, X.; Chen, C.; Dong, Z.; Dong, Y.; Du, C.; Peng, Y. Analysis of the Impact of Meteorological Factors on Particle Size Distribution and Its Characteristic over Guanzhong Basin. *Meteorol. Mon.* **2018**, *44*, 929–935. [CrossRef]

31. Chen, B.; Jin, Q.; Chai, H.; Guo, F. Spatiotemporal distribution and correlation factors of PM2.5concentrations in Zhejiang Province. *Acta Sci. Circumst.* **2021**, *41*, 817–829. [CrossRef]

32. Zhang, Z.; Zheng, M.; Zhang, Y.; Zhou, J.; Liu, H. The Survey and Influence Factors of Air Pollution in Ningbo. *Environ. Monit. China* **2020**, *36*, 96–103. [CrossRef]

43. Li, L.; Li, H.; Peng, L.; Li, Y.; Zhou, Y.; Chai, F.; Mo, Z.; Chen, Z.; Mao, J.; Wang, W. Characterization of precipitation in the background of atmospheric pollutants reduction in Guilin: Temporal variation and source apportionment. *J. Environ. Sci.* **2020**, *98*, 1–13. [CrossRef] [PubMed]
44. Boleti, E.; Hueglin, C.; Grange, S.K.; Prévôt, A.S.H.; Takahama, S. Temporal and spatial analysis of ozone concentrations in Europe based on timescale decomposition and a multi-clustering approach. *Atmos. Chem. Phys. Discuss.* **2020**, *20*, 9051–9066. [CrossRef]
45. Ji, M.; Jiang, Y.; Han, X.; Liu, L.; Xu, X.; Qiao, Z.; Sun, W. Spatiotemporal Relationships between Air Quality and Multiple Meteorological Parameters in 221 Chinese Cities. *Complexity* **2020**, *2020*, 6829142. [CrossRef]
46. Wang, Z.-B.; Li, J.-X.; Liang, L.-W. Spatio-temporal evolution of ozone pollution and its influencing factors in the Beijing-Tianjin-Hebei Urban Agglomeration. *Environ. Pollut.* **2020**, *256*, 113419. [CrossRef] [PubMed]

 *sustainability*

Article

Competitiveness of Industrial Companies Forming the Value Chain of Wind Energy Components: The Case of Lithuania

Akvile Cibinskiene [1,*], Daiva Dumciuviene [1], Viktorija Bobinaite [2] and Egidijus Dragašius [3]

1 School of Economics and Business, Kaunas University of Technology, Gedimino str. 50, LT-44239 Kaunas, Lithuania; daiva.dumciuviene@ktu.lt
2 Lithuanian Energy Institute, Laboratory for Energy System Research, Breslaujos str. 3, LT-44403 Kaunas, Lithuania; viktorija.bobinaite@lei.lt
3 Faculty of Mechanical Engineering and Design, Kaunas University of Technology, Studentu str. 56, LT-51424 Kaunas, Lithuania; egidijus.dragasius@ktu.lt
* Correspondence: akvile.cibinskiene@ktu.lt; Tel.: +370-687-77771

Abstract: Sustainable energy development has attracted attention worldwide, partly because of the value chain of the wind energy industry that focuses on the overall value creation and innovation. In order to achieve not only ambitious goals in the fight against climate change, but also to create significant economic benefits for European Union citizens, it is necessary to ensure the production of renewable energy components in Europe itself and in Lithuania at the same time. This paper aims to evaluate the competitiveness of Lithuanian companies that manufacture wind energy components. The research was conducted applying methods such as a survey of manufacturers of wind energy components, expert assessment and descriptive analysis. The results of the competitiveness assessment revealed that the existing conditions and trends are favourable for the development of their performance and strengthening of their competitiveness. The government solutions to promote industry could facilitate the performance of companies operating in the value chain of wind energy components and encourage new companies to join it. This would encourage the Lithuanian industry to expand its participation in the value chain of the European Union's renewable energy industry, create more jobs, and increase the added value.

Keywords: wind power industry; manufacturers of wind energy components; value chain of wind energy components; competitiveness

Citation: Cibinskiene, A.; Dumciuviene, D.; Bobinaite, V.; Dragašius, E. Competitiveness of Industrial Companies Forming the Value Chain of Wind Energy Components: The Case of Lithuania. *Sustainability* **2021**, *13*, 9255. https://doi.org/10.3390/su13169255

Academic Editors: Baojie He, Ayyoob Sharifi, Chi Feng and Jun Yang

Received: 21 July 2021
Accepted: 13 August 2021
Published: 18 August 2021

Publisher's Note: MDPI stays neutral with regard to jurisdictional claims in published maps and institutional affiliations.

1. Introduction

Sustainable energy development has attracted attention around the world, partly because of the value chain of the wind energy industry that focuses on overall value creation and innovation. The global wind energy industry is going through major changes. The development of renewable and sustainable energy technologies makes the wind energy industry more intense, efficient, greener and more complex in terms of value creation. It is therefore intended to minimize value-added operations and maximize value creation as much as possible. The value chain model of the wind energy industry allows for a comprehensive analysis of the factors forming the value chain of wind energy components (VCWEC), to determine the overall value creation process.

In order to achieve not only ambitious goals in the fight against climate change, but also to create significant economic benefits for European Union (EU) citizens, it is necessary to ensure the production of renewable energy components in Europe itself and in Lithuania at the same time. Lithuania proposes to include wind production and energy storage technologies in the EU's strategic value chain. The EU needs a stable and strong industrial sector and research and development (R&D) efforts to become a leader in the developing renewable energy technologies.

It is strategically important to invest rapidly and actively in the recovery and growth of the Lithuanian economy after the COVID-19 crisis. Firstly, it is vital to exploit the situation and ensure a sustainable, innovative and high-value-added economy. It is also vital to attract investment, reduce dependence on the Asian market and shorten and diversify the value chains. Finally, it is important to increase their resilience, create new jobs and ensure the extraction of green energy and the supply of this energy to other sectors.

Wind energy is one of the types of "green" energy which can address the aforementioned issues. In detail, due to the development of offshore and onshore wind energy, people are supplied with jobs, which accounted for 300,000 jobs in EU in 2019, and the economy is supplied with growth, which was generated from value-added contributions of EUR 37.2 billion and exports in goods and services of EUR 8 billion [1]. Furthermore, wind energy made relevant contributions to the local economy through taxes, which were of EUR 1.3 billion (not linked to corporate profit) and were designated to local governments and communities [1], and income for farmers in a form of lease payments to landlords of the area [2]. Indeed, in the presence of rising carbon and fuel prices, energy demand and more [3], the low operating cost of wind energy [2] kept wholesale electricity prices low and decreasing [3,4]. Being a local industry, wind energy reduced the country's dependence on polluting fossil fuels [5], helped stabilize the cost of electricity and reduced vulnerability to price spikes and supply disruptions [2]. Beyond the economic benefits, wind energy is relevant for environmental purposes as it reduces greenhouse gas (GHG) emissions [6], including lifecycle GHG emissions [7]. The latest research demonstrates that the known environmental, economic and other benefits of wind energy might even be larger in monetary terms if novel solutions are adapted, including wind farm layout optimization [5,8].

Changes, trends, forecasts, and investments of strategic importance to the country require detailed analysis and substantiation. In order for Lithuania to successfully solve the tasks mentioned above and join the VCWEC manufacturing, research of Lithuania's opportunities and perspectives in this field is essential and relevant, not only in the country's context, but also in the context of the EU.

The World Economic Forum (WEF) defines competitiveness as the ability of a country or firm to create more wealth than its competitors under global market equilibrium [9]. The Organization for Economic Co-operation and Development (OECD) argues that competitiveness under favourable market conditions shows the extent to which a firm can produce goods and services for an internationally competitive market while synchronizing real domestic income and growth in living standards [10]. The International Institute for Management Development (IMD) defines competitiveness as an effective mean of achieving growth in living standards and social well-being [11].

Although the concept of competitiveness has been used and applied in research for a relatively long time, there are many studies of the competitiveness of various types of companies, industries and countries. In addition, the regular calculation and publication of national competitiveness indices are promoted (Global Competitiveness Index (WEF), World Competitiveness Rating (IMD), the European Regional Competitiveness Index (European Commission)). In general, competitiveness shows the ability of a company to make business decisions using its available resources, which allows it to occupy a higher position than its competitors do. Therefore, the competitiveness assessment can provide valuable insights in evaluating the development opportunities, perspectives and resilience of the companies forming the VCWEC.

This paper aims to evaluate the competitiveness of Lithuanian companies that manufacture wind energy components (WEC).

The results of this research can provide insights for the Lithuanian government to take actions that will encourage the Lithuanian industry to participate more widely in the value chain of the EU's renewable energy industry, including wind energy, create more jobs and increase added value.

2. Theoretical Background of Competitiveness Assessment

The origins of competitiveness are linked to the ability of individual countries to accumulate more wealth. This concept was later developed and adapted to various levels, not just countries. It was taken into account that the country's assets depend on the industries and individual companies operating within it. In this way, the factors that may affect competitiveness are starting to be examined, and models for competitiveness assessment are created.

2.1. The Concept of Competitiveness and the Levels of Its Assessment

The representatives of the mercantilist worldview may be considered the pioneers of the country's competitiveness and the analysis of its factors. In the 14th century, they held the position that a country that is able to accumulate more wealth by promoting exports and restricting imports is in a better position. A. Smith's [12] concept of absolute advantage and D. Ricardo's [13] theory of comparative advantage, which states that a country has a comparative advantage in the production of a certain product when the opportunity cost of this product is lower than in other countries, can also be considered as the basics of a competitiveness assessment. The representatives of classical and neoclassical economic theory paid great importance to trade, which ensures the country's competitiveness. According to them, competitiveness is gained by being able to profit from international trade by exporting expensive products or services produced cheaply and using inexpensive raw materials. Competitiveness was later developed within the framework of Shumpeter's theory of entrepreneurship and innovation, the Porter Diamond model and even later in Krugman's [14] theory of new economic geography criticizing competitiveness [15].

The term "competition" is derived from the Latin word *concurrentia*, which means collision and embarrassment [16]. The concept of "competitiveness" is treated slightly differently by many authors. Porter [17] can be considered as the originator of the modern concept of competitiveness, and he argued that competitive advantage could be understood as "a country's ability to create an environment that enables companies to develop and innovate faster than foreign competitors". According to Fang et al. [18], the definition of competitiveness varies depending on the context, scale and purpose of its application.

A review of the concepts of competitiveness presented by various authors shows that this category can be examined in very different ways. Summarizing the research of many authors, Travkina and Tvaronavičienė [19] state that a country's competitiveness can be analysed by levels (micro, meso, macro), areas (economics, politics, society and technology) and time perspectives (medium or long term). Only the same levels of competitiveness can be compared. However, we can observe that the competitiveness of the lower level (e.g., companies) forms the competitiveness of the higher level (industry), and this also affects the overall competitiveness of the country [20]. This dependence of the country's economic competitiveness on the competitiveness of its constituent industries, which in turn depend on the competitiveness of enterprise that is determined by the competitiveness of employees, was also revealed by Reiljan, Hinrik and Ivanov [21].

2.2. The Assessment of Competitiveness

Many different models can be found to assess competitiveness. Balkytė and Tvaronavičienė [22] state that many economists develop competitiveness models to adapt them to different factors that affect competitiveness. The broader the analysis of competitive advantages, the more complex the model is, leading to differences in the views of researchers, even when assessing at the same sectors or industries.

One of the most commonly used models for competitiveness assessment is the Porter Diamond Model [17]. It covers four main groups of factors that affect national competitiveness: factor conditions, demand conditions, related and supporting industries and firm strategy, structure and rivalry.

Factor conditions. This group of factors includes the factors of production needed to compete in a given industry. These factors include human resources (quantity of labour, skills, costs, etc.), physical resources (natural resources or raw materials used in production, their quantity, quality, price, availability; they may also include the country's natural or climatic conditions), knowledge resources (scientific, technical knowledge resources, capital (amount of capital, cost) and infrastructure (types of infrastructure, quality and cost of use).

Demand conditions. These are the conditions that shape the demand for manufactured products or services in domestic and foreign markets. In the absence of demand, it is hardly possible to achieve competitiveness.

Related and supportive industries. Nowadays, this is often referred to as a cluster of a particular industry or activity. These are industries or companies that are geographically close to each other, interact, collaborate or complement each other with their specific activities. These may include local suppliers or scientific and other organizations.

Firm strategy, structure and rivalry. It is the internal organization, management and strategy of a company that can affect its competitiveness in one way or another.

In addition to these four main groups of factors, Porter [17] adds government and a chance that can affect all four main groups of factors. Direct intervention, significant technological discoveries, jumps in factor prices, significant changes in financial markets, jumps in exchange rates, fluctuations in global and regional demand, political decisions of foreign governments, wars and pandemics are considered as a chance. Porter treats the government not as a factor of the Diamond model but rather as an influencer of its structure and efficiency.

Although the Porter Diamond model has been supplemented and refined in various ways, the main groups of factors identified have become the basis for the various models used in the research. It has become popular for competitiveness assessment because it covers the key elements that shape competitiveness, and the factors that affect each key element can be chosen flexibly depending on the object of the research.

3. Porter Diamond Competitiveness Assessment Model for Assessing the Competitiveness of Enterprises Forming the VCWEC

Zhao et al. [23] used the Diamond model to assess the performance and competitiveness of the Chinese wind energy industry, and Liu et al. [24] applied it to a review of the value chain of the Chinese wind energy industry. Irfan et al. [25], with the help of this model, examined the main factors affecting the Indian wind energy industry. Fang et al. [18] applied this model to assess the competitiveness of national renewable energy in the G20. They took only the four main factor groups of the Porter Diamond model for the assessment of national renewable energy competitiveness but combined it with the indicator system. Zhao et al. [23] applied the improved dynamic Diamond model, which includes government as the fifth determinant that directly affects all main groups of factors except for factor conditions. However, one of the factors under the group of factor conditions discussed in that study includes programs and projects that might be initiated by the government. Therefore, the determinant government should affect the factor conditions as well. Technology in that model is incorporated as an intermediate variable, affecting only demand conditions and firm strategy, structure and rivalry. Under the technology, Zhao et al. [23] discuss technical R&D; R&D capacity building and standards and norms building. These elements may have an impact not only on demand conditions and firm strategy, structure and rivalry but also on factor conditions and related and support industries. The chance in this model has influence only on factor conditions and firm strategy, structure and rivalry but in fact may have influence on the other main Diamond model factor groups. Irfan et al. [25] applied the Diamond model with the core four groups of factors with government and chance both affecting all main elements. Technology in this case was not included as an element of the model but showed up in the analysis under the chance and firm strategy, structure and rivalry. Although Zhang et al. [26] discuss Porter's Diamond model for the international competitiveness of China's wind turbine manufacturing industry, they did not apply it for evaluation.

Summarizing the performed research of wind energy competitiveness, the adapted Diamond model to assess the competitiveness of Lithuanian companies forming the VCWEC was applied (Figure 1).

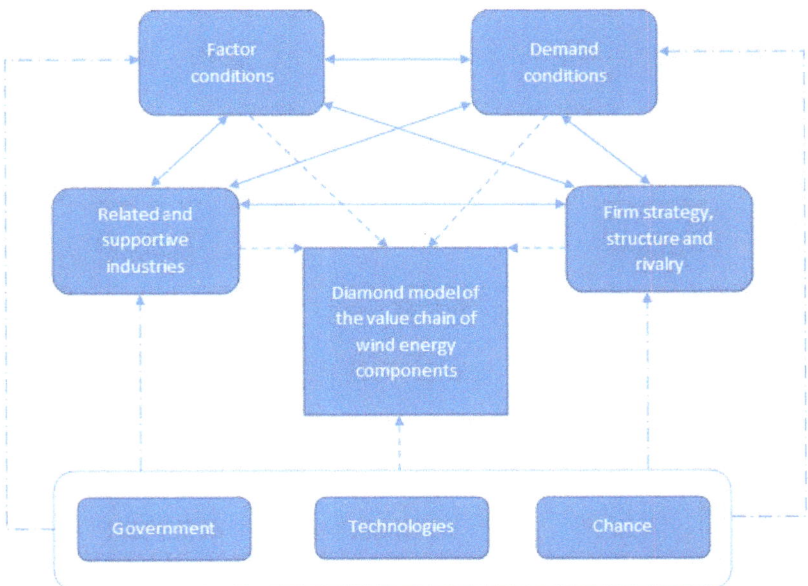

Figure 1. Diamond model of the VCWEC (adapted according to Liu et al. [24]).

In assessing the competitiveness of the wind energy industry according to the Diamond model, foreign authors have assigned to Porter's selected factor conditions the appropriately selected factors relevant to each activity. The authors studying the competitiveness of the wind energy industry, renewable energy and wind energy industry value chain have assigned the following factors to the respective groups of factors in the Porter's Diamond model:

- **Factor conditions:** availability of resources; capital injections; level of technology; skilled labour; wind resource potential; changes in energy structure; programs and projects; production costs and electricity price in the grid.
- **Demand conditions:** market size; substitution costs; environmental requirements; policy incentives; energy demand; installed wind capacity; energy supply conditions; wind energy rejection and off-grid wind power.
- **Firm strategy, structure and rivalry:** local renewable energy companies; specialization; technological innovations; differentiation strategies; vertical integration and social capital investments.
- **Government:** laws and regulations; political decisions; taxes and government support.
- **Technologies:** energy storage; construction of ultra-high voltage electricity transmission networks; technical R&D; increase of R&D capacity; development of standards and norms.
- **Chance:** wind industry opportunities (rich wind resources; high energy demand; government support; decrease of wind energy production costs; increasing share of wind energy in the country's total energy balance; local economic development), wind industry challenges (unbalanced distribution of wind energy resources; inefficient and obsolete wind farms, lack of financial mechanism and economic incentives, resistance to wind energy projects, need for capital, dependence on foreign technologies).

As can be seen, some of the above-mentioned factors of the assessment of the competitiveness of Lithuanian VCWEC companies cannot be applied since they were applied at another level of research and become irrelevant in this case. Examples of these factors include local renewable energy companies as a factor in the firm strategy, structure and rivalry of companies. In the case of this study, the element must include the companies that form the VCWEC.

Fang et al. [18] singled out the substitution costs under the demand conditions when examining the competitiveness of renewable energy, which was considered as a possible change in energy prices due to the changed energy structure (an increasing share of energy from renewable sources may lead to an increase in the price to the final consumer, leading to fewer consumers choosing to buy energy from renewable sources). This study chooses the domestic and foreign market demand for "green" products as factors with similar meaning. Fang et al. [18] also included in the demand conditions a factor of political incentives that may encourage consumers to abandon fossil fuel energy and choose energy from renewable energy sources, which may affect demand. The Diamond model used in this study distinguishes a separate group of "government" factors that affect all the major groups of the other factors. Therefore, the factors related to government decisions were assigned to the group of "government" factors in this study. The wind energy rejection factor singled out by Liu et al. [24] is associated with possible restrictions on the supply of wind energy due to the limited capacity of China's transmission networks. The off-grid energy factor examined by Zhao et al. [23] is also specifically relevant to the Chinese market, as it relates to wind energy that is generated in rural areas and consumed locally which cannot be supplied to the broader market as it would be economically inefficient to build power transmission lines to these areas. Due to the specifics of these factors which are relevant to the Chinese market and not relevant in Lithuania, they were not included in this study. The manufacturers of WEC are classified as a group of related and supporting industries by Fang et al. [18], Irfan et al. [25] and Zhao [23], as they examined the competitiveness of wind (renewable) energy. In the case of this study, the factor is attributed to the firm strategy, structure, and rivalry because the manufacturers of WEC are competitors to the firms in question. Typically, the output of manufacturers of WEC is very specific and only in rare cases can one manufacturer of WEC be a supplier or customer to the other component manufacturers. For this reason, they are not included in the group of related and supporting industries in this study. Irfan et al. [25] assigned a specialization factor to firm strategy, structure, and rivalry, emphasizing on firm specialization to achieve competitive advantage. In this study, Lithuanian manufacturers of WEC usually produce a wider range of products, only one part of which is WECs; therefore, the inclusion of this factor in the study becomes meaningless. The opportunities and challenges of a chance for foreign authors are related to the conditions of demand and factors, as well as government and technology elements, and therefore, these factors are not applied in this study. To describe a chance in this study, price spike, shortage of raw materials, supply disruptions and other factors are used.

Irfan et al. [25] attributed the technological innovation factor to the element of firm strategy, structure and rivalry, but in the case of this study, it is more appropriate to attribute it to the element of technology. The factors "energy storage" and "construction of ultra-high voltage electricity transmission networks" that were attributed to the group of technological factors by Liu et al. [24] are not relevant for manufacturers of WEC in this study, therefore, they were further omitted.

Summarizing the factors applied in the research of foreign authors, as well as the factors of the business environment, statistics and analysis of the current situation, the following factors in assessing the competitiveness of Lithuanian VCWEC were singled out.

Factor conditions:

- skilled labour costs;
- cost of energy used in the production process;
- price of other production inputs;

- origin of raw materials;
- cost of capital borrowing;
- changes in the investment environment;
- potential of wind energy resources;
- changes in energy structure;
- EU and national business support;
- diversity of business financing sources and access to them;
- changes in the exchange rate;
- inflation;
- status of production facilities;
- lack of requested qualification staff.

Demand conditions:

- foreign market demand for "green" products;
- domestic market demand for "green" products;
- energy demand;
- demand structure;
- demand dynamics;
- installed wind energy capacity.

Related and supporting industries:

- wind turbine manufacturers;
- grid constructors;
- raw material suppliers;
- sales intermediaries.

Firm strategy, structure and rivalry:

- differentiation strategies;
- export;
- competitors.

Government:

- targets and support measures for increasing the production and consumption of energy from renewable sources in Lithuania;
- EU and Lithuanian emission reduction targets;
- energy efficiency targets and support measures;
- regional development policy, its objectives and support measures;
- requirements for the management and utilization of waste generated in the production process;
- phase out of subsidies for fossil fuels;
- taxes.

Technologies:

- implementation of the latest technologies;
- dynamics of technological change and innovation implementation;
- expenditure on R&D.

Chance: price spike, shortage of raw materials, supply disruptions.

4. Methods of Assessing the Competitiveness of Enterprises Forming a VCWEC

Various quantitative and qualitative methods are used to assess competitiveness according to the Porter's Diamond model. The applied method is determined by the selected competitiveness assessment factors and data describing them. It is difficult to limit oneself to quantitative methods, as not all of the selected factors can be quantified or there is a lack of quantitative data. Some of identified competitiveness factors for this study describe the business environment, another part is the analysis of statistical data and the third part is the answers to purposefully formulated questions for companies in the questionnaire. Irfan et al. [25], Liu et al. [24] and Zhao et al. [23] performed a competitive assessment using the method of descriptive analysis. Zhang [26] and Fang et al. [18] used quantitative methods to assess competitiveness. However, in their case, the number of competitiveness factors was lower, and they performed an international competitiveness assessment comparing the competitiveness of renewable energy producers and international wind power producers in different countries, respectively.

In order to cover as many factors of competitiveness of the companies forming the VCWEC as possible, a descriptive logical analysis, a questionnaire survey and an expert assessment were used. The results of the survey of manufacturers of WEC allow for the simultaneous assessment of the business environment and the factors shaping the competitiveness. Based on these, it is possible to draw conclusions about the influence of individual groups of competitiveness factors on the competitiveness of the companies in question. The expert assessment was applied to assess the business environment and the significance of various individual groups of competitiveness factors. In our case, the experts selected were the representatives of energy and industry associations.

The general aim of *the survey of manufacturers of WEC* was to identify and describe the peculiarities of these companies' performance in terms of the group of products they produce and could produce in future, the share of WEC in the total production value, the historical development of sales volumes of WEC, the export markets and the shares of exports, the customers they deal with, the trade intermediaries they have, the competitors in domestic and foreign markets, the purchases of raw materials, the post-sale services they provide, the participation in R&D, the factors of business environment they face, the issues they meet and the role of government improving the conditions for performance. The survey was constructed based on the questionnaire, which consisted of 24 questions covering the activities mentioned above. This paper focuses mainly on responses of the manufacturers of WEC to questions about the factors of the business environment. The identified factors were grouped considering the PEST analysis method which allowed for an analysis of the Political, Economic, Sociodemographic and Technological determinants of the business environment. The manufacturers of WEC were asked to assess whether the factors provide opportunities or threats to their participation in the VCWEC and, later on, to assess the strength of this impact on a scale from 1 of 5, where:

- 1—very small opportunity or threat;
- 2—small opportunity or threat;
- 3—medium opportunity or threat;
- 4—high opportunity or threat;
- 5—very high opportunity or threat.

In addition, if manufacturers of WEC supposed that the factor of business environment is irrelevant to their participation in the VCWEC, they could scale it as of impact 0, which means it is insignificant factor.

From the practical perspective, the scaled strength of the impact on opportunities should be understood as the opportunity to increase production value, earn net profit, grow, increase exports, enter new markets or achieve other benefits, which are adjusted to the core aim of that company. In contrast, the scaled strength of the impact on threats means that the factor of the business environment contradicts the main aim of that company's performance and results. A rating of 0 means that the factor of business environment is insignificant as it does not impact the performance and results of manufacturer. No numerical values were attributed to measures of opportunities and threats; therefore, manufacturers were left with personalized understanding. Such a decision was made by the Authors of this paper with the purpose to normalize the responses.

The expert assessment aims to assess how strongly the factors of the political, economic, sociodemographic and technological environment impact the participation of companies in VCWEC and evaluate the importance of the factors determining the competitiveness of these companies. The expert assessment questionnaire consisted of seven questions. Of these, four questions were designed to assess the factors of the business environment according to the opportunities and threats (on a scale from 1 to 5, see above), one question on the assessment of the importance of groups of competitiveness factors (on a scale from 1 of 5) and two open-ended questions. One open-ended question was asked to find out the reasons that prevent companies from getting involved in the VCWEC, as this may not have been reflected in the assessment of business environment. The second open question sought to find even more companies involved in the manufacture of WEC that could be interviewed.

The assessment of the competitiveness of the companies forming the Lithuanian VCWEC was carried out in three stages, as shown in Figure 2.

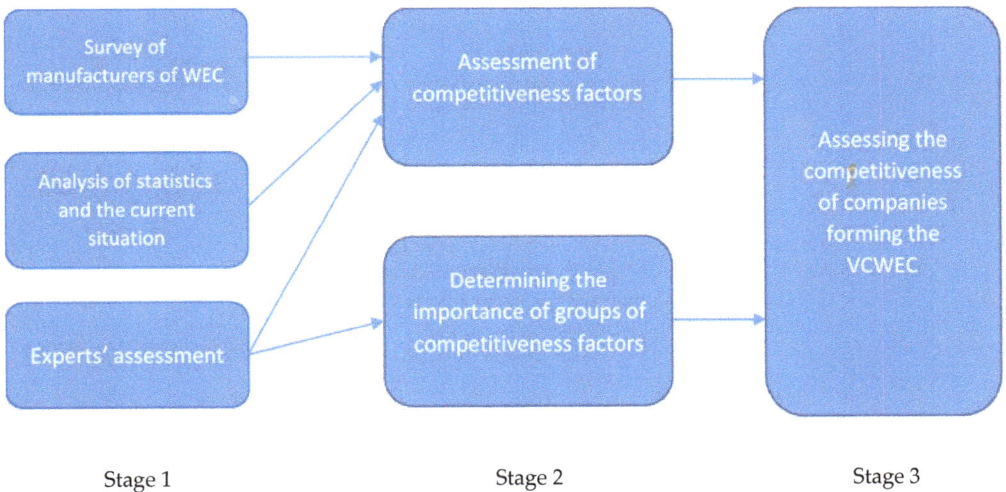

Figure 2. The assessment of the competitiveness of the companies forming the Lithuanian VCWEC.

Factors that describe both the business environment and the results of the analysis of statistics and the current situation were selected to assess the competitiveness of the companies forming the VCWEC. Thus, in the first stage of the research, a survey of companies, experts' assessment, analysis of statistics and the current situation were carried out. In the second stage of the study an analysis of the data collected during the first stage is performed, i.e., assessing competitiveness factors and determining the importance of groups of competitiveness factors. By combining the evaluations carried out in the second stage, their results are summarized in the third stage.

During the research, 25 manufacturers of WEC were identified and contacted electronically by sending the questionnaires or the references to questionnaires during 1 March 2021 and 5 July 2021. In total, seven manufacturers of WEC sent their responses. The responses about business environment factors were summarized considering the weighted average method, which was applied for assessments of factors providing both opportunities and threats. Four experts participated in the survey. Since there have not been many manufacturers forming Lithuanian VCWEC, many industrial associations could not be experts. An expert assessment was organized electronically between 28 May 2021 and 14 June 2021 by sending the questionnaires or the references to experts.

The Kendall concordance coefficient was used to assess the level of coincide (similarity) of experts' opinions [27], as follows:

$$W = \frac{12S}{m^2(n^3 - n)} \tag{1}$$

where m—number of experts, n—number of assessed (rated) factors and S—sum of squares of deviations from the mean:

$$S = \sum_{i=1}^{n} \left(R_i - \overline{R} \right)^2 \tag{2}$$

$$R_i = \sum_{j=1}^{m} r_{ij} \tag{3}$$

where $\overline{R} - R_i$—the average and r_{ij}—assessment of i-th factor of j-th expert.

The Kendall concordance coefficient can take values from 0 to 1. The closer the value of the coefficient to 1 is, the more similar the opinions of the experts are.

5. Results of Research on Competitiveness of Manufacturers Forming VCWEC

5.1. Assessment of Competitiveness Factors of VCWEC

The calculated overall Kendall concordance coefficient of all experts' responses indicates a mean concordance of opinions, as it is 0.556. The greatest agreement is reached on the impact of the factors of the sociodemographic and economic environments on the participation of Lithuanian manufacturers in the VCWEC. The greatest difference of opinion, on the other hand, is in the assessment of the importance of the factors determining the competitiveness of manufacturers participating in the VCWEC and the impact of the factors of technological environment (Table 1).

Table 1. Values of the Kendall concordance coefficient by groups of factors: case of research of Lithuanian VCWEC.

Group of Factors	Assessment of Positive Impact (Opportunities)	Assessment of Negative Impact (Threats)
Factors of political environment	0.430	0.490
Factors of economic environment	0.651	0.580
Factors of sociodemographic environment		0.816
Factors of technological environment	0.308	
Factors impacting on competitiveness	0.272	
Total (based on all groups of factors) value of Kendall concordance coefficient	0.556	

According to experts, within the group of factors of the political environment, the EU and Lithuanian emission reduction targets, regional development policies, aims and supporting policies provide large opportunities to manufacturers of WEC. The first factor creates assumptions to increase demand for WEC, but the second factor creates assumptions to develop WEC business in regions having wind energy potential and relevant infrastructure, such as seaports. The medium threats come from the requirements for the management and utilization of waste generated in the production process. In the group of factors of the economic environment, increasing foreign and domestic market demand for "green" products, active EU and national support for business, as well as access to variety of business financing sources bring the WEC business medium to large opportunities; while increasing prices of goods and services, growth of production cost, as well changes of exchange rate are recognized as unfavourable conditions. The threats they cause are assessed as medium to very high. The experts singled out the shortage of requested qualification staff and emigration as a source of threat within the group of factors of the sociodemographic environment but agreed that immigration, education and society's improving attitude towards "green" products provide opportunities. The status of production facilities, implementation of new technologies and innovation, development of general infrastructure and expenditure on R&D are assessed as providing opportunities to manufacturers of WEC under the technological environment.

Based on the factors singled out in the literature review, eight factors of the economic environment, one factor of technological and sociodemographic environment, two factors describing the external environment and one factor to the question about raw materials used in the production of WEC were prescribed to a group of factor conditions. The impact of environmental factors was estimated as an average of manufacturers of WEC and expert assessment. The descriptive analysis to assess other factors was used (Table 2).

Table 2. Assessment of competitiveness factors: a group of factor conditions.

Factor Conditions	Description	Assessment
Status of production facilities	Factor of technological environment	Opportunities 4.0
EU and national business support	Factor of economic environment	Opportunities 3.4
Changes in the investment environment	Factor of political environment	Opportunities 3.1
Diversity of business financing sources and access to them	Factor of economic environment	Opportunities 2.9
Lack of requested qualification staff	Factor of sociodemographic environment	Threats 4.0
Skilled labour cost	Factor of economic environment	Threats 2.9
Cost of capital borrowing	Factor of economic environment	Threats 2.9
Price of other production inputs	Factor of economic environment	Threats 2.8
Changes in exchange rate	Factor of economic environment	Threats 2.2
Inflation	Factor of economic environment	Threats 2.0
Cost of energy used in production process	Factor of economic environment	Threats 1.9
Origin of raw materials	Question in the questionnaire: 1 Where does the company procure materials and raw materials for the production of wind energy components?	56% of manufacturing companies purchase raw materials in EU markets, 31% in non-EU markets and 13% in the Lithuanian market.
Potential of wind energy resources	Analysis of current situation	In the Litgrid Network Development Plan for 2020–2029 [28], it is foreseen that total capacity of wind power plants will increase to 2.206 MW in 2029, including 700 MW off-shore wind farm. The onshore wind capacity will increase by approximately 1.000 MW, i.e., from current 540 MW to 1506 MW.
Changes in energy structure	Statistical analysis	Wind energy in electricity generation from renewable energy sources accounted for 48% in 2015 and 61% in 2019 in Lithuania [29].

With reference to the information in Table 2, in a group of factor conditions, the lack of requested qualification staff is the most serious source of threat. Respondents noted that the business faces both: a shortage of skilled engineers and low-skilled staff. Other threats come from increasing cost of key production factors, including skilled labour, capital borrowing and energy. Most manufacturers of WEC procure raw materials on EU markets, but a third of them import raw materials from outside the EU and settle for raw materials using exchange rates other than EUR. Therefore, changes in exchange rates bring uncertainty. Respondents assessed the threat coming from the changes in exchange rate as low. Historically, the inflation rate was moderate in Lithuania (up to 10% a year), and during some years, deflation was fixed (during 2014–2015, inflation was 0.1–0.3% a year). Respondents assessed that threat (for example, reducing purchasing power of customers or orders of WEC) coming from increasing prices is low. The current status of production facilities assures the execution of orders. Manufacturers of WEC invest in update and development of production facilities. Since 2016, manufacturers of WEC have invested EUR 24.4 million a year when production volume has been EUR 250–300 million a year. Lithuania implements its Programme for Investment Promotion and Industry Development [30] by using EU Structural Funds and other resources which focus on linking industry and services to networks and industrial cooperation, increasing production of advanced technologies and using raw materials and energy efficiently. Respondents assessed that changes in investment environment are positive and promising. Moreover, within the Programme, the EU and national support measures are provided to address the aforementioned issues. Respondents assessed this as a medium to high opportunity. The market provides various financing sources to business. However, respondents assessed that the opportunities from this factor are medium, as the access conditions are strict. The share of wind energy in electricity production from renewable sources has been increasing in recent years in the country. The potential of wind energy resources and changes in the structure allow for the assumption that there are favourable conditions for the development of wind energy in Lithuania. The assessment of a group of condition factors and their tendencies show that these conditions are moderately favourable for manufacturers of WEC.

Two factors of the economic environment and two factors focusing on customers of WEC and the dynamics of sales of those components over the last five years included in the manufacturers of WEC survey are assigned to a group of demand conditions. The assessment of three factors described by statistical analysis is additionally added to a group of demand conditions (Table 3).

Table 3. Assessment of competitiveness factors: a group of demand conditions.

Factors of Demand Conditions	Description	Assessment
Foreign market demand for "green" products	Factor of economic environment	Opportunities 4.2
Domestic market demand for "green" products	Factor of economic environment	Opportunities 3.1
Energy demand	Statistical analysis	The final electricity consumption changed insignificantly in Lithuania during 2016–2020. Consumption increased by 6.2% from 2016–2020 but decreased by 2% from 2019–2020. Since 2018, electricity consumption has been declining in industry and households, but household consumption has increased in 2020 [29].
Demand structure	Question: Who are the customers of wind energy components produced by the company?	Manufacturers of wind power plants and their components (58% of respondents), wind power plant construction (installation) companies (33%) and companies for wind power plant maintenance services (8%).
Demand dynamics	Question: How have the sales volumes of wind energy components produced in the company changed during the last 5 years?	Decreased (40% of Respondents), increased (30%) and unchanged (30%).
Installed wind energy capacity	Statistical analysis	The installed capacity of wind power plants was 0.5 GW in Lithuania in 2019. It increased by 0.2 GW in 5 years. The installed capacity of wind power plants was 167.1 GW in EU in 2019. It increased by 51.5 GW in 5 years [31]. Wind power installations have fallen by 6% to 14.7 GW in 2020 in EU [32].

The changing attitudes of consumers searching for sustainable, environment-friendly solutions increase the demand for "green" products. As it is seen from Table 3, due to expressed demand for "green" products, opportunities arise for the manufacturers of

WEC. Subject to the fact that the Lithuanian market is small for conducting business, opportunities provided by foreign markets demand for "green" products were assessed higher than those provided by the domestic market. Demand structure is not homogeneous as it is formed by the manufacturers of wind power plants and their components, the companies constructing wind power plants and the companies providing maintenance services to wind power plants. The demand structure allows for the assumption that the WEC for wind power plants installed in Lithuania are applied in different stages of the production and maintenance of wind power plants and may have the potential to increase by expanding the construction of wind power plants. The dynamics of demand over the last 5 years show that WECs sales have changed slightly, as 40% of the respondents said that sales were declining, and 30% said that they were increasing or did not change. The installed wind energy capacity has been growing in Lithuania since 2008 but has remained constant since 2016 (0.5 GW). The installed wind power capacity in the EU market is constantly growing, only decreasing by 5.8% in 2020 due to the COVID-19 pandemic. The International Renewable Energy Agency (IRENA) predicts that wind energy will meet 35% of total energy demand by 2050 in order to meet the climate change targets of the Paris Agreement. In this way, the installed wind power capacity must triple by 2030 and increase nine times by 2050 compared to 2018. Asia (mainly China) will dominate in installed onshore wind power capacities (more than 50% in 2050), while in Europe, 10% will be installed. Accordingly, with 60% of installed capacity, Asia will be the leader in offshore wind power capacities, too; in Europe 22% of offshore wind power capacities will be installed [33]. In this way, the demand conditions to produce WEC show positive trends and can be welcomed.

Four factors are assigned to a group of factors of related and supporting industries. They are included in the survey to manufacturing companies and refer to the questions about the customers of WEC, the suppliers of raw materials and the sales intermediaries (Table 4).

Table 4. Assessment of factors of a group of related and supporting industries.

Factors of A Group of Related and Supporting Industries	Description	Assessment
Wind turbine manufacturers	Question: Who are the customers of wind energy components produced in the company?	Manufacturers of wind power plants and their components (58% of respondents), wind power plant construction (installation) companies (33%), companies for wind power plant maintenance services (8%) and other (to be identified) (0%).
Grid constructors		
Raw material suppliers	Question: How many suppliers of materials and raw materials does the company have for the production of wind energy components?	3–10 raw material suppliers (60% of respondents), more than 10 (30%) and a single supplier (10%).
Sales intermediaries	Question: How many sales intermediaries are involved in the distribution of wind energy components produced by the company?	No sales intermediaries at all (55% of respondents), one (27%), and more than 3 (18%).

Assessing the factors of a group of related and supporting industries, it is found that the manufacturers of wind power plants and their components, the companies constructing wind power plants, and the companies providing wind power plant maintenance services are the key related and supporting industries. Grid constructors were not identified as related and supporting industries for Lithuanian manufacturers of WEC. The suppliers of raw materials are related and supporting industries uphold the competitiveness of manufacturers of WEC. It was found that the majority of manufacturers of WEC have 3–10 raw material suppliers; one-third of those companies demand raw materials from more than 10 suppliers, but up to 10% of manufacturers of WEC are dependent on a single supplier. The sales intermediaries are less important in this group of factors, as a significant proportion (55%) of manufacturers of WEC answered that they do not have sales intermediaries. According to the results of the survey of manufacturers of WEC, it is seen that they usually communicate with their customers without intermediaries, have regular suppliers of raw materials; therefore, there is some synergy between related and supporting industries.

Three factors (Table 5) were prescribed to a group of firm strategy, structure and rivalry. Seeking to analyse them, six questions were formed in the questionnaire for the survey of manufacturers of WEC. The content of those questions referred to the groups of WEC that the manufacturers produce, the shares of WEC in the production structures, the shares of WEC exports, the export markets and the domestic and foreign competitors.

Table 5. Assessment of firm strategy, structures and rivalry.

Factors of Firm Strategy, Structure and Rivalry	Description	Assessment
Differentiation strategies	Questions: 1. Which groups of WEC does the company produce and / or could produce in future? 2. What is the share of WEC in the company's total production volume?	Mechanical, electrical, electronic, thermal, composite and other components. WEC make 15% in the company's production volume (in 70% of surveyed WEC companies), 70% (in 10% of surveyed WEC companies), 95–100% (in 20% of surveyed WEC companies)
Export	Questions: 1. What is the share of WEC in the total production volume that the company exports? 2. To which countries does the company export it's WEC?	More than 50% (70% of surveyed WEC companies) EU member states are key export markets (62% of surveyed WEC companies); 38% of surveyed WEC companies export to other countries than EU.
Competitors	Questions: 1. How many companies in the Lithuanian market are the company's competitors in the production of WEC? 2. How many companies in foreign markets are the company's competitors in the production of WEC?	60% of surveyed WEC companies have 1–3 competitors and 40% of companies do not have competitors at all in Lithuania. WEC companies compete with more than 3 competitors, except for 1 WEC company, which has up to 3 competitors in foreign markets.

According to the factor of differentiation strategy, it was found that the manufacturers of WEC produce various components, but usually they are not the only products they manufacture. Several companies specialize in WEC production; therefore, WEC make up to 95–100% of the production structure. Exports of WEC are quite widely developed among the surveyed manufacturers. They mainly export to EU countries, but 38% of the surveyed manufacturers of WEC export their products to other countries of the world, too. Lithuanian companies are too small to attract the largest wind power manufacturers. Competition in the Lithuanian market is not high, and some manufacturers of WEC have no competitors at all (40%). Competition in the foreign markets is high, but due to the specifics of the products, it is not very intense. In any case, international competition is not beneficial for Lithuanian producers due to the low costs of producers from other countries. Most of the surveyed companies indicated that they are subdivisions of global enterprise groups. Therefore, they face the following difficulties:

- they cannot apply for Lithuanian and EU financial support;
- it is difficult to become a direct supplier due to the complex organization with the customer;
- by managing risks, customers regulate all processes at their direct suppliers;
- unattractive obligations when working with large groups—binding warranty conditions for products, strict requirements for quality and supply indicators.
- high penalties are provided for delays or non-compliance with other obligations.

As opportunities to address issues related to firm strategy, structure and rivalry factors, companies see opportunities to look for partners who are already working with power plant manufacturers and aspire to be second or third in the supply chain.

Six factors of political environment and one factor of the economic environment were prescribed to a group of governmental factors. They were assessed as an average of results from the survey of manufacturers of WEC and expert assessment (Table 6).

Table 6. Assessment of governmental factors.

Governmental Factors	Description	Assessment
EU and Lithuanian emissions' reduction targets	Factor of political environment	Opportunities 3.7
Targets and support measures for increasing the production and consumption of energy from renewable sources in Lithuania	Factor of political environment	Opportunities 3.3
Regional development policy, its objectives and support measures	Factor of political environment	Opportunities 3.0
Phase out of subsidies for fossil fuels	Factor of political environment	Opportunities 2.9
Energy efficiency targets and support measures	Factor of political environment	Opportunities 2.3
Taxes	Factor of economic environment	Threats 3.1
Requirements for the management and utilization of waste generated in the production process	Factor of political environment	Threats 2.4

The group of governmental factors is dominated by the factors of political environment. One factor of the economic environment was included. According to the Respondents, governmental decisions provide companies with more opportunities than threats. The most immense opportunities can be provided by the EU and Lithuanian emissions' reduction targets, followed by targets and support measures for increasing the production and consumption of energy from renewable sources in Lithuania, as well as regional development policy, its objectives and support measures as the factors create demand for WEC, facilitate business doing in places favourable to development of wind energy or to relevant infrastructure (for example, seaports).

Companies, naming opportunities for the development of their activities, indicate that the governmental decisions could stimulate it:

- taxation of profits only by withdrawing funds from the company would encourage shareholders to invest more in Lithuania by increasing the value of the company;
- the form of business promotion and taxation used in free economic zones' territories would be adapted for entire regions;
- providing equal opportunities to foreign and Lithuanian capital companies; promotion of Lithuanian capital investments in manufacturing companies;
- promotion of the establishment of joint Lithuanian and foreign capital companies with know-how experience.

Three factors of technological environment were prescribed to a group of technological factors (Table 7).

Table 7. Assessment of technological factors.

Technological Factors	Description	Assessment
Implementation of the latest technologies	Factor of technological environment	Opportunities 3.8
Dynamics of technological change and innovation implementation	Factor of technological environment	Opportunities 3.3
Expenditure on R & D	Factor of technological environment	Opportunities 2.5

As it is seen from Table 7, the introduction of the latest technologies, dynamics of technological changes and innovations and R&D expenditures may provide medium to high opportunities to manufacturers of WEC.

The impact of a chance on the competitiveness of the manufacturers of WEC is assessed only based on the results of an experts' survey. Its importance is scored as 3.25 on a 5-point rating scale.

5.2. Assessment of the Importance of Groups of Competitiveness Factors and the Competitiveness of WEC Companies

Assessment of the competitiveness factors of manufacturers of WEC was performed in three stages. The first stage examined the factors of business environment (analysis of responses of surveyed manufacturers of WEC and experts). In the second stage, the factors of business environment were summarized and supplemented by the factors of statistical and current situation analysis. In the third stage, the assessment results of the importance of the groups of competitiveness factors were summarized according to the results of the surveyed experts (Table 8).

Table 8. Assessment of the importance of the factors determining the competitiveness of manufacturers of WEC.

Groups of Factors	Average of Assessment Score	Minimum Assessment Score	Maximum Assessment Score
Factor conditions	5	5	5
Demand conditions	3.3	1	5
Related and supported industries	3.5	2	5
Firm strategy, structure and rivalry	4	3	5
Government	4	2	5
Chance	3.3	1	5

Assessing the importance of the factors determining the competitiveness of manufacturers of WEC, experts unanimously agree that the factor conditions are the most important. However, the other factors mentioned were also considered to be quite important.

The firm strategy, structure and rivalry and the government were assessed as very important by experts too (the average score is 4.0). The analysis of individual factors of these groups revealed that Lithuanian manufacturers of WEC are characterized by their activity differentiation, the development of their exports to the EU and other countries and that they do not face the intense competition.

The importance of related and supporting industries for the competitiveness of manufacturers of WEC was assessed by experts at the average score of 3.5. The analysis of individual factors showed that companies engaged in these activities sell their products without the intermediaries, have regular suppliers of raw materials and a list of different customers, including manufacturers of wind turbines and their components, companies constructing and installing wind power plants and companies providing maintenance services to wind power plants.

The demand conditions and the impact of the chance were assessed as the least important in comparison to other discussed factors. Their average score is 3.3. Nevertheless, the analysis of demand factors has revealed that they are favourable and positively affect the competitiveness of WEC companies.

The methodology developed and applied for the research of the competitiveness of the industrial companies forming the VCWEC in Lithuanian is universal and may be applied to other research in other countries'. The manufacturer surveys and expert assessment may reveal completely different results in larger countries with more developed industries. It would be useful to carry out such a study in the context of EU countries. In this case, it might be challenging to assess government factors as they can vary significantly from one EU country to another. This may be a further area of research.

Competitiveness assessment studies often face difficulties in conducting quantitative assessments, as competitiveness models cover a wide range of factors, but most of them cannot be quantified, so qualitative assessment methods have to be applied.

A short (half a year) project implementation deadline and the quarantine made data collection more complex. With more time for companies to be interviewed and opportunities to meet, the study results might reveal slightly different results.

6. Conclusions

The research was conducted applying methods such as a survey of manufacturers of WEC, expert assessment and descriptive analysis

Summarizing the factors assigned to the groups of factors assessing the competitiveness of Lithuanian industrial companies forming VCWEC assessed according to the results of companies' survey and experts' assessment, it is seen that the greatest opportunities are provided by demand conditions (3.6), followed by factor conditions (3.4), government (3.3) and technology (3.3). The threats are posed by factor conditions (3.7) and government (2.8). The analysis disclosed that the existing conditions are sufficiently favourable for the competitiveness of Lithuanian industrial companies forming the VCWEC.

According to experts, financial reasons and insufficient state legal regulation are the main obstacles that make it difficult for companies to join and/or operate in the VCWEC in Lithuania.

The interviewed Lithuanian industrial companies forming the VCWEC identified the following difficulties for their performance:

- The size of Lithuania and its industry is too small to attract large wind power manufacturers;
- Competition with producers and service providers in low-cost countries is high;
- There is a lack of both low-skilled and skilled workers with engineering education.

The survey of manufacturers revealed that most of them are entities of global enterprise groups. For this reason, they face the following difficulties:

- They are not eligible for Lithuanian and EU financial support;
- It is difficult to become a direct supplier;
- Complex qualification process: managing risks customer companies regulate all processes at their direct suppliers;
- Unattractive obligations when working with large corporations: binding warranty conditions for products, strict requirements for quality and supply indicators. High fines for delays or non-compliance with other obligations.

The interviewed manufacturers see the following opportunities for improvement or development of their performance and the measures that would help to implement them:

- Finding partners who are already working with the power plant manufacturers, be second or third in the supply chain and thus avoid much of bureaucracy working directly;
- Applying corporate income tax only by withdrawing funds from the company would encourage shareholders to invest more in Lithuania by increasing the company's value;
- Applying enterprise promotion and taxation that are used in free economic zones territories for entire regions;
- Ensuring the education of more engineering professionals and creating better conditions for the employment of missing professionals from third countries;
- Promoting Lithuanian capital investments in manufacturing companies;
- Encouraging the establishment of joint Lithuanian and foreign capital companies with know-how.

The research results reveal that government solutions to promote industry could facilitate the activities of Lithuanian industrial companies forming VCWEC and encourage new companies to join it. This could encourage the Lithuanian industry to participate in the value chain of the EU's renewable energy industry, create more jobs and increase the value-added created.

Author Contributions: Conceptualization: D.D. and A.C.; methodology: D.D., A.C. and V.B.; software: A.C.; validation: A.C. and V.B.; investigation: D.D., A.C. and V.B.; supervision: E.D.; writing

and review: D.D., A.C. and V.B. All authors have read and agreed to the published version o
the manuscript.

Funding: This research "Feasibility study of Lithuanian industry participation in the value chair
of wind energy components. (VEGRAND)" is funded by the Research Council of Lithuania unde
Grant No. S-REP-21-5.

Institutional Review Board Statement: Not applicable.

Informed Consent Statement: Not applicable.

Conflicts of Interest: The authors declare no conflict of interest. The funders had no role in the desigr
of the study; in the collection, analyses, or interpretation of data; in the writing of the manuscript; o
in the decision to publish the results.

References

1. Wind Europe. Wind Energy and Economic Recovery in Europe. How Wind Energy Will Put Communities at the Heart o
the Green Recovery. Available online: https://windeurope.org/intelligence-platform/product/wind-energy-and-economic
recovery-in-europe/ (accessed on 4 August 2021).
2. U.S. Department of Energy: Energy Efficiency & Renewable Energy. PowerPoint Version of the Wind Energy Integratior
Slideshow. Available online: https://windexchange.energy.gov/slideshows (accessed on 4 August 2021).
3. European Commission. Quarterly Report on European Electricity Markets with Focus on the Developments in Annual Wholesale
Prices. Available online: https://ec.europa.eu/energy/sites/default/files/quarterly_report_on_european_electricity_markets_
q4_2020.pdf (accessed on 4 August 2021).
4. Bobinaite, V.; Priedite, I. RES-E support policies in the Baltic States: Price aspect (Part 2). *Latv. J. Phys. Tech. Sci.* **2015**, *52*, 13–25.
5. Ju, X.; Liu, F. Wind farm layout optimization using self-informed genetic algorithm with information guided exploitation. *Appl
Energy* **2019**, *248*, 429–445. [CrossRef]
6. Bobinaite, V.; Priedite, I. Assessment of impacts of wind electricity generation sector development: Latvian case. *Procedia Soc
Behav. Sci.* **2015**, *213*, 18–24. [CrossRef]
7. Li, J.; Li, S.; Wu, F. Research on carbon emission reduction benefit of wind power project based on life cycle assessment theory
Renew. Energy **2020**, *155*, 456–468. [CrossRef]
8. Ju, X.; Liu, F.; Wang, L.; Lee, W.-J. Wind farm layout optimization based on support vector regression guided genetic algorithm
with consideration of participation among landowners. *Energy Convers. Manag.* **2019**, *196*, 1267–1281. [CrossRef]
9. World Economic Forum (WEF). Global Competitiveness Report 2014–2015. Geneva. Available online: http://reports.weforum
org/global-competitiveness-report-2014-2015/ (accessed on 18 May 2021).
10. Vivien, W. Technology and the economy—The key relationships: (Organisation for economic co-operation and development
Paris, 1992) pp. 328, 260 FF. *Res. Policy* **1994**, *23*, 473–475.
11. International Institute for Management Development (IMD). World Competitiveness Yearbook 2014. Available online: https
//www.imd.org/research-knowledge/articles/com-may-2014/ (accessed on 15 May 2021).
12. Smith, A. The Wealth of Nations: An Inquiry into the Nature and Causes of the Wealth of Nations. Harriman House Limited
2007. Available online: https://www.ibiblio.org/ml/libri/s/SmithA_WealthNations_p.pdf (accessed on 20 May 2021).
13. Ricardo, D. On the Principles of Political Economy and Taxation. (Original Work Published 1817, Third Edition 1821). London
Dent 2001. Available online: https://socialsciences.mcmaster.ca (accessed on 20 May 2021).
14. Krugman, P.R. Making sense of the competitiveness debate. *Oxf. Rev. Econ. Policy* **1996**, *12*, 17–25. [CrossRef]
15. Siudek, T.; Zawojska, A. Competitiveness in the economic concepts, theories and empirical research. *Acta Sci. Pol. Oeconomia
* **2014**, *13*, 91–108.
16. Beniušienė, I.; Svirskienė, G. Konkurencingumas: Teorinis aspektas. *Ekon. Vadyb. Aktual. Perspekt.* **2008**, *4*, 32–40.
17. Porter, M.E. The competitive advantage of nations. *Compet. Intell. Mag.* **1990**, *1*, 14.
18. Fang, K.; Zhou, Y.; Wang, S.; Ye, R.; Guo, S. Assessing national renewable energy competitiveness of the G20: A revised Porter's
Diamond Model. *Renew. Sustain. Energy Rev.* **2018**, *93*, 719–731. [CrossRef]
19. Travkina, I.; Tvaronavičienė, M. An Investigation into Relative Competitiveness of International Trade: The Case of Lithuania
Available online: http://dspace.vgtu.lt/bitstream/1/597/1/504-510_Travkina_Tvaronaviciene.pdf (accessed on 18 May 2021).
20. Lietuvos Pramonės Eksporto Konkurencingumo Vertinimas. Available online: https://www.vdu.lt/cris/bitstream/20.500.12259
/1221/1/ISSN1822-7996_2012_T_6_N_2.PG_49_72.pdf (accessed on 18 May 2021).
21. Key Issues in Defining and Analysing the Competitiveness of a Country. Available online: https://papers.ssrn.com/sol3/papers
cfm?abstract_id=418540 (accessed on 18 May 2021).
22. Balkytė, A.; Tvaronavičienė, M. Perception of competitiveness in the context of sustainable development: Facets of "sustainable
competitiveness". *J. Bus. Econ. Manag.* **2010**, *11*, 341–365. [CrossRef]
23. Zhao, Z.Y.; Hu, J.; Zuo, J. Performance of wind power industry development in China: A DiamondModel study. *Renew. Energ
* **2009**, *34*, 2883–2891. [CrossRef]

4. Liu, J.; Wei, Q.; Dai, Q.; Liang, C. Overview of wind power industry value chain using diamond model: A case study from China. *Appl. Sci.* **2018**, *8*, 1900. [CrossRef]

5. Irfan, M.; Zhao, Z.Y.; Ahmad, M.; Mukeshimana, M.C. Critical factors influencing wind power industry: A diamond model based study of India. *Energy Rep.* **2019**, *5*, 1222–1235. [CrossRef]

6. Zhang, S. International competitiveness of China's wind turbine manufacturing industry and implications for future development. *Renew. Sust. Energ. Rev.* **2012**, *16*, 3903–3909. [CrossRef]

7. Legendre, P. Coefficient of concordance. *Encycl. Res. Des.* **2010**, *1*, 164–169.

8. Litgrid. Lietuvos Elektros Energetikos Sistemos 400–100 kv Tinklų Plėtros Planas 2020–2029 m. Vilnius 2020. Available online: https://www.vert.lt/SiteAssets/posedziai/2021-02-19/litgrid_planas_priedas_1.pdf (accessed on 12 June 2021).

9. Lietuvos Statistika. Rodiklių Domenų Bazė-Oficialiosios Statistikos Portalas. Available online: https://osp.stat.gov.lt/statistiniu-rodikliu-analize#/ (accessed on 13 July 2021).

10. Programme for Investment Promotion and Development of Industry during 2014–2020. Available online: https://e-seimas.lrs.lt/portal/legalAct/lt/TAD/731c6e80457911e4ba35bf67d0e3215e (accessed on 11 August 2021).

11. Eurostat. 2021. Available online: https://ec.europa.eu/eurostat/data/database (accessed on 6 June 2021).

12. Wind Europe. Wind Energy in Europe 2020 Statistics and the Outlook for 2021–2025. Available online: https://windeurope.org/intelligence-platform/product/wind-energy-in-europe-in-2020-trends-and-statistics/ (accessed on 14 July 2021).

13. IRENA. Future of Wind: Deployment, Investment, Technology, Grid Integration and Socio-Economic Aspects. A Global Energy Transformation Paper. International Renewable Energy Agency 2019, Abu Dhabi. Available online: https://www.irena.org/-/media/files/irena/agency/publication/2019/oct/irena_future_of_wind_2019.pdf (accessed on 14 July 2021).

 *sustainability*

Article

Bus Load Forecasting Method of Power System Based on VMD and Bi-LSTM

Jiajie Tang [1], Jie Zhao [1,*], Hongliang Zou [2], Gaoyuan Ma [1], Jun Wu [1], Xu Jiang [2] and Huaixun Zhang [1]

1   School of Electrical Engineering and Automation, Wuhan University, Wuhan 430072, China;
    2020282070144@whu.edu.cn (J.T.); mgaoyuan@whu.edu.cn (G.M.); byronwu@whu.edu.cn (J.W.);
    2021282070156@whu.edu.cn (H.Z.)
2   Taizhou Power Supply Company of State Grid Zhejiang Electric Power Co., Ltd., Taizhou 318000, China;
    jamie942706435@126.com (H.Z.); superbacking@163.com (X.J.)
*   Correspondence: jiez_whu@whu.edu.cn

Abstract: The effective prediction of bus load can provide an important basis for power system dispatching and planning and energy consumption to promote environmental sustainable development. A bus load forecasting method based on variational modal decomposition (VMD) and bidirectional long short-term memory (Bi-LSTM) network was proposed in this article. Firstly, the bus load series was decomposed into a group of relatively stable subsequence components by VMD to reduce the interaction between different trend information. Then, a time series prediction model based on Bi-LSTM was constructed for each sub sequence, and Bayesian theory was used to optimize the sub sequence-related hyperparameters and judge whether the sequence uses Bi-LSTM to improve the prediction accuracy of a single model. Finally, the bus load prediction value was obtained by superimposing the prediction results of each subsequence. The example results show that compared with the traditional prediction algorithm, the proposed method can better track the change trend of bus load, and has higher prediction accuracy and stability.

Keywords: variational mode decomposition (VMD); Bayesian optimization; bidirectional long short-term memory (Bi-LSTM); power system bus load forecasting

Citation: Tang, J.; Zhao, J.; Zou, H.; Ma, G.; Wu, J.; Jiang, X.; Zhang, H. Bus Load Forecasting Method of Power System Based on VMD and Bi-LSTM. *Sustainability* **2021**, *13*, 10526. https://doi.org/10.3390/su131910526

Academic Editors: Nicu Bizon and Baojie He

Received: 2 August 2021
Accepted: 14 September 2021
Published: 23 September 2021

Publisher's Note: MDPI stays neutral with regard to jurisdictional claims in published maps and institutional affiliations.

1. Introduction

With the development of energy environment, the proportion of power consumption on the user side in energy consumption is gradually increasing. The large-scale introduction of distributed generation and the diversification of user behavior characteristics pose new challenges to power dispatching planning. Because the power on the user side is mostly collected from the bus side, accurate prediction of bus load is of great reference significance to power planning. The bus load can be predicted by exploring the internal connection and development law between the load influencing factors and the bus load. Accurate load forecasting can be applied to many fields such as power system planning, market transactions, dispatching, etc., and serves as an important basis for the work of related departments. Electric energy consumption has become one of the most important modes of energy consumption in the world. The accurate prediction of electric energy consumption can also predict the future energy development trend, provide a basis for the development of renewable energy, and help the sustainable development of humans and the environment.

In recent years, the research on the theory of bus load forecasting has become increasingly mature, and its forecasting methods can be divided into three categories: statistical forecasting methods, intelligent forecasting methods, and combined forecasting methods. The statistical forecasting algorithm analyzes the time series based on the implicit time dependence and recursive relationship between the bus loads at different times, and then obtains the short-term bus load forecast values, including time series models, gray models,

147

etc. However, the time series model has a suitable prediction effect when the relationship between load and time is linear or exponential, but the relationship between bus load and time is irregular and does not have obvious linear or exponential relationship; thus, the application scenarios have certain limitations. The intelligent prediction method realizes prediction by extracting the features of the data. Typical representatives are Deep Neural Network (DNN) [1], Support Vector Machine (SVM) [2], long and short-term memory (LSTM) Network [3–5], DeepAR [6], N-BEATS [7], Transformer [8], etc. These methods have the ability to model the nonlinear load process, can better adapt to nonlinear spikes and more accurately model the data characteristics of the load, and have better forecasting accuracy. They have become the main research direction of short-term bus load forecasting in recent years.

With the increasing application of data preprocessing theories such as wavelet transform, empirical mode decomposition (EMD) [9,10], ensemble empirical mode decomposition (EEMD) [11], and variational mode decomposition (VMD) [12,13], in view of the nonlinear and non-stationary characteristics of the load sequence, the data preprocessing method is used to decompose the original sequence, and each sub-sequence is predicted separately, and the prediction result is obtained by superimposing and reconstructing. With the progress of data preprocessing methods, bus load forecasting has more processing methods, and the combined forecasting method has been developed. Combination forecasting methods can be divided into two categories. One is to weight and synthesize the forecast results of different forecasting methods to obtain the combined forecast. The forecast results are easily affected by weight distribution. The other is based on the non linearity of the bus load sequence non-stationary characteristics, using signal processing methods such as wavelet transform, EMD, and VMD to decompose the original sequence separately model each sub-sequence component, and reconstruct its prediction results through superposition to obtain the combined prediction results that meet the accuracy re quirements. Reference [14] uses EMD to decompose the bus load for multi-step prediction which has achieved ideal prediction results, but EMD is prone to mode aliasing and cannot choose the number of decomposed components. VMD can find the optimal solution of the natural modal function model through repeated iterations within a limited number of times, which can effectively avoid modal aliasing and improve robustness. Reference [15] proposed four hybrid models based on four decomposition methods, EMD, VMD, wavelet packet transform, and intrinsic time-scale decomposition to forecast agricultural commod ity futures prices. The results showed that VMD contributed the most in improving the forecasting ability.

In intelligent prediction methods, the forecasting model based on LSTM in the field of deep learning has great prediction performance. Reference [16] improved the multi-level gated LSTM prediction model to effectively improve the accuracy of bus load prediction. However, the hyperparameters of LSTM are artificially set before the machine learning model starts the learning process, rather than parameters such as weights and biases obtained by training. The choice of hyperparameters plays a vital role in the improvement of model performance. Therefore, it is necessary to carry out parameter adjustment work according to different application scenarios. Hyperparameter optimization methods mainly include grid search method, random search method, Bayesian optimization algorithm, and so on. Among them, the grid search method is an exhaustive search method that traverses the hyperparameter space. It has the disadvantages of being time-consuming and low efficiency when searching in the high-dimensional space. The random search algorithm avoids the above to a certain extent through sparse and simple sampling. However, it still has the disadvantage of not being able to use prior knowledge to select the next set of hyperparameters. The basic idea of Bayesian optimization algorithm is to use prior knowledge to approximate the posterior distribution of the unknown objective function and then adjust the hyperparameters, which significantly improve the efficiency and accuracy of search in high-dimensional space. The data preprocessing method is selected for noise reduction preprocessing before bus load curve prediction in order to obtain more stable prediction results. In reference [17], the bidirectional long-term and short-term memory (Bi-LSTM) network is used to predict the bus load sequence, which effectively improves the learning ability of the network to historical data. However, with the combination prediction model dividing the complete sequence into multiple subsequences, not all sequences are suitable for bidirectional training. Therefore, it is necessary to use the optimization method to select the Bi-LSTM.

Based on this, a bus load forecasting method based on VMD and Bayesian Optimization Bi-LSTM is proposed in this paper. Firstly, VMD is used to stabilize the original bus load series, which is decomposed into a group of subsequence components with different frequencies. Then, the LSTM neural network prediction model of each subsequence component is constructed, the network related super parameters are optimized by Bayesian theory, and whether Bi-LSTM is used is judged to improve the prediction accuracy of a single model. Finally, the prediction results of each subsequence are superimposed to obtain the predicted value of bus load. In the second part of this paper, we expound the theoretical part of the proposed method and compare and verify the method through an example in the third part. The example results show that compared with the traditional algorithm, the prediction model constructed in this paper has a significant improvement in the accuracy of single-step prediction and multi-step prediction, and can better track the change trend of bus load.

The rest of the paper is organized as follows. Section 2 introduces the theories of methods that are used in the bus load forecasting method of power system, which includes VMD, LSTM, BiLSTM, and Bayesian optimization method, and establishes the VMD-BiLSTM combined prediction model. There is a case study in Section 3 to test the model and compare the model proposed in this paper with SVM, LSTM, Bayesian-LSTM, Bayesian-BiLSTM, and VMD-BiLSTM methods to prove its effectiveness. The conclusion and future studies are presented in Section 4.

2. Theoretical Framework

Here, we mainly explain the theoretical framework of the bus load forecasting in this article.

2.1. Variational Mode Decomposition

2.1.1. Construction of Variational Mode Decomposition Function

Variational modal decomposition (VMD) is an adaptive signal processing method proposed by Dragomiretskiy, which can be effectively applied to the smoothing processing of nonlinear and non-stationary time series [13]. It iteratively searches for the optimal solution of the variational mode, continuously updates each mode function and center frequency, and obtains a number of Intrinsic Mode Functions (IMF) with a certain bandwidth.

In the process of VMD, each natural mode is a finite bandwidth with a center frequency, so the variational problem can be defined as seeking k natural mode functions $u_k(t)$ and making the bandwidth of each mode is the smallest, and the sum of each mode is equal to the input signal f. The specific construction steps are as follows:

(1) Through the Hilbert transform, the analytical signal of the modal function $u_k(t)$ is obtained:

$$[\delta(t) + \frac{j}{\pi t}] * u_k(t) \tag{1}$$

Among them, $\delta(t)$ is the Dirichlet function, $*$ is the convolution symbol.

(2) Perform frequency mixing on the analytical signal to transform the frequency spectrum of each mode to the fundamental frequency band:

$$\left[\left(\delta(t) + \frac{j}{\pi t}\right) * u_k(t)\right] e^{-j\omega_k t} \tag{2}$$

(3) The constraints of the optimized variational model are:

$$\begin{cases} \min\left\{ \sum\limits_{k=1}^{K} \left\| \partial_t \left[\left(\delta(t) + \frac{j}{\pi t} \right) * u_k(t) \right] e^{-j\omega_k t} \right\|_2^2 \right\} \\ s.t. \sum\limits_{k=1}^{K} u_k(t) = X(t) \end{cases}$$ (3)

In the formula, K is the number of IMFs, $\{u_k\} = \{u_1, u_2,..., u_k\}$ is IMFs, and $\{\omega_k\} = \{\omega_1, \omega_2, ..., \omega_k\}$ is the center frequency of u_k.

2.1.2. Solution of Variational Mode Decomposition Function

(1) Using the quadratic penalty factor α and the Lagrangian multiplication operator $\lambda\,(t)$, the constrained problem is turned into a non-constrained problem.

Extended Lagrangian function expression:

$$L(\{u_k\}, \{\omega_k\}, \lambda) = \sum\limits_{k=1}^{K} \left\| \partial_t \left[\left(\delta(t) + \frac{j}{\pi t} \right) * u_k(t) \right] e^{-j\omega_k t} \right\|_2^2$$
$$+ \left\| X(t) - \sum\limits_{k=1}^{K} u_k(t) \right\|_2^2 + \left\langle \lambda(t), X(t) - \sum\limits_{k=1}^{K} u_k(t) \right\rangle$$ (4)

(2) Initialize $\hat{u}_k^1$, ω_k, $\hat{\lambda}^1$:

Iteratively update $\hat{u}_k$, ω_k, $\hat{\lambda}^n$ under the condition of $\omega \geq 0$:

$$u_k^{n+1}(\omega) = \frac{\hat{f}(\omega) - \sum\limits_{i \neq k} \hat{u}_i^{n+1}(\omega) + \frac{\hat{\lambda}^n(\omega)}{2}}{1 + 2\alpha\left(\omega - \omega_k^n\right)^2}$$ (5)

$$u_k^{n+1}(\omega) = \frac{\int_0^\infty \omega \left| \hat{u}_k^{n+1}(\omega) \right|^2 d\omega}{\int_0^\infty \left| \hat{u}_k^{n+1}(\omega) \right|^2 d\omega}$$ (6)

$$\hat{\lambda}^{n+1}(\omega) = \hat{\lambda}^n(\omega) + \left(\hat{f}(\omega) - \sum\limits_k \hat{u}_k^{n+1}(\omega) \right)$$ (7)

Until $\sum_k \left\| \hat{u}_k^{n+1} - \hat{u}_k^n \right\|_2^2 / \left\| \hat{u}_k^n \right\|_2^2 < \varepsilon$.

By using the VMD, different scales or trend components can be decomposed from the bus load sequence step by step to form a series of sub-sequence components with different time scales. The sub-sequences have stronger stationarity than the original series, and regularity help to improve forecast accuracy.

2.2. Long Short-Term Memory Neural Network

2.2.1. LSTM Operation Rules

The long and short-term memory (LSTM) network is a special modified version of the cyclic neural network. While retaining the cyclic feedback mechanism, the topology of the LSTM network controls the accumulation speed of information by introducing a gating unit, selectively adding new information, and selectively forgetting. The previously accumulated information solves the long-term dependence problem in sequence modeling.

Compared with ordinary RNN, LSTM neural network is also composed of an input layer, output layer, and hidden layer, but its hidden layer is replaced by ordinary neurons with memory modules containing gating mechanism. Its internal structure is shown in the Figure 1.

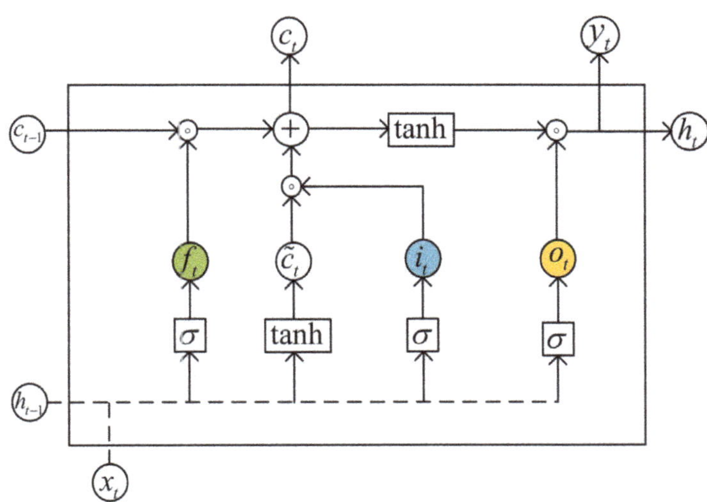

Figure 1. Internal structure of LSTM.

The memory unit c_t is the core component of the LSTM memory module. It is realized by controlling the forget gate, input gate, and output gate. It contains the long-term memory information of the sequence. The hidden layer state h_t contains the short-term memory information of the sequence, which is updated faster than memory unit update speed [16].

Suppose a total of k time steps of the input vector sequence are divided into x_1, x_2, ..., x_k according to the input time sequence, and the t-th time step is taken for analysis.

The LSTM operation rules are as follows:

(1) Update the output of the forget gate, select the historical information that the memory unit needs to keep, and control the degree of influence of c_{t-1} on c_t:

$$f_t = \sigma(W_f x_t + U_f h_{t-1} + b_f) \tag{8}$$

(2) Update the two parts of the output of the input gate, select the current input information that the memory unit needs to retain, and control the degree of influence of x_t on c_t:

$$f_t = \sigma(W_f x_t + U_f h_{t-1} + b_f) \tag{9}$$

$$\tilde{c}_t = \tanh(W_c x_t + U_c h_{t-1} + b_c) \tag{10}$$

(3) Update the cell status according to the input gate and forget gate:

$$c_t = f_t \circ c_{t-1} + i_t \circ c_t{}' \tag{11}$$

(4) Update the output gate output, select the output information that the memory unit needs to retain, and control the degree of influence of c_t on h_t:

$$o_t = \sigma(W_o x_t + U_o h_{t-1} + b_o) \tag{12}$$

$$h_t = o_t \circ \tanh(c_t) \tag{13}$$

(5) Update the forecast output at the current moment:

$$\hat{y}_t = \sigma(V h_t + c) \tag{14}$$

Among them, f_t, i_t, and o_t represent the calculation results of the forget gate, input gate, and output gate at time t; W_f, Wi, and Wo represent the weight matrix of the forget gate, input gate, and output gate, respectively; b_f, b_i, and b_o represent the weight matrix of the forget gate, input gate, and output gate, respectively. The bias term of the forget gate, input gate, and output gate; o is the dot product symbol of the matrix element; $\sigma(x)$ is the Sigmoid activation function of each gate:

$$\sigma(x) = \frac{1}{1 + e^{-1}} \tag{15}$$

Its value range is (0, 1). Through the conversion of Sigmoid function, the input can be converted into probability values, so it is widely used as the activation function of artificial neural network transmission.

When x_t is input to the network, it will be processed by a tanh neural layer and three gates at the same time as the hidden layer vector h_{t-1} of the previous time step. The tanh function is a hyperbolic tangent activation function, and its expression is:

$$\tanh(x) = \frac{\sinh(x)}{\cosh(x)} = \frac{e^{2x} - 1}{e^{2x} + 1} \tag{16}$$

Its value range is $(-1, 1)$, the output is centered at the origin, and the convergence speed is faster than that of Sigmoid. It is usually used as the activation function of the output gate.

The tanh layer will create a new candidate state vector $c_t{}'$. The forgetting gate f_t determines what information to discard and retain from the cell state c_{t-1} of the previous time step. The input gate it determines how to update the candidate state vector. After the cell state is updated, the output gate o_t decides how to filter the new state vector c_t into output information h_t.

2.2.2. Training Process of LSTM

The training algorithm of LSTM neural network mainly includes two categories: back propagation algorithm over time and real-time cyclic learning algorithm. The concept of backpropagation algorithm is clear and it has advantages in computational efficiency. This paper selects this method to train LSTM neural network.

The back propagation method expands the LSTM into a deep feedforward neural network in time sequence, and then further calculates the relevant parameter gradients according to the error back propagation algorithm of the feedforward network, and trains the model [18]. The LSTM network sequence expansion diagram is shown in Figure 2.

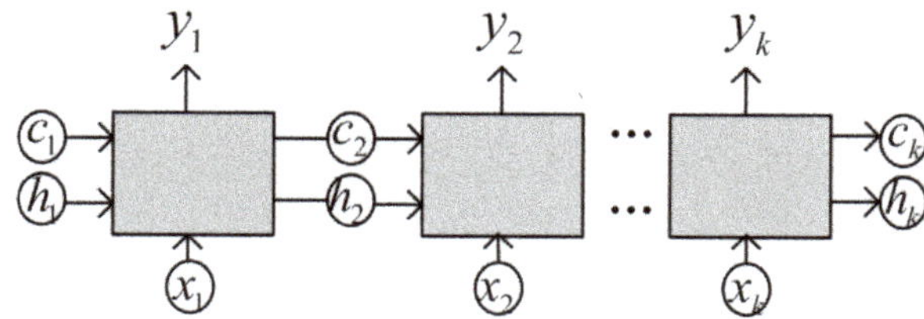

Figure 2. Sequence structure diagram of LSTM.

The specific training steps are as follows in Figure 3.

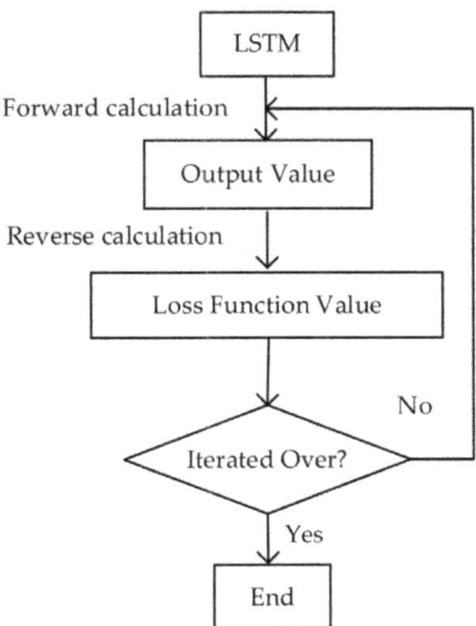

Figure 3. Training steps of LSTM.

(1) Forward calculation of the output value of the LSTM memory module.

The c_t and h_t of the current time step are calculated by the gating mechanism of LSTM and are retained for the calculation of the next time step. When the calculation of the last step is completed, the hidden layer vector h_k will be used as the output and the prediction corresponding to this set of sequences' value (tag value) to compare.

(2) Reverse calculation of the error item value of each memory module, including two reverse propagation directions in chronological order and at network level.

Calculate the loss function value according to the comparison result of step 1, and select the square sum error function as the loss function of LSTM; the expression is:

$$L = \frac{1}{2} \sum_{i \in outputs} (\hat{y}_i - y_i)^2 \tag{17}$$

(3) According to the corresponding error term, calculate the gradient of each weight, and the gradient descent method iterates the parameters including W, U, V, c_t, and h_t:

$$\hat{\theta}_j = \theta_j - \alpha \frac{\partial}{\partial \theta_j} L(\theta_j) \tag{18}$$

Through the gating mechanism and perfect parameter update rules, LSTM realizes the selection and screening of the input information flow, and improves the processing ability of the recurrent neural network for long sequences.

The back-propagation algorithm can compare the predicted value with the real value by using the error function after the forward calculation, and then optimize the network parameters. Therefore, the back-propagation algorithm can back calculate and optimize many parameters of LSTM. The cyclic learning algorithm can deal with some sequence problems, but it has serious long-term dependence problems after multi-stage propagation—the gradient tends to disappear or explode, and it is difficult to continue to optimize within the number of iterations in many cases. However, the introduction of gating mechanism in LSTM solves the gradient disappearance problem and performs well in sequence processing.

2.2.3. Bidirectional LSTM

The bidirectional long and short-term memory (Bi-LSTM) network is derived from the bidirectional cyclic neural network [19]. Its main feature is to increase the learning function of the neural network for future information, thereby overcoming the defect that the unidirectional LSTM network can only process historical information. The Bi-LSTM mainly splits the ordinary LSTM into two directions, but the two LSTMs are connected to the same output layer. Such a structure can provide complete upper and lower sequence information for the input sequence of the output layer. Using a Bi-LSTM network to model bus load forecasting is to input historical data into a forward LSTM network and a reverse LSTM network at the same time, so as to capture a complete time series global information. The Bi-LSTM neural network structure diagram is as follows in Figure 4.

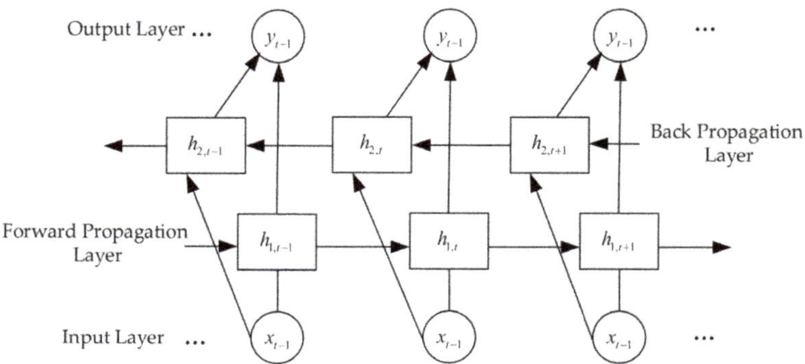

Figure 4. Bi-LSTM Neural Network Structure Diagram.

Because the bus load has certain regularity and planning, the joint grasp of historical and future information can better learn the characteristics and laws of the load curve.

The update formula of the back-to-forward cycle neural network layer is:

$$h_{1,t} = f\left(W_{h_1,t}x_t + W_{h_1}h_{1,t-1} + b_{h_1}\right) \tag{19}$$

The update formula of the looping neural network layer from front to back is:

$$h_{2,t} = f\left(W_{h_2,t}x_t + W_{h_2}h_{2,t+1} + b_{h_2}\right) \tag{20}$$

The two layers of recurrent neural networks are superimposed and input to the hidden layer:

$$y_t = g\left(U_{h_1}h_{1,t} + U_{h_2}h_{2,t} + b_y\right) \tag{21}$$

Among them, h_1, t, and h_2, t are the hidden units of the front pass layer and the back pass layer at time t, respectively; y_t is the model output at time t; $f(*)$, $g(*)$ are optional activation functions; $W_{h1,t}$, W_{h2}, t, W_{h1}, W_{h2}, U_{h1}, and U_{h2} are the weight matrices of the corresponding objects; and b_{h1}, b_{h2}, and b_y are the bias terms of the corresponding objects.

2.3. Bayesian Optimization of LSTM

2.3.1. Bayesian Optimization Theory

Gaussian Regression Process

In the feasible region, uniformly select points that obey the multi-dimensional normal distribution as candidate solutions to establish a Gaussian regression model [20]:

$$\begin{bmatrix} y_1 \\ \vdots \\ y_n \end{bmatrix} \sim N\left(0, \begin{bmatrix} k(x_1, x_1) & \cdots & k(x_1, x_n) \\ \vdots & & \vdots \\ k(x_n, x_1) & \cdots & k(x_n, x_n) \end{bmatrix}\right) \tag{22}$$

Among them, K is the covariance matrix, x is the input value, and y is the response output value.

Through the training set, the updated value y^* is obtained according to the posterior formula:

$$P\left(y_* \big| y \sim N\left(K_* K^{-1} y, K_{**} - K^{-1} K_*^T\right)\right) \tag{23}$$

Among them, K_* is the covariance of the training set, and K_{**} is the covariance of the newly added sample.

Update the Gaussian regression model according to the updated value:

$$\begin{bmatrix} y \\ y_* \end{bmatrix} \sim N\left(0, \begin{bmatrix} K & K_*^T \\ K_* & K_{**} \end{bmatrix}\right) \tag{24}$$

The Gaussian regression model considers the relationship between y_N and y_{N+1} and establishes input and output functions to provide a search basis for parameter optimization.

Acquisition Function Process

On the basis of the Gaussian regression model, it is necessary to use the collection function to solve the optimal solution of the function. This paper selects the expectation improvement method to use the mathematical expectation to solve the optimal solution.

$$x_{n+1} = \mathrm{argmax} EI_n(x) \tag{25}$$

$$EI_n(x) = E_n\left[\left[f(x) - f_n^*\right]^+\right] \tag{26}$$

$$= (\mu - f_n^*)\left(1 - \phi\left(\frac{f_n^* - \mu}{\sigma}\right)\right) + \sigma\varphi\left(\frac{f_n^* - \mu}{\sigma}\right) \tag{27}$$

Among them, $p(x)$ is the probability density of the normal distribution, $\varphi(x)$ is the standard normal distribution about x, μ is the mean value of the input value x, and σ is the variance of the input value x.

The expected value is used for parameter optimization, and the next optimal value is effectively sought within a certain range, so as to find the optimal parameter within the number of iterations [20].

2.3.2. Hyperparameter Optimization of LSTM

The hyperparameters of the LSTM neural network used for bus load prediction can be divided into two categories: structural hyperparameters and training hyperparameters. The structural hyperparameters mainly include the number of hidden neurons in the network, etc [21]. The number of hidden layer neurons determines the expressive ability of the network, but also determines whether the network is over-fitting and the network is time-consuming. Reasonable hidden layer neurons help improve the performance of the network's predictive ability. The use of the bidirectional long-term memory network improves the network's ability to learn historical data, but it also leads to slow network prediction and different effects. Choosing different curves to predict whether to use a Bi-LSTM can save prediction time while ensuring prediction accuracy [22].

The training hyperparameters of the LSTM neural network mainly include learning rate, L2 regularization parameters, etc. A suitable initial learning rate helps to significantly improve the iterative convergence speed and prediction accuracy of deep learning models. L2 regularization parameters are additional items of the network loss function and can prevent the over-fitting problem to a certain extent.

In summary, this paper will use Bayesian optimization algorithm to optimize and debug the number of hidden layer neurons of LSTM neural network, whether to use Bi-LSTM, initial learning rate, and L2 regularization parameters.

2.4. VMD-Bi-LSTM Combined Prediction Model

The power system bus load itself has fluctuating characteristics and is affected by distributed power dispatch and user-side behavior characteristics. Its curve has a certain degree of non-linear and non-stationary characteristics. The use of conventional learning forecasting methods to improve the forecasting accuracy is relatively limited. Considering the outstanding advantages of variational modal decomposition technology in sequence smoothing processing and the excellent performance of LSTM networks in time series data modeling, this paper proposes a VMD-Bi-LSTM combined forecasting model. The specific modeling process is as follows, as shown in the Figure 5.

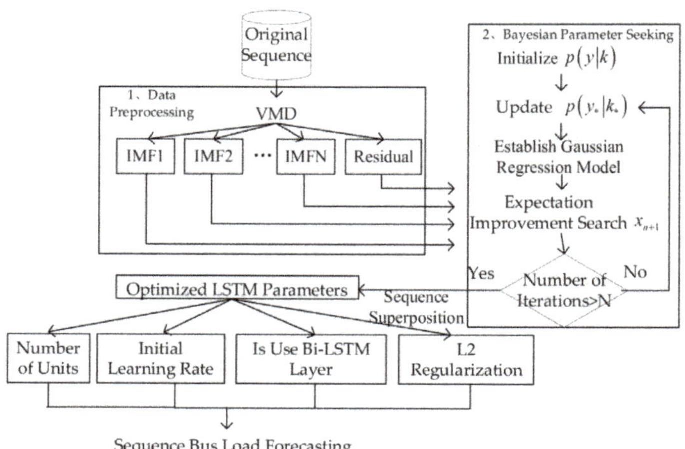

Figure 5. Flowchart of VMD-BiLSTM combined prediction.

(1) In view of the non-stationary characteristics of the bus load sequence, the VMD method is used to decompose, and each IMF component and residual component are obtained

(2) Normalize each sub-sequence component separately, and divide the training sample and the test sample according to the same ratio;

(3) Construct an LSTM neural network prediction model for each sub-sequence component, and use Bayesian optimization algorithm to optimize the hyperparameters of a single model to obtain the most suitable hyperparameter combination for decomposing the sequence and determine whether to use Bi-LSTM in sub-sequences;

(4) Train the prediction model after hyperparameter optimization, use the trained prediction model to perform multi-step extension prediction, and superimpose the reconstruction to obtain the multi-step prediction value of bus load;

(5) Compared with actual data, the multi-step prediction performance of the prediction model is evaluated by calculating error indicators.

3. Case Study

Here, we analyze the example of the method proposed in this paper. The example takes the bus load data of Canberra, Australia, from 16 January to 22 January 2016 as the data set, including 270 time steps (30 min as a time step). The first 244 time steps of the data are used as the training sequence and the last 36 time steps are used as the verification sequence.

3.1. Temporal Data Decomposition

The original bus load sequence is decomposed by VMD, and six groups of IMF components and one group of residual components are separated step by step. The decomposition results are shown in Figure 6.

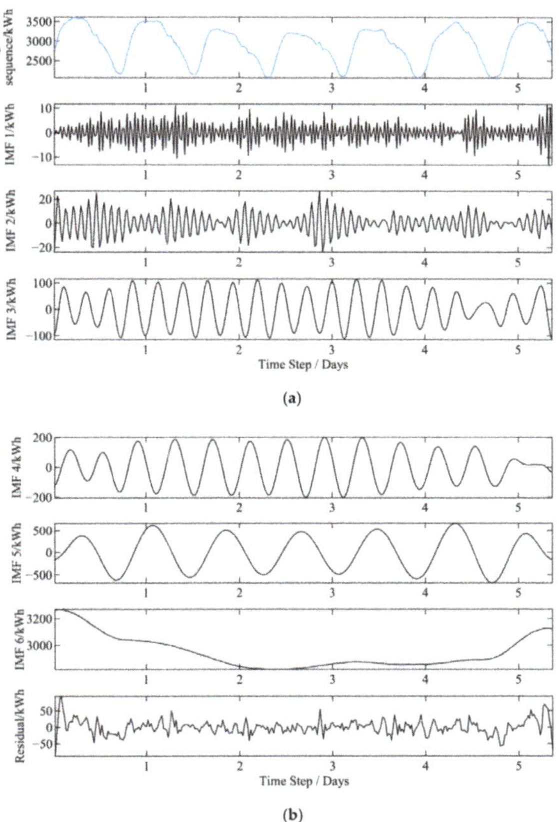

Figure 6. (**a**) Original sequence and IMFs; (**b**) IMFs and residual.

Decomposing the bus load sequence through variational modal decomposition shows that the natural modal function consists of multiple sub-sequences with small amplitude and stable frequency. Although the frequency of the remainder is relatively unstable, the amplitude is small, which affects the overall bus load. The forecast trend change of the load has less impact.

3.2. Hyperparameter Optinization of VMD-LSTM

On the basis of smoothing the original sequence, it is necessary to construct the LSTM network prediction model of the sub-sequence components, and to optimize the related structure hyperparameters and training hyperparameters. The hyperparameter results obtained by using Bayesian optimization algorithm in this paper are shown in the Table 1.

Table 1. Hyperparameters of each subsequence.

Sequence	Number of Units	Using Bi-LSTM Layer	Initial Learning Rate	L2 Regularization
IMF 1	197	1	0.01	0.0013
IMF 2	73	2	0.011	0.00018
IMF 3	66	1	0.011	0.000021
IMF 4	64	2	0.016	0.0024
IMF 5	199	1	0.01	1.16
IMF 6	140	1	0.021	0.00018
Residual	130	1	0.01	0.00097

3.3. Model Evaluation Index

In order to visually analyze the predicted error value, the curve correlation coefficient is introduced to evaluate the shape difference between the predicted curve and the real curve, and the root mean square error (RMSE) is introduced to visually analyze the deviation between the observed value and the true value. The accuracy is analyzed, and the standard error is selected as the criterion for predicting the dispersion level.

In order to evaluate the forecasting effect of the proposed bus load forecasting method, the RMSE, NRMSE, standard error, and correlation coefficient are selected as the overall forecasting results evaluation index of the short-term bus load forecasting method.

$$e_{RMSE} = \sqrt{\frac{1}{n}\sum_{i=1}^{n}(y_i - \hat{y}_i)^2} \tag{28}$$

$$e_{NRMSE} = \sqrt{\frac{RMSE}{mean\left(\sum_{i=1}^{n} y_i\right)}} \tag{29}$$

$$e_{STD} = \sqrt{\frac{\sum_{i=1}^{n}(y_i - \hat{y}_i)^2}{n}} \tag{30}$$

$$\rho(A, B) = \frac{1}{N-1}\sum_{i=1}^{N}\left(\frac{A_i - \mu_A}{\sigma_A}\right)\left(\frac{B_i - \mu_B}{\sigma_B}\right) \tag{31}$$

where σ_A and σ_B are the standard deviation of dataset A and dataset B, respectively, and the standard deviation formula is as follows:

$$\sigma_x = \sqrt{\frac{\sum_{i=1}^{n}(x_i - \overline{x})^2}{n - 1}} \tag{32}$$

where y_i is the true value of bus load and $\hat{y}_i$ is the predicted value of bus load. The correlation coefficient r is used to characterize the accuracy of the prediction curve and the calibration curve. The closer the correlation coefficient is to 1, the higher the prediction accuracy.

3.4. Model Evaluation Index

After optimizing the Bayesian parameters of each sub-sequence of the bus load, the long and short-term memory network is trained, and the next time step is predicted in Figure 7, and the single-step prediction results of each sub-sequence component are integrated into the historical monitoring data. The new input sequence of the single-step forecasting model can realize the multi-step rolling forecast of each component, which can realize the rolling forecast load value for a period of time in the future. The rolling prediction method is used to predict multiple time steps in the future, and the prediction results are shown in Figure 8.

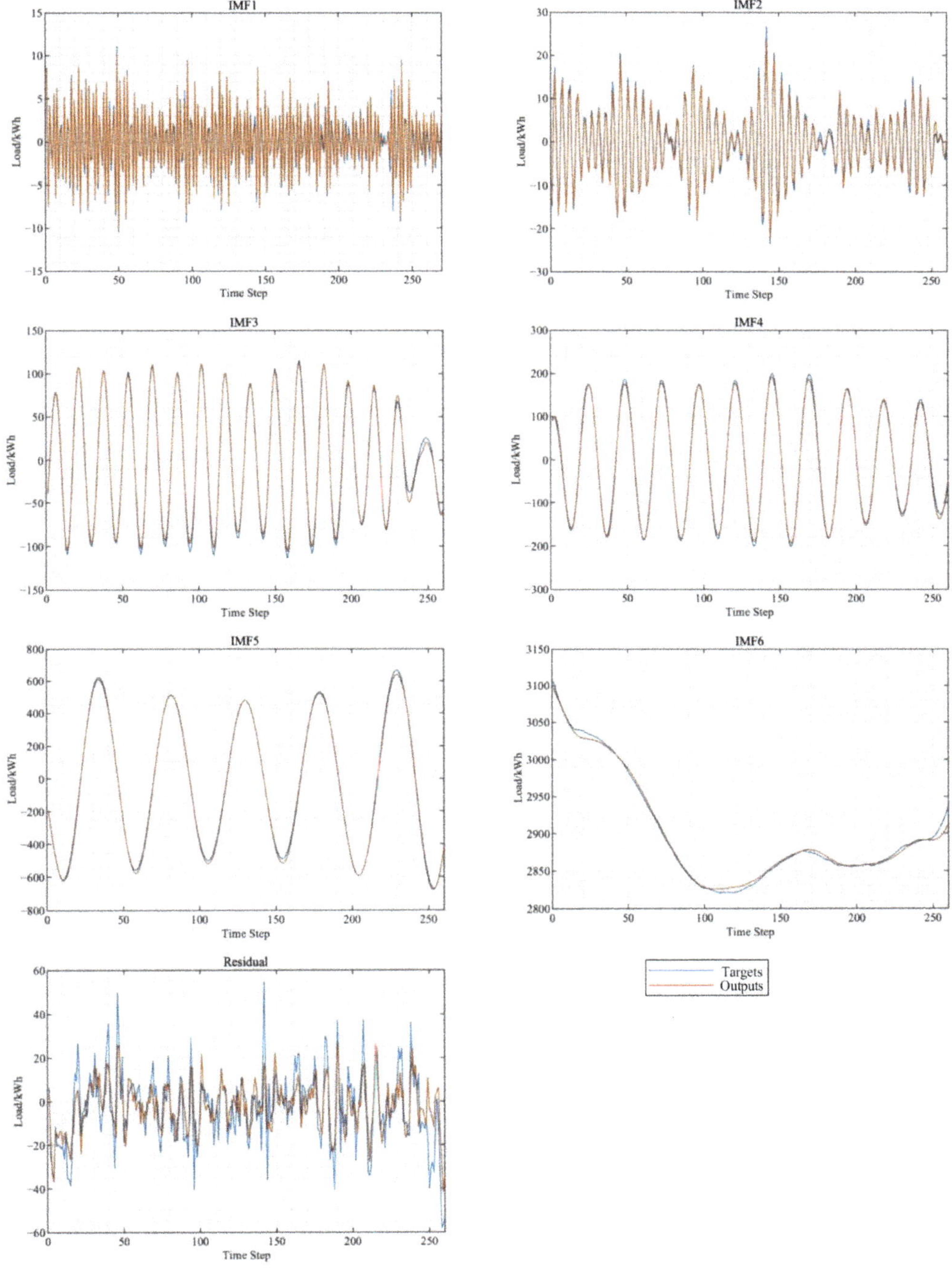

Figure 7. Subsequence training prediction results.

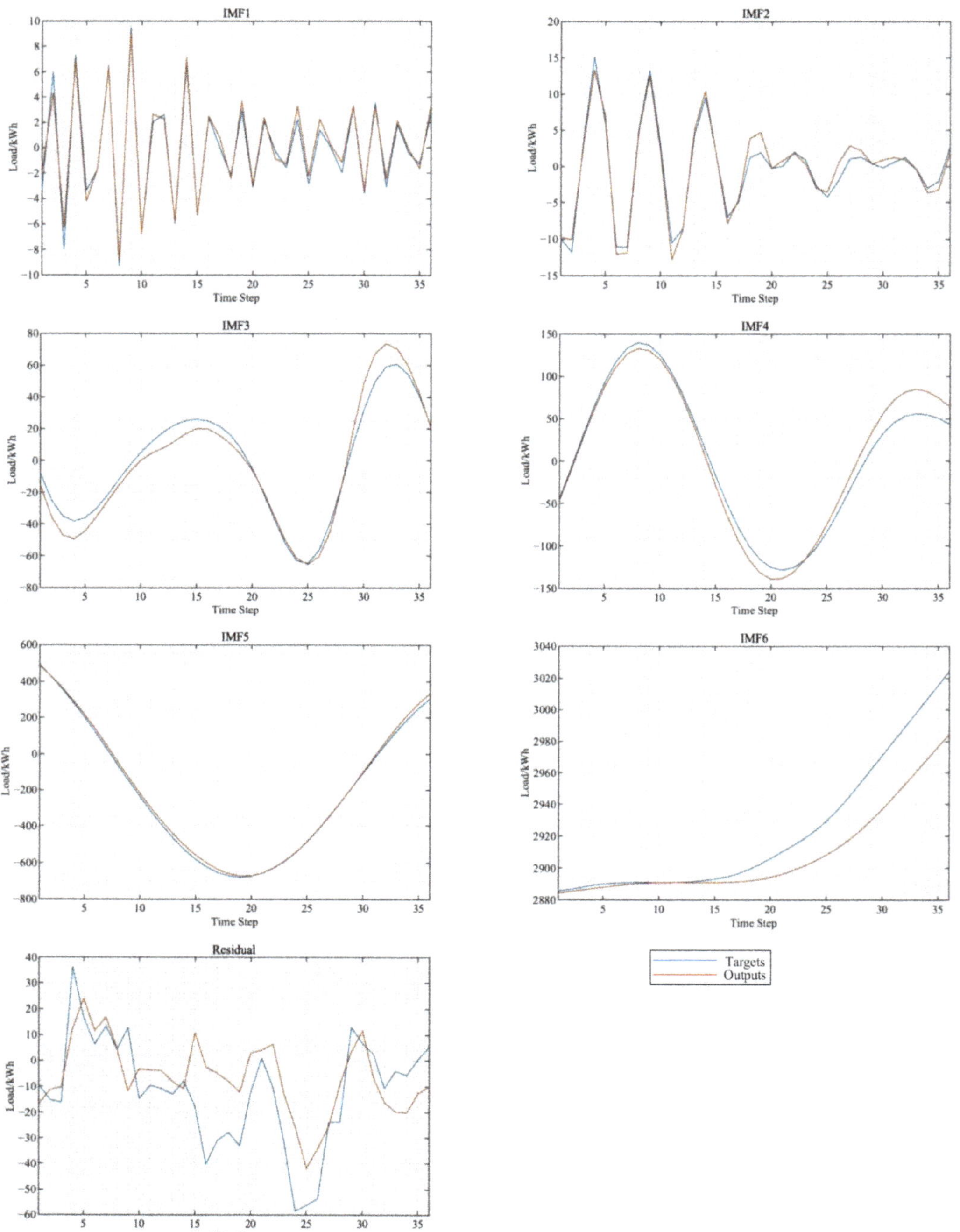

Figure 8. Subsequence multi-step prediction results.

The multi-step prediction results of bus load can be obtained by superimposing the predicted values of each subsequence, and the error analysis of each subsequence is carried out by using the prediction evaluation index. The error results are shown in Table 2. From the training prediction results and multi-step prediction results, it can be seen that the e_{RMSE} of IMF6, which accounts for a large proportion of the original sequence of bus load, is 1.3635 and e_{STD} is 0.8233 in the next 12 time steps, with small prediction error. With the increase in prediction time steps, the prediction e_{RMSE} in the next 36 time steps will increase to 20.2677, e_{STD} and e_{NRMSE} will also increase accordingly, but the r will increase from 0.9521 to 0.992. The shape of the prediction curve will gradually stabilize and be closer to the shape of the original curve. With the increase in time steps, the prediction accuracy will decline, but the prediction curve will gradually remain stable; the shape gradually approaches the original curve, which has suitable prediction stability.

Table 2. Performance comparison of multi-step prediction.

Sequence	Time Step	e_{RMSE}	r	e_{NRMSE}	e_{STD}
	12	0.8909	0.9917	7.6779	0.9299
IMF 1	24	0.7262	0.9904	−39.3426	0.7265
	36	0.6631	0.9899	−33.9655	0.6387
	12	1.1507	0.9941	−0.7562	1.0931
IMF 2	24	1.1929	0.9877	−2.3908	1.2184
	36	1.1635	0.9834	−2.5087	1.1713
	12	7.9527	0.9904	−0.5446	2.9920
IMF 3	24	6.7540	0.9882	−0.7751	4.2085
	36	7.7266	0.9840	−4.8082	7.6166
	12	5.0755	0.9999	0.0630	1.5155
IMF 4	24	8.7905	0.9987	2.8258	5.2502
	36	14.6416	0.9869	2.2186	14.6664
	12	15.7076	0.9999	0.2571	8.4288
IMF 5	24	15.2865	0.9997	−0.0559	9.7277
	36	15.3647	0.9996	−0.0740	10.4295
	12	1.3635	0.9521	0.000472	0.8233
IMF 6	24	7.6753	0.9585	0.0026	5.9782
	36	20.2677	0.992	0.0069	14.6026
	12	11.4701	0.6938	9.4522	11.9700
Residual	24	17.4227	0.6755	−1.5639	15.2513
	36	15.9639	0.6827	−1.3800	15.3315

The prediction e_{NRMSE} of other subsequences is lower than 16, the r is higher than 0.98, the prediction error is small, and the prediction shape remains suitable. The prediction results of a single stable natural mode function subsequence meet expectations.

The decomposition remainder of bus load has many burrs and unstable frequency, so the prediction correlation coefficient is low. However, due to its small amplitude, its prediction error is also small, and its impact on the overall prediction result of bus load is relatively small.

The multi-step prediction results of each subsequence and remainder are superimposed to obtain the multi-step prediction curve of bus load, as shown in Figure 9. The prediction curve fits with the real curve and has accurate prediction results. In order to verify the prediction performance of the proposed method, different models in various cases are selected for comparative analysis. The SVM that uses radial basis function as kernel function for prediction is compared with LSTM to verify the advantages of LSTM in time series prediction. The VMD-Bayesian-BiLSTM model is compared with EEMD-Bayesian-BiLSTM model and EMD-Bayesian-BiLSTM to verify the advantages of VMD in short-term combination forecasting. The VMD-LSTM combined prediction model and LSTM model are selected for prediction to verify the prediction accuracy and stability of the combined prediction model. The VMD-Bayesian-LSTM model is compared with VMD-LSTM model, and the Bayesian-BiLSTM model is compared with the Bayesian-LSTM model and LSTM to verify the improvement effect of Bayesian optimization theory on LSTM prediction accuracy and the necessity to consider the applicability of BiLSTM in sequence prediction. The comparison results are shown in Table 3 and Figure 10.

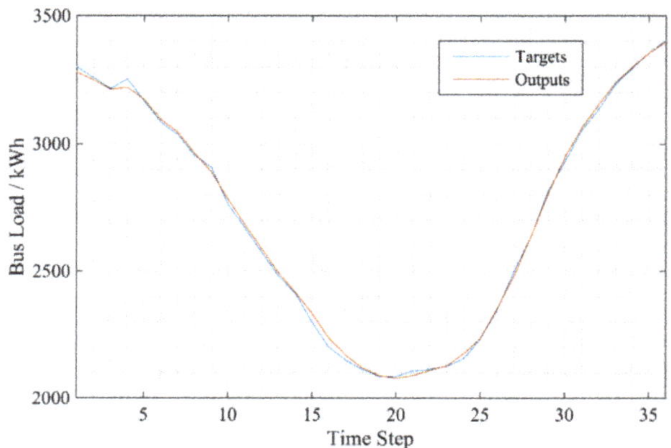

Figure 9. Prediction results of VMD-BiLSTM.

Table 3. Performance comparison of different models.

Title 1	Time Step	e_{RMSE}	r	e_{NRMSE}	e_{STD}
VMD-Bayesian-BiLSTM	12	15.7524	0.9986	0.0052	16.4296
	24	16.6947	0.9994	0.0064	16.6392
	36	14.9219	0.9995	0.0055	14.8022
EEMD-Bayesian-BiLSTM	12	19.6386	0.9963	0.0073	16.4825
	24	22.3739	0.9971	0.0088	17.9276
	36	26.8462	0.9951	0.0091	21.9372
EMD-Bayesian-BiLSTM	12	21.8372	0.9968	0.0083	17.8362
	24	25.8376	0.9946	0.0096	19.8372
	36	28.3826	0.9938	0.0105	23.8261
VMD-BiLSTM	12	15.8235	0.9978	0.0052	16.4784
	24	22.9478	0.9974	0.0073	20.4936
	36	23.7018	0.9968	0.0088	20.9806
Bayesian-BiLSTM	12	25.6357	0.9965	0.0092	21.9365
	24	30.6387	0.9956	0.0105	25.3794
	36	32.7487	0.9947	0.0124	28.3748
Bayesian-LSTM	12	49.3128	0.9865	0.0163	43.5910
	24	38.2263	0.9971	0.0147	35.6199
	36	44.1296	0.9953	0.0163	44.7550
LSTM	12	66.9500	0.9867	0.0222	40.1171
	24	80.9277	0.9933	0.0311	53.9405
	36	71.9871	0.9937	0.0266	55.8161
SVM	12	86.4025	0.9777	0.0286	51.8872
	24	97.3993	0.9894	0.0374	67.1863
	36	89.0407	0.9891	0.0329	71.4600

The comparison of different models shows that the prediction accuracy of each time step of LSTM is greatly improved compared with SVM. After decomposing the bus load sequence, the e_{RMSE}, r and other indices of VMD-LSTM are greatly improved on the basis of the LSTM model, which greatly improves the prediction accuracy and stability. The comparison of VMD-Bayesian-BiLSTM and EEMD-Bayesian-BiLSTM and the comparison of VMD-Bayesian-BiLSTM and EMD-Bayesian-BiLSTM shows that the prediction error of the prediction model considering VMD in multiple time steps is lower than that of EMD and EEMD. With the proposal of Bayesian optimization theory, the e_{RMSE} of 36 time steps of combined prediction is reduced from 23.9219 to 14.9219, the e_{NRMSE} is reduced from 0.0088 to 0.0055, the e_{STD} is reduced from 20.9806 to 14.8022, and the results of LSTM are also greatly improved after Bayesian optimization. The results of the Bayesian-BiLSTM model are improved compared with those without considering Bi-LSTM. The comparison of various data shown in Figure 10 verifies that the VMD-LSTM combined prediction model based on Bayesian optimization proposed in this paper has more accurate prediction results and more stable multi-step prediction results.

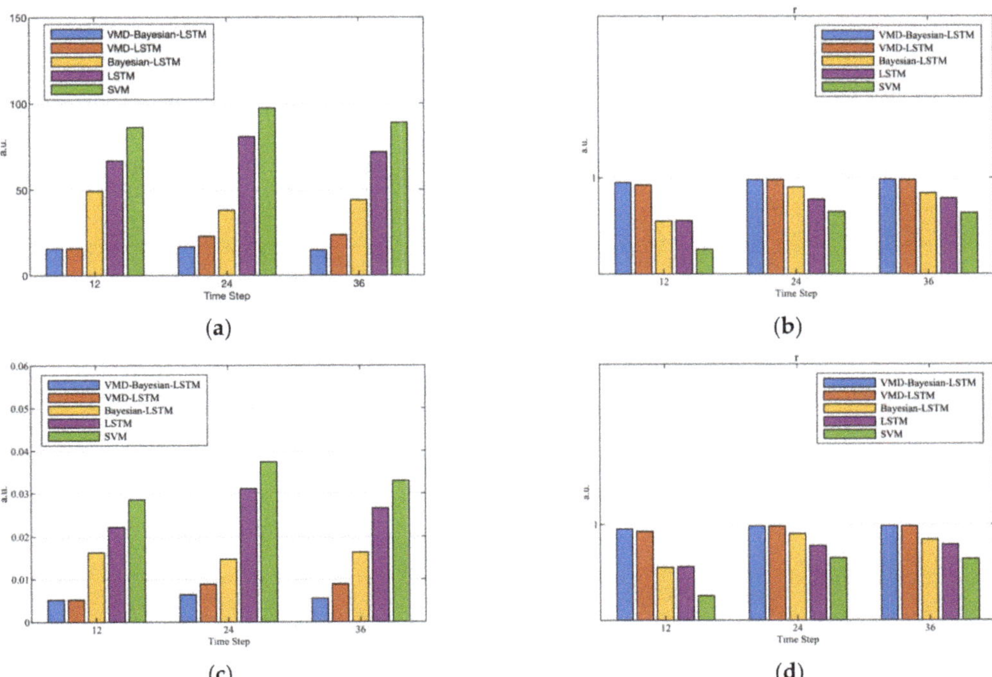

Figure 10. (**a**) Comparison of e_{RMSE}; (**b**) comparison of r; (**c**) comparison of e_{NRMSE}; (**d**) comparison of e_{STD}.

In order to verify the applicability of the model proposed in this paper, this paper randomly selects two groups of 270 time step data sets in the bus load data set in Canberra, Australia, in 2016, which are divided into data set 2 and data set 3 for verification. In order to eliminate the regional influence, three groups of 270 time step data sets are randomly selected from the 2014 Beijing bus load data set, namely data set 4, data set 5, and data set 6, and their errors are tested in Table 4.

Table 4. Test of different data sets.

Title 1	Time Step	e_{RMSE}	r	e_{NRMSE}	e_{STD}
	12	15.7524	0.9986	0.0052	16.4296
Data set 1	24	16.6947	0.9994	0.0064	16.6392
	36	14.9219	0.9995	0.0055	14.8022
	12	16.8362	0.9993	0.0058	16.4784
Data set 2	24	17.9378	0.9987	0.0078	17.9387
	36	17.9373	0.9989	0.0083	17.9272
	12	15.9327	0.9991	0.0055	16.9365
Data set 3	24	16.9372	0.9989	0.0067	18.3794
	36	18.8372	0.9988	0.0093	19.3748
	12	17.3128	0.9995	0.0163	18.5910
Data set 4	24	18.2263	0.9988	0.0147	19.6199
	36	19.1296	0.9985	0.0163	21.7550
	12	17.8362	0.9993	0.0087	18.8362
Data set 5	24	19.8367	0.9983	0.0093	20.9837
	36	21.8272	0.9969	0.0128	22.9472

Table 4. *Cont.*

Title 1	Time Step	e_{RMSE}	r	e_{NRMSE}	e_{STD}
Data set 6	12	18.8367	0.9991	0.0115	19.8362
	24	18.9272	0.9992	0.0134	20.9372
	36	20.8367	0.9983	0.0176	21.9272
Average Value	12	17.0845	0.9992	0.0088	17.8513
	24	18.0933	0.9989	0.0097	19.0830
	36	18.9149	0.9985	0.0116	19.7889

It can be seen from Table 4 that through the test of multiple groups of randomly selected data in the same time step, the error value is always controlled in a low range, and the average error of six groups of data has suitable prediction results. This shows that the proposed method has appropriate prediction stability and adaptability.

4. Conclusions and Future Studies

Aiming at the current research hotspot in the field of deep learning, this paper studies the bus load forecasting, establishes the bus load forecasting method based on VMD BiLSTM, and draws the following conclusions:

(1) The VMD method is used to deal with the non-stationary characteristics of bus load series and reduce the interaction between different time scale information, which is conducive to further mining the characteristics of original series and improving the prediction performance of the model;

(2) The cyclic network structure and gating mechanism of LSTM neural network are used to capture the temporal correlation of each sub sequence component, so as to track the change trend of bus load more effectively. Compared with other models, the VMD-LSTM combined prediction model has a significant improvement in multi-step prediction accuracy;

(3) Bayesian optimization algorithm is used to optimize the super parameter combination of LSTM neural network to overcome the adverse effect of empirical selection on the improvement of model prediction performance;

(4) Bayesian optimization considers the applicability of the sequence to the bidirectional neural network. While using the Bi-LSTM network to enhance the training ability considering the applicability of the sequence, the prediction performance of the network has been optimized and improved.

This method has certain feasibility in the field of bus load forecasting, can be applied in the direction of energy consumption forecasting and power production planning, and is conducive to the planning and development of clean energy in the future and the sustainable development of energy.

The model proposed in this paper has achieved suitable results in short-term prediction, but it is uncertain whether it can ensure the prediction accuracy and stability in long-term prediction when the data support is sufficient, and whether it is adaptable in other areas. Due to time constraints, this paper does not compare it with ETS, ARIMA, and Thet to reflect the optimization performance. In the future, this proposed model can be implemented in different areas to validate its effectiveness and compared to alternative approaches and stronger baselines. Moreover, by taking a new variety of data input for sustainability studies, the model can be implemented for carbon emission forecasting. The author will study and optimize the excellent prediction methods in the field of machine learning in order to establish an accurate prediction model that can be applied in the field of sustainability.

Author Contributions: Conceptualization, methodology, software, writing—original draft preparation, validation, formal analysis, investigation, J.T.; resources, H.Z. (Hongliang Zou), X.J., H.Z. (Huaixun Zhang); data curation, J.Z., G.M.; writing—review and editing, J.Z., J.W. All authors have read and agreed to the published version of the manuscript.

Funding: This work was supported by State Grid Zhejiang Electric Power Co., Ltd. Science and technology project: Research on bus load forecasting method in spot market (5211TZ1900S4).

Institutional Review Board Statement: Not applicable.

Informed Consent Statement: Not applicable.

Data Availability Statement: Data sharing not applicable.

Acknowledgments: This work was supported by State Grid Zhejiang Electric Power Co., Ltd. Science and technology project: Research on bus load forecasting method in spot market (5211TZ1900S4). Thanks for the information and data provided by power system company.

Conflicts of Interest: The authors declare that there is no conflict of interests regarding the publication of this paper.

Nomenclature

$\delta(t)$	The Dirichlet function
K	The number of IMFs/the covariance matrix
u_k	IMF
ω_k	The center frequency of u_k
α	The quadratic penalty factor
$\lambda(t)$	The Lagrangian multiplication operator
c_t	The memory unit
h_t	The hidden layer state
x_k	The input vector sequence
f_t	The calculation results of the forget gate
i_t	The calculation results of the input gate
o_t	The calculation results of the output gate
W_f	The weight matrix of the forget gate
W_i	The weight matrix of the input gate
W_o	The weight matrix of the output gate
b_f	The weight matrix of the forget gate
b_i	The weight matrix of the input gate
b_o	The weight matrix of the output gate
o	The dot product symbol of the matrix element
$\sigma(x)$	The Sigmoid activation function of each gate
x	The input value
y	The response output value
y^*	The updated value through the training set
K^*	The covariance of the training set
K^{**}	The covariance of the newly added sample
$\phi(x)$	The probability density of the normal distribution
$\varphi(x)$	The standard normal distribution about x
μ	The mean value of the input value
σ	The variance of the input value x

Abbreviations

DNN	Deep Neural Network
SVM	Support Vector Machine
IMF	Intrinsic Mode Functions
LSTM	Long Short-term Memory
Bi-LSTM	Bidirectional Long Short-term Memory
EMD	Empirical mode decomposition
VMD	Variational mode decomposition
RMSE	Root mean square error
NRMSE	Normalized root mean square error

References

1. Deutsch, J.; He, D. Using Deep Learning-Based Approach to Predict Remaining Useful Life of Rotating Components. *IEEE Tran Syst. Man Cybern. Syst.* **2017**, *48*, 11–20. [CrossRef]
2. Zhao, P.; Dai, Y. Power load forecasting of SVM based on real-time price and weighted grey relational projection algorithm. *Powe Syst. Technol.* **2020**, *44*, 1325–1332.
3. Zang, H.; Xu, R.; Cheng, L.; Ding, T.; Liu, L.; Wei, Z.; Sun, G. Residential load forecasting based on LSTM fusing self-attentio mechanism with pooling. *Energy* **2021**, *229*, 120682. [CrossRef]
4. Memarzadeh, G.; Keynia, F. Short-term electricity load and price forecasting by a new optimal LSTM-NN based predictio algorithm. *Electr. Power Syst. Res.* **2020**, *192*, 106995. [CrossRef]
5. Ning, J.; Hao, S.; Zeng, A.; Chen, B.; Tang, Y. Research on Multi-Timescale Coordinated Method for Source-Grid-Load wit Uncertain Renewable Energy Considering Demand Response. *Sustainability* **2021**, *13*, 3400. [CrossRef]
6. Salinas, D.; Flunkert, V.; Gasthaus, J.; Januschowski, T. DeepAR: Probabilistic forecasting with autoregressive recurrent network *Int. J. Forecast.* **2019**, *36*, 1181–1191. [CrossRef]
7. Oreshkin, B.N.; Carpov, D.; Chapados, N.; Bengio, Y. N-BEATS: Neural Basis Expansion Analysis for Interpretable Time Serie Forecasting. In Proceedings of the ICLR 2020, Addis Ababa, Ethiopia, 30 April 2020.
8. Li, S.; Jin, X.; Xuan, Y.; Zhou, X.; Chen, W.; Wang, Y.; Yan, X. Enhancing the Locality and Breaking the Memory Bottleneck o Transformer on Time Series Forecasting. In Proceedings of the NeurlPS 2019, Vancouver, Canada, 8–14 December 2019.
9. Zheng, J.; Su, M.; Ying, W.; Tong, J.; Pan, Z. Improved uniform phase empirical mode decomposition and its application i machinery fault diagnosis. *Measurement* **2021**, *179*, 109425. [CrossRef]
10. Wang, J.; Athanasopoulos, G.J.; Hyndman, R.; Wang, S. Crude oil price forecasting based on internet concern using an extrem learning machine. *Int. J. Forecast.* **2018**, *34*, 665–677. [CrossRef]
11. Lin, G.; Lin, A.; Cao, J. Multidimensional KNN algorithm based on EEMD and complexity measures in financial time serie forecasting. *Expert Syst. Appl.* **2021**, *168*, 114443. [CrossRef]
12. Zhang, Z.; Hong, W. Application of variational mode decomposition and chaotic grey wolf optimizer with support vecto regression for forecasting electric loads. *Knowl. Based Syst.* **2021**, *228*, 107297. [CrossRef]
13. Zhu, Q.; Zhang, F.; Liu, S.; Wu, Y.; Wang, L. A hybrid VMD–BiGRU model for rubber futures time series forecasting. *Appl. So Comput.* **2019**, *84*, 105739. [CrossRef]
14. Liu, Y.; Xu, Z.; Dong, W.; Li, Z.; Gao, S. Concentration prediction of dissolved gases in transformer oil based on empirical mod decomposition and long short-term memory neural networks. *J. Chin. Electr. Eng. Sci.* **2019**, *39*, 3998–4008.
15. Wang, C.; Yue, S.; Wei, S.; Lv, J. Performance analysis of four decomposition-ensemble models for one-day-ahead agricultura commodity futures price forecasting. *Algorithms.* **2017**, *10*, 108. [CrossRef]
16. Zhao, Y.; Wang, X.; Jiang, C.; Zhang, J.; Zhou, Z. A novel short-term electricity price forecasting method based on correlatio analysis with the maximal information coefficient and modified multi-hierachy gated LSTM. *J. Chin. Electr. Eng. Sci.* **2021**, *41* 135–146.
17. Zhen, H.; Niu, D.; Yu, M.; Wang, K.; Liang, Y.; Xu, X. A Hybrid Deep Learning Model and Comparison for Wind Powe Forecasting Considering Temporal-Spatial Feature Extraction. *Sustainability* **2020**, *12*, 9490. [CrossRef]
18. Kraft, E.; Keles, D.; Fichtner, W. Modeling of frequency containment reserve prices with econometrics and artificial intelligence. *Forecast.* **2020**, *39*, 1179–1197. [CrossRef]
19. Mughees, N.; Mohsin, S.; Mughees, A.; Mughees, A. Deep sequence to sequence Bi-LSTM neural networks for day-ahead pea load forecasting. *Expert Syst. Appl.* **2021**, *175*, 0957–4174. [CrossRef]
20. Han, Y.; Lam, J.C.; Li, V.O.; Reiner, D. A Bayesian LSTM model to evaluate the effects of air pollution control regulations i Beijing, China. *Environ. Sci. Policy* **2020**, *115*, 26–34. [CrossRef]
21. Jamilloux, Y.; Romain-Scelle, N.; Rabilloud, M.; Morel, C.; Kodjikian, L.; Maucort-Boulch, D.; Bielefeld, P.; Sève, P. Developmen and Validation of a Bayesian Network for Supporting the Etiological Diagnosis of Uveitis. *J. Clin. Med.* **2021**, *10*, 3398. [CrossRef [PubMed]
22. Wu, K.; Wu, J.; Feng, L.; Yang, B.; Liang, R.; Yang, S.; Zhao, R. An attention-based CNN-LSTM-BiLSTM model for short-term electric load forecasting in integrated energy system. *Int. Trans. Electr. Energy Syst.* **2021**, *31*, e12637. [CrossRef]

Article

The Good, the Bad and the Future: A SWOT Analysis of the Ecosystem Approach to Governance in the Baltic Sea Region

Savitri Jetoo * and Varvara Lahtinen

Faculty of Social Sciences, Business and Economics, Åbo Akademi University, Tuomiokirkontori 3, 20500 Turku, Finland; varvara.lahtinen@abo.fi
* Correspondence: savitri.jetoo@abo.fi

Abstract: The ecosystem approach has been used extensively as a guiding principle in water policies of the Baltic Sea Region since the 1970s. In addition to its operationalization as a management framework in this region, it also has expansive theoretical underpinnings. However, despite extensive literature on this approach, there has not yet been any systematic assessment of the internal and external factors that influence its implementation. This kind of assessment could form the basis for improved thinking around the concept and better implementation actions. As such, this article presents a Strengths, Weaknesses, Opportunities and Threats (SWOT) analysis of the ecosystem approach in the Baltic Sea Region by using content analysis on Baltic Sea documents. This study found that key strengths of the principle are its interdisciplinary focus and its acceptance as a framework for conservation, whilst resource intensiveness and its operational complexity are key weaknesses. The SWOT analysis revealed that a key opportunity in the external environment is the ease of alignment with other policies whilst the key external threat is the difficulty integrating disciplines. This study showed that with a streamlined allocation of resources, more stakeholder engagement through capacity building and political leadership, the ecosystem approach could facilitate interdisciplinary knowledge pooling to achieve a good ecological status of the Baltic Sea.

Keywords: ecosystem approach; Baltic Sea; SWOT analysis; implementation; water policy; water governance; barriers

Citation: Jetoo, S.; Lahtinen, V. The Good, the Bad and the Future: A SWOT Analysis of the Ecosystem Approach to Governance in the Baltic Sea Region. *Sustainability* **2021**, *13*, 10539. https://doi.org/10.3390/su131910539

Academic Editors: Baojie He, Ayyoob Sharifi, Chi Feng and Jun Yang

Received: 22 August 2021
Accepted: 18 September 2021
Published: 23 September 2021

Publisher's Note: MDPI stays neutral with regard to jurisdictional claims in published maps and institutional affiliations.

1. Introduction

Humans constantly increase the stressors against the global marine ecosystems that, together with climate change, lead to severe loss of biodiversity, habitats and resources. It has already become apparent from the middle of the 20th century that traditional management practices could not address the ever-increasing loss of biodiversity [1–3]. The Ecosystem Approach (EA) is a concept that arose by the end of the 1980s, initially endorsed for the management of terrestrial ecosystems [1,2] to counteract the increasing loss of biodiversity. It emphasized the integration of information into management practices, the involvement of stakeholders in the decision-making process and the consideration of environmental relationships in different spatial, organizational and biological scales [2].

The term Ecosystem Management, conceptually similar to the ecosystem approach, was used to signify an ecosystem approach to resource management at the end of the 1980s [2]. Another similar concept, ecosystem-based management, emerged in the 1950s for managing terrestrial ecosystems [3–5]. The usage of many different terms (e.g., ecosystem approach, ecosystem-based management, ecosystem management, integrated management, etc.) in reference to similar concepts created confusion among scientists and practitioners [6,7]. Many researchers use ecosystem (-based) management and ecosystem approach interchangeably [5,6,8]. It is argued that there is a difference between the ecosystem approach and ecosystem management with the first being a vision and simultaneously a working plan for its realization, while the second is judged in human terms and may be altered [9].

The internationalization of the ecosystem approach occurred through the Convention of Biological Diversity (CBD), which adopted it as the framework for the implementation of its objectives [10]. According to the COP V/6 of the CBD, the ecosystem approach is "a strategy for the integrated management of land, water and living resources and promotes conservation and sustainable use in an equitable way" [11]. Furthermore, the CBD provided a list of 12 interlinked principles (Malawi principles) that characterize the ecosystem approach, and five points were proposed for guiding the operationalization of EA [11].

Regional Sea Conventions such as the Commission of the Convention on the Protection of the Marine Environment of the North-East Atlantic (OSPAR) and the Helsinki Commission (HELCOM) adapted the definition to the needs of their regional marine ecosystems. They jointly adopted "a vision of an ecosystem approach to managing human activities impacting on the marine environment (an "ecosystem approach") in their maritime areas" [6,12]. They defined the ecosystem approach as "the comprehensive integrated management of human activities based on the best available scientific knowledge about the ecosystem and its dynamics, in order to identify and take action on influences which are critical to the health of marine ecosystems, thereby achieving sustainable use of ecosystem goods and services and maintenance of ecosystem integrity" [12]. In 2007, HELCOM adopted the ecosystem approach in its Baltic Sea Action Plan (BSAP) [13] as it was defined in the first joint ministerial meeting of the Helsinki and OSPAR Commissions in 2003, and it calls for its implementation in the management of the Baltic Sea Region.

EA was also adopted by European Union (EU) policy documents. The Water Framework Directive (WFD) of 2000 did not explicitly mention the ecosystem approach but calls for the administrative reorganization of the water institutions to the appropriate scales according to the ecological borders [14]. The Marine Strategy Framework Directive (MSFD) adopted in 2008, explicitly refers to the application of the ecosystem-based approach to the management of human activities in the marine strategies of EU members for attaining good environmental status in the European marine regions and sub-regions [15]. However, the MSFD does not provide a definition for the ecosystem-based approach. The same term is also to be found in the EU Maritime Spatial Planning Directive (MSPD) of 2014, where the ecosystem-based approach will "contribute in the sustainable development and growth of the maritime and coastal economies and the sustainable use of marine and coastal resources" [16]. Furthermore, it identifies the maritime spatial planning as a tool for the application of the ecosystem-based approach in attaining good environmental status [16].

The ecosystem approach has been studied extensively globally but also within the Baltic Sea Region, which is this research's focus. One line of research focuses on the ecosystem approach to fisheries (EAF) or ecosystem-based fisheries management (EBFM) [17–21] that is not a fully cross-sectoral ecosystem approach but instead applies ecosystem-based policies within an individual sector [5]. Another line of research discusses appropriate ecosystem indicators to be developed for the implementation of ecosystem-based management [22–25]. A large part of the literature discusses frameworks as well as guidelines and the sequence of steps for the implementation of the ecosystem approach/management [5,26–28].

Against this backdrop of extensive literature discussing different aspects of EA, there has not yet been any systematic assessment of the internal and external factors that influence EA implementation in the Baltic Sea Region. This article intends to fill this gap by presenting a Strengths, Weaknesses, Opportunities and Threats (SWOT) analysis of the ecosystem approach in the Baltic Sea Region. The analytical framework itself guides the research questions of this study: what are the internal strengths and weakness of the ecosystem approach and what are the external opportunities and threats for the ecosystem approach in the Baltic Sea Region?

The purpose of this article is to provide a clear view of the key features that can promote or hinder the implementation of EA in the Baltic Sea Region and the achievement of good environmental status but also to inform the ongoing policy processes such as the updating of HELCOM's Baltic Sea Action Plan.

2. Materials and Methods

This is a qualitative research process as illustrated in Figure 1. The SWOT and content analysis and further details of the research method are elaborated in the following sections.

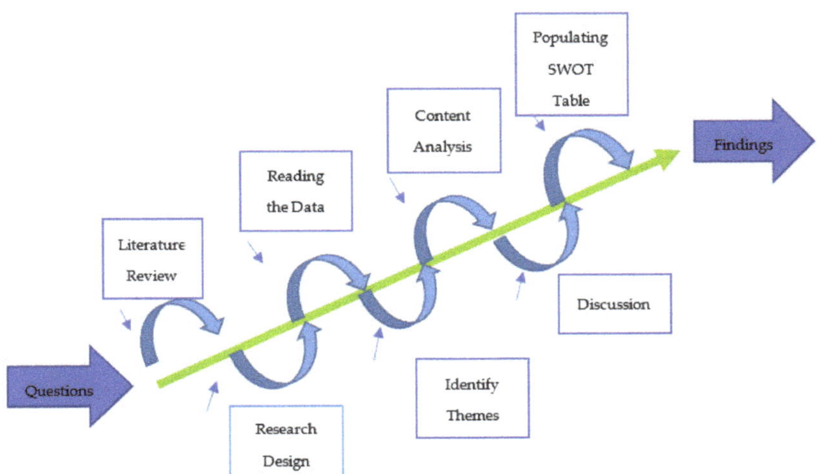

Figure 1. The research process.

2.1. The SWOT Analysis

The term SWOT derives from business literature and refers to Strengths, Weaknesses, Opportunities and Threats (SWOT) analysis. It is used as an assessment tool for strategic planning to position organizations to achieve their stated goals. The SWOT analysis is an effective tool for taking reconnaissance of the organization's resources and capabilities, its deficiencies, market opportunities and threats to its future [29]. The SWOT analysis views the performance of a given organization with respect to its goals and varies with the internal characteristics of that organization and the broader external context in which that organization must operate [30]. Strengths are internal characteristics that position the organization favorably towards the achievement of its goals, whereas weaknesses are internal factors that can detract from goal realization. Conversely, opportunities and threats are factors in the external environment that can facilitate or thwart the achievement of goals, respectively. These factors can include political, economic, social, technological, legal and environmental factors.

The SWOT analysis is a useful framework for analysis of the use of the ecosystem approach in the Baltic Sea Region as its value stems from the way the features of the concept interact in the external and internal environments to achieve its success. Its usefulness also arises from its ability to be used in the identification of barriers to success and the subsequent implementation of long-term strategies to achieve its objectives [31]. When applied to the ecosystem approach, the SWOT analysis can be applied to the features of the concept of the ecosystem approach that allow for the realization of its goals [32], to:

➢ Consider the ecosystem as an integrated whole;
➢ Use sustainability and people as the heart of achieving good environmental status; and
➢ Value the services we obtain from nature.

In this context, strengths can be seen as the features of the ecosystem approach that allow for the achievement of the above stated goals, whilst weaknesses are characteristics of the approach that compromise the realization of these goals. As such, strengths and weaknesses are internal attributes of the ecosystem approach itself whilst opportunities and threats are external features that retard the realization of the above goals. In this case, external features are examined with respect to the Baltic Sea environment and the achievement of the good ecological status of the Baltic Sea. More specifically, the political economic, sociological, technological, legal and environmental (later the PESTLE framework) factors within the implementation framework of the ecosystem approach in the context of the Baltic Sea Action Plan are considered. This is illustrated in Figure 2, with research questions included.

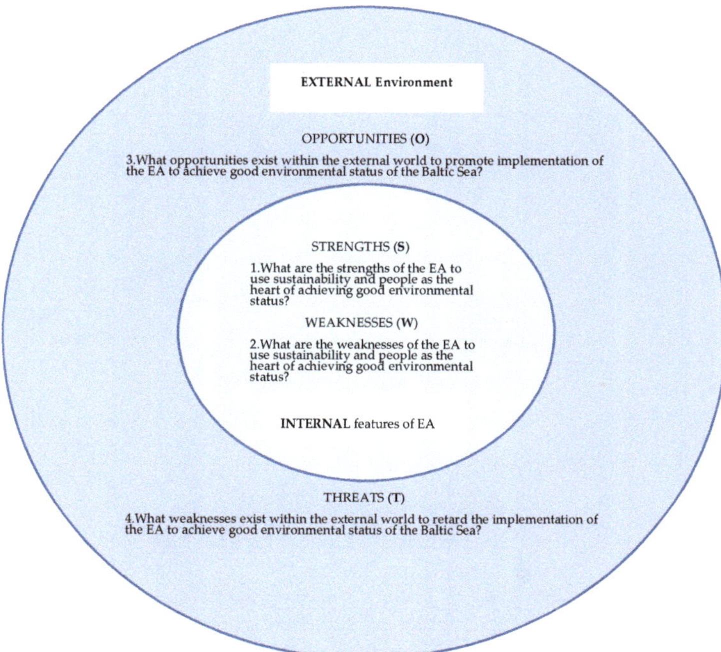

Figure 2. Research questions showing the internal and external SWOT environments.

2.2. The Content Analysis

The qualitative content analysis interprets the text's data content subjectively by using the systematic classification method of coding in the identification of patterns or themes [33]. It has an interpretive character that enables discussion on the points and themes that are essential to the study's research goal [34]. As a qualitative method, it has many advantages such as flexibility and applicability in a plethora of text types, being rather "safe" [35] and having "less rules to follow" by not being strictly linked to any specific science [36]. The disadvantages, on the other hand, are the inability to study other types of communication (e.g., body language, etc.) but the recorded communication-text, the potential loss of useful data due to restricted access or altogether missed data, its complexity and the fact that it is labor intensive and time consuming [35].

Qualitative content analysis was deemed the most appropriate method to be used in this research for a number of reasons. Firstly, due to the data used being existing documents and communications; secondly, due to their relatively limited numbers that renders their study attainable through qualitative content analysis and thirdly, due to qualitative content analysis being the most appropriate research method in answering the posed research questions.

The documents studied in this research were mainly the minutes from the meetings of the Group for the Implementation of the Ecosystem Approach (GEAR) [37]. GEAR is one of HELCOM's working groups, established in 2012 for coordinating the actions in implementing both the Baltic Sea Action Plan (BSAP) and the Marine Strategy Framework Directive (for the Contracting Parties also EU Member States). It also has oversight of pertinent global processes with the overarching goal being the implementation of EA and the achievement of good environmental status in the marine waters of the Baltic Sea [38]. By August 2021, the GEAR Group had 24 meetings. All 24 GEAR minutes and accompanying documents were studied but four of these minutes-documents were chosen specifically for coding analysis. These were the minutes of the GEAR 1 meeting (2012) [39], GEAR 13 (2016) [37], GEAR 18 (2018) [37] and GEAR 24 (2021) [37]. The GEAR 1 [39] and 24 [37] minutes were chosen to observe the changes the Group has undergone within the 9 years of operating. The GEAR 13 of 2016 [37] was chosen as the year when the EU Member States (also Contracting Parties) had to establish and implement Programmes of Measures (PoMs) in order to achieve Good Environmental Status (GES), set in Article 13 of MSFD (MSFD 2008); it was therefore interesting to observe how this mandate was met through the GEAR meeting. Finally, the GEAR 18 of 2018 [37] was chosen as the year of the beginning of the second cycle of MSFD implementation and therefore interesting to note how this process was reflected in the GEAR meeting.

The coding process began after having collected the documents. It is acknowledged that a successful qualitative analysis is the outcome of a well-thought out and well-executed coding process [33]. Therefore, the focus was kept on the research questions in order to avoid too broad or too narrow coding that would not answer them. The coding process did not aim at generating any theory, and therefore a deductive use of existing theoretical framework took place. The theoretical/conceptual framework utilized in this study was the 12 Malawi Principles of EA of the CBD [32]. After studying and discussing the Malawi Principles, some of the principles were somewhat overlapping and therefore condensed into themes that were relevant to the primary content of the principles and relevant to the Baltic Sea Region. The themes that emerged are as follows, with principle number in brackets: stakeholder participation (1), governance scale (2,7,8), institutional arrangements (3,6,9,10), economic considerations (4,5) and science-policy interface (11,12).

For the coding process, an excel spreadsheet was utilized for each GEAR minutes separately, where paragraphs from the minutes were inserted with their assigned theme as a code. Additionally, a column "remarks" was inserted where each minutes' paragraph was classified as internal strength or weakness, or as external threat or opportunity along with the justification for the classification. Many topics discussed in the minutes touched upon more than one theme and therefore assigned more than one code accordingly.

3. Results

This section presents the results of the SWOT analysis, focusing on each part of the SWOT analysis in turn.

3.1. Strengths

One of the key strengths of the ecosystem approach that was evident in all the themes and came across in the document analysis is the integrated approach in practice. This was evident in the meeting minutes, for example, in the Gear 24 min, paragraphs 5.51–5.56, which noted the "plans for assessments for marine litter, integrated eutrophication assessment tool, underwater noise, integrated hazardous substances assessment tool" [37]. These paragraphs describe the latest developments on the topic of the assessments for marine litter, underwater noise, eutrophication, etc. HELCOM has developed the necessary tools for assessing all these stressors, and their efficient utilization can lead to accurate assessments that can facilitate the implementation of EA in management. It exemplifies an internal strength of the approach, which causes seemingly disparate areas to connect with each other in pursuit of one common overarching goal of good environmental status.

Another strength of the approach is its simplicity in communicating the necessity of humans to protect their environment, in this case, the good ecological status of the Baltic Sea. One theme throughout the GEAR meeting portal is that there is consensus that the ecosystem approach to management is the accepted guiding framework for the achievement of good ecological status of the Baltic Sea. The meeting minutes [37] demonstrate that there is agreement that everyone understands the guiding principles of the ecosystem approach and that humans are an integral part of ecosystems.

Stakeholder participation is one of the key features of EA and one of its internal strengths. The practice of stakeholder participation can be traced during different processes described in the GEAR minutes. The adequacy of participation or the breadth of stakeholders of course can be argued but nonetheless it is present in different occasions. For example, in GEAR 13 min, the paragraphs from 4.14 to 4.17 discuss the national consultation process regarding the first results of HOLAS II. The comments that would occur from this consultation will feed the HOLAS II updating. Additionally, it was discussed the "option to have a regional consultation on the HOLAS II report ... oriented in particular towards an international audience in the Baltic Sea Region ... " [37]. A few more examples of stakeholder participation can be found in the minutes of GEAR 24, in paragraph 5.35, "a project meeting ... to enable interested experts and CPs to contribute more concretely to the development of driver indicators ... " and in paragraph 6.2, "outcome of the HELCOM Stakeholder Conference in 2021 ... " [37]. Finally, already from its first meeting, GEAR discussed "how in principle the projects (steered by GEAR) should be carried out ... good involvement of national experts in the projects and project preparation ... better results than purely scientific ... of mainly external scientific institutions" [39].

Another internal strength of EA is the constant generation and updating of scientific data and indicators for better monitoring and assessing of different components of marine ecosystem. For example, the GEAR 24 min in paragraphs 5.17 to 5.23 discuss the establishment of a list of commercial species for the HELCOM area and (5.21) the "possible need to include species such as perch ... commercially important in some areas of the Baltic but for which data is not reported under the CFP." [37]. Furthermore, in 5.22, it was "agreed that all commercially exploited species ... should be included on the list ... only those species for which landings data is available can be included in the assessments ... collect initial information on the availability of landings data for regionally important commercial species for which data is not reported under the CFP." [37]. Though no scientific break through is communicated through this example, the constant effort to improve the existing indicators for better monitoring is however evident.

One strength closely related to the constant generation of data and stakeholder participation is adaptation; an inherent feature of EA. Change is inevitable and is it recognized by EA, which encourages for the right adjustments and adaptations to it. This is exemplified in the minutes of GEAR 24, paragraph 6.6: the "current set up of the HELCOM Roadmap on EA is no longer optimal and that as part of updating the Roadmap in 2022 ... an approach more along the line of a meta-analysis would be more appropriate ... " [37]. The Group recognized the need for improvement and adjustments to fit the input from feedback loops for facilitating further EA implementation.

3.2. Weaknesses

The scientific and interdisciplinary basis of the ecosystem approach requires comprehensive monitoring data which may be difficult to implement due to resources limitations. This incomplete scientific information was a theme in the meeting minutes of the GEAR group. For example, in the first meeting minutes [39], Section 4.4 states that "new approaches for eutrophication targets ... the report should be considered as input from science to the HELCOM work ... acknowledging the views of Germany and Poland that it has its deficiencies in terms of data use ... HELCOM work continues to be based on science and its technical work should follow the best available science" [39]. This statement illustrates that there is incomplete data which hampers the achievement of eutrophication targets, as now HELCOM is forced to use the incomplete information, or best available science. Another inherent weakness is the inconsistency in the application of EA, due to the diverse interpretation of what it means and how it lends itself to the local context. For example, the Gear 1 meeting minutes demonstrate in Sections 4.2 and 4.3: "Poland ... requested information on sampling stations ... Poland to observe that only data from the 1980s and 1990s had been used … most useful for the project to have data also from the end of the 1990s and 2000s … proposals by the project for further sub-regionalized targets" [39]. These comments were made by delegates considering the final report of the HELCOM Approaches and methods for eutrophication target setting in the Baltic Sea Region (TARGREV) project [39]. Since Poland discharges one of the largest nutrient loads in the Baltic Sea Region, meeting participants were pointing out for the need for a more comprehensive data set to be able to set sub-regional targets.

An additional weakness of EA is the vastness of the marine ecosystem itself as well as a number of factors that cannot be monitored easily or issues of national jurisdictions or natural phenomena and climate change that cannot be fully predicted or controlled. This aspect is reflected in the minutes of GEAR 13 in paragraph 3.13 where "took note of the information on the national plans to use exceptions under MSFD Article 14 … " [37] and it continues in presenting the exceptions used by the different Member States-Contracting Parties (CPs). Article 14 of MSFD [15] in essence gives the opportunity to the Member States (MS) to determine areas within their national marine waters that cannot meet the environmental targets or the GES within the time schedule due to a number of listed reasons. Those reasons include natural causes, unforeseeable circumstances that prevent MS to fulfill their obligation, natural conditions that hinder the prompt improvement of marine waters' status, etc.

3.3. Opportunities

One of the big opportunities of EA was recognized in the very first GEAR meeting minutes, which is the potential for alignment with existing policies and strategies and finding collaborative synergies with other countries and agencies. For example, the GEAR 1 meeting, Section 2.3 recognizes "The importance of starting early on with the regional coordination … also cooperate with other organizations such as OSPAR and Black Sea Commission" [39]. This was further reiterated in Section 2.4, which went on to say that members "supported the alignment of the delivery of the relevant HELCOM products with timetables under other international frameworks" [39]. This theme continued forward in the GEAR 24 meeting minutes, where Section 6.7 looks into "use GEAR as a platform to share national experiences on implementing ecosystem based management (EBM), in the long term develop joint guidance on EBM, consider how to strengthen interactions with other sectors" [37].

These synergies can have further benefits apart from sharing knowledge, scientific data and expertise. These benefits can be the saving of resources, both in terms of funding and working force, avoiding duplication of work and unnecessary workload. This is clearly stated in GEAR 18 min, paragraph 3.7: "global fora are good channels to present … high quality and best practices developed in the Baltic Sea Region … the well-structured assessments … from EU and RSC (Regional Sea Convention) work should be exported … to avoid duplication of work … " [37].

Another opportunity that surfaced in the GEAR 24 meeting minutes is the possibility to combine the potential for the alignment of policies with funding support. For example, Section 5.64 notes that "Sweden is looking into the possibility of providing additional resources to support the work under HOLAS III" [37] and 5.69 "Finland and Sweden investigate potential funding opportunities to support closer participation by Russia" [37]. The EU also provides funding support in the operationalization of projects that promote the implementation of MSFD and EA such as the BALSAM project, which provided recommendations for marine monitoring in the Baltic Sea. Such reference is found in GEAR 6 paragraph 3.26: "EC that is providing the main source of funding for BALSAM, that project goals and deliverables could in discussion with the EC be adjusted in order to make best use of resources" [40]. More supportive actions, even if their nature is not specified, we find in GEAR 13, paragraph 4.1, where Germany will continue supporting "activities that were not possible to finalize within the timeframe of the project (EUTRO-OPER) e.g., development of GES boundaries for new pre-core indicators … " [37].

An additional opportunity in the implementation of EA in the Baltic Sea Region i performing both national and regional assessments for better management and implemen tation of EA. Such a statement is found in GEAR 24, paragraph 3.7: "regional and nationa assessment results . . . valuable information even if these differ, with regional assessmen . . . a transboundary overview . . . national level assessments enables results to be used fo direct management needs which often take place at a local scale" [37].

Finally, the active participation of observers in GEAR meetings but also in action taking place within HELCOM can be considered as an external opportunity in EA imple mentation in the Baltic Sea Region. Such an example is to be found in GEAR 18, paragraph 4.19: "the need to include those observers which have been taking part in the HOLAS II work" [37].

3.4. Threats

The top threats to the implementation of EA in the Baltic Sea Region relate to the themes already discussed. For example, inadequate funding is a possible threat to the third HELCOM Holistic assessments of the Baltic Sea environment (HOLAS III). This wa captured in the GEAR 24 meeting minutes, Section 5.63: "resources allocated for HOLAS II . . . further consideration on resources needs of the planned work . . . very important tha there are dedicated resources for drafting the HOLAS III texts in a coherent and readabl manner" [37]. Planning for the funding is very important and it must be secured before the planned work began. Amidst the lack of resources, wasting them is an additional threa for EA implementation. Such a fear was discussed in GEAR 24 paragraph, 3.24: "carefu consideration . . . the added value of the resources required for the pre-filling . . . to wha degree countries will utilize this information" where a careful planning is required to avoi using resources to unutilized information" [37].

Another threat that was clearly articulated in the meeting minutes of GEAR 24 is the challenge of integrating the different disciplines. For example, Appendix 3 of the minutes speak of a future action by 2023 to "integrate economic and social analyses in HELCOM work . . . for assessment of the linkages between the marine environment and human wellbeing, including carrying out regionally coordinated economic and social analysis o the marine environment" [37]. This difficulty of coordinating and integrating differen approaches was tabled at the minutes of the GEAR meeting nine years earlier in 2012 when the GEAR 1st meeting minutes Section 3.15 stated that "the different approaches lack of sufficient coordination and harmonization . . . will be no possibility to coordinat and harmonize the targets between the neighboring countries' sea areas during the firs implementation round of the MSFD" [39]. This quotation also points to the threat o inadequate political leadership to ensure collaborative governance for the meeting of the goals of the ecosystem approach.

An additional threat is the differences in national agendas, which can have variou origins (differences in political will, in prioritizations, in resources, in capabilities, in dat etc.). These differences can be reflected in GEAR, a group where mutual agreement is required for decisions to be made. Such an example can be found in the minutes o GEAR 13 in paragraphs 3.3 to 3.8 where study reservations on the "Consolidated versio of the Joint documentation of regional coordination of programmes of measures in the Baltic Sea area" [37] by Denmark and Germany are discussed. In paragraph 3.6 is state "the strict approach by Germany . . . expected a number of its comments and propose amendments . . . to be agreed by the other CPs as a prerequisite to Germany clarifying it study reservation . . . harsh action . . . effectively blocks constructive work of the grou . . . against the good practice of HELCOM work" [37]. These kind of incidents may not b frequent but they certainly can have a disruptive effect on the implementation of EA.

The lack of consistency in the delegates of the CPs' presence in the GEAR meeting can pose an external threat for EA in the Baltic Sea Region. This issue was raised in GEAR 14 min, paragraph 2.3: "only four Contracting Parties were represented . . . stressed that a better representation of Contracting Parties is needed Group to steer the implementation of the HELCOM BSAP and to facilitate the regional coordination for the implementation of the EA and the MSFD" [37].

Finally, partly ineffective tools can be considered as an external threat. Such an example can be found in GEAR 24, paragraph 4.9: "the information on the updated HELCOM Explorer" and 4.11 "could be refined to be more easily accessible for the general public … proposals for simplifying the language" [37]. HELCOM Explorer is a very important tool because it gives the general public the opportunity to be informed about the different joint and national actions in the Region for the implementation of EA and to be part of this process, which mainly takes place among experts. Therefore, it should cater to the needs of the public and the language simplification is a step towards that direction.

4. Discussion

4.1. Results of the SWOT Analsyis and Strategies Going Forward

The results of the SWOT analysis as conducted for the Baltic Sea Region are summarized in Table 1. The research questions that this study posed were: what are the internal strengths and weaknesses of ecosystem approach to use sustainability and people as the heart of achieving good environmental status, and what are the external opportunities and threats to promote or retard the implementation of EA in achieving these goals in the Baltic Sea Region? The Baltic Sea Action Plan (BSAP) is HELCOM's key tool for achieving good ecological status of the Baltic Sea by 2021, while EU MSFD established a framework for achieving or maintaining GES by 2020, both having EA as their guiding principle. However, none of these realized their goal because GES has not yet been achieved in the Baltic Sea Region. Nonetheless, progress towards EA implementation has been made under these frameworks in the Region, a fact that should be not overlooked and understated and that was evident in the minutes of the GEAR meetings examined in this study. However, as Table 1 shows, there are key threats such as the difficulty integrating disciplines, inadequate funding, lack of consistent stakeholder participation and lack of political will that potentially undermine the strengths of the approach.

Table 1. Summary of SWOT results.

	The GOOD	The BAD
INTERNAL to EA	STRENGTHS -Integrated approach -Interdisciplinary approach -Simplicity -Accepted framework for conservation -Stakeholder participation -Generation of scientific data, update of indicators and creation of new ones -Adaptation -Brining persons closer to the environment	WEAKNESSES -Requires comprehensive data -Resource intensive -Incomplete scientific information -Complex and difficult to apply in practice -Unpredictability in ecosystem processes, natural causes and phenomena and climate change.
EXTERNAL including the Baltic Sea Region	OPPORTUNITIES -Widely accepted framework for Baltic Sea protection -Can be easily aligned with existing policies and strategies such as the UN sustainable development goals (SDGs). -Funding from EU pools for MSFD -Synergies in reporting and monitoring saving funds and avoiding duplication of workload -Regional and national assessments -Observers' participation -Technological and scientific resource availability	THREATS -Difficulty integrating different disciplines -Inadequate funding -Inadequate resources e.g. personnel -Lack of political will -Different national agendas signifying different approaches -Lack of consistent participation of delegates -Wasting resources -Tools that may not fulfill their purpose completely -Lack of institutional capacity in some Baltic Sea countries

This study has found confirmation of the usefulness of the ecosystem approach as a framework towards the achievement of the good ecological status of the Baltic Sea. It has extended previous studies on the ecosystem approach as it has applied the SWOT framework and is using it in the discussion of strategies for taking the approach forward into the current update of the Baltic Sea Action plan for more effective implementation. The strategies that will be discussed below flow out of a systematic examination of the threats and weaknesses and the utilization of the strengths and opportunities to mitigate those impediments.

4.2. The Future—Strategies Going Forward

These results are now used to devise strategies for moving forward with EA in the Baltic Sea Region. The SWOT analysis is not only a simple framework used to highlight strengths, weaknesses, opportunities and threats, but its additional value can be found in combining these elements in the development of strategies toward goal achievement. There are many ways to link the strategies from a SWOT analysis, based on the internal and external factors. According to the literature [41,42], four strategies can be developed, as follows: SO (strengths-opportunities), ST (strengths-threats), WO (weakness-opportunities) and WT (weakness-threats). This can be combined in a SWOT matrix, as follows [43]:

- SO strategies: taking advantage of opportunities
- ST strategies: avoiding threats
- WO strategies: introducing new opportunities by reducing weakness
- WT strategies: avoiding threats by minimizing weakness

4.2.1. SO Strategies

SO strategies are those combinations that allow internal strengths of the ecosystem approach to utilize external opportunities in the Baltic Sea environment. This combination of internal strengths of the ecosystem approach with the external opportunities in the Baltic Sea Region and other areas of the world could help to mitigate threats in the environment. Blending the strengths of the ecosystem approach, being an inherently interdisciplinary one with the opportunity of it being a widely held and accepted framework for Baltic Sea protection, could help to mitigate the threat of the difficulty of integrating different disciplines. A strategy recommendation would be to hold facilitated events such as workshops or conferences that bring together different disciplines to discuss the ecosystem approach in the Baltic Sea Region. Such workshops and conferences would also help to raise awareness and create a common understanding of the approach.

Stakeholder participation is an inherent strength of EA that helps to determine the objectives of ecosystem-based management. The Baltic Sea Region presents the great opportunity of broad participation of stakeholders in the process of developing EA. These stakeholders can be national experts participating in GEAR, experts working on EA actions in national contexts or observers from different organizations such as foundations, NGOs etc. An example of strategy recommendation would be that by highlighting the significance of stakeholder participation in the core of EA, participating experts in the Baltic Sea Region can promote action that would further cultivate and elevate the participatory culture addressing especially the countries that have less experience of it in their political systems. Furthermore, observers could also promote their experience in participating in Baltic Sea Region actions and the positive impact and effect such participation can have in protecting the environment and the achievement of GES. They could use the websites of their organization as platforms to promote such experiences to gain visibility and raise awareness. This combination of strengths and opportunities can also be used to combat the threat of the loss of political will, as showcasing the framework in public events would help to garner political support.

One of the EA cornerstones is the production of data and scientific knowledge, which is an important strength of the principle. The production of scientific data in the Baltic Sea Region is voluminous as it is a high priority in the region. More tools are developed, and more data are collected. One strategy action could be for HELCOM to export the expertise of creating tools and frameworks that have been effectively utilized in the Baltic Sea Region not only to other regions in Europe but also at an international level. The poor ecological status of the Baltic Sea due to its pollution, adding to its natural vulnerability, has led to a high scientific expertise in the Region that is constantly looking for more effective tools and framework. Such expertise in the Baltic Sea Region should become visible, recognized and exported globally to harness financial resources.

The opportunity of technological advances in the region along with the inherent strength of the scientific focus of the approach could help to mitigate the threat of lack of institutional capacity in some Baltic Sea countries. Environmental forerunners such as Finland, Sweden and Denmark could help in training staff of the environmentally weaker countries such as the Baltic States through mobility staff exchanges.

4.2.2. ST Strategies

ST strategies use internal assets to help lessen the impact of external threats in the Baltic Sea and wider world. The integrated, interdisciplinary focus of the ecosystem approach could be combined with the threat of integrating disciplines, leading to a strategy of increasing collaborations among disciplines. It is noted that HELCOM has a natural sciences focus, so collaborations should be extended with the humanities and the social sciences to gather local knowledge to combat stressors such as climate change.

The internal strength of stakeholder participation could be coupled with the threat of the voluntary participation of national delegates in HELCOM processes. A strategy recommendation may be to detect the reasons behind the occasional lack of participation and trying to address those reasons if possible. For example, if delegates cannot be present to a certain venue due to lack of funds or excessive workload, then actions could be taken to find some funds or hold the meetings in a virtual manner, in a hybrid mode, in order for more delegates and observers to participate in the meetings.

Another possible combination may be the internal strength of data collection and knowledge production in conjunction with the threat of developed tools not completely fulfilling their purpose. The constant development of methods and tools for collecting information about the marine ecosystems in the Baltic Sea and their monitoring has been a very important component of EA implementation in the Region. A strategic recommendation would be a more careful design of tools and possibly their pilot use to determine their performance in their designated task before their release to broad use. The constant scientific research with a broadening interdisciplinary focus will lead to gains in knowledge on EA implementation. Therefore, it can inform the process of developing tools suited to purpose, in order to avoid wasting resources by creating impractical tools.

4.2.3. WO Strategies

WO strategies are those that use external opportunities in the Baltic Sea environment and wider world to mitigate the internal flaws of the ecosystem approach. One opportunity that recurs in HELCOM's meeting minutes is the alignment of the ecosystem approach with other frameworks such as the sustainable development goals (SDGs). This can be harnessed in overcoming the weakness of the approach in being complex and difficult to apply in practice. A strategy to do this will see experts of EA helping managers and non-specialists to understand the approach and to apply the tools of the approach. They can be part of project teams that help in the implementation of the approach on the ground in the Baltic Sea Region. This could result in greater understanding, uptake and local ownership of the ecosystem approach.

One important external opportunity in the Baltic Sea Region is conducting assessments at both a national and regional level for fulfilling the requirements of the MSFD, among other things, since eight out of the nine countries-CPs of HELCOM are also member states of the EU. This opportunity can be combined with the inherent weakness of EA, which is the incomplete scientific data for EA implementation. A strategic recommendation could call for intensifying the combination of the data collected at a national and regional level in the Region in order to fill in the gaps of scientific knowledge. Furthermore, national assessments can provide data for smaller geographical areas and be more specific. This information should be utilized in a manner that would enhance knowledge of more components of the marine ecosystems of the Baltic Sea and it would promote further the implementation of EA.

An often-reoccurring opportunity in the GEAR meetings is the synergies in processes such as reporting and monitoring that can lead to saving funds as well as less workload. It can be combined with the internal weakness of EA of being resource-intensive and require many comprehensive data. There is already a broad network of collaboration between HELCOM and other organizations in the Baltic Sea Region and at an international level. A strategic action could be to enhance the existing collaboration and to assign tasks in a coordinated manner so as to share different monitoring and reporting processes and costs. The collaborations aimed at producing data in the Baltic Sea Region could be strengthened in order to produce solid scientific data to export for international assessments in ameliorating the weakness of increased resources and data requirements for EA implementation.

4.2.4. WT Strategies

WT is the double whammy strategy, as it consists of strategies aimed at lessening internal weaknesses and external threats at the same time. Lessening the negative effect of threats and weaknesses simultaneously can be confounding. The weakness of the approach being resource-intensive combined with the threat of inadequate funding could be addressed by utilizing funding pools available from the EU. As such, it would be pertinent to identify the funding gaps and their causes and to combine this with matching these gaps to the pools available.

The important internal weakness of EA, the incomplete scientific data information and the external threat of inadequate resources, e.g., personnel, could be addressed together. A strategic recommendation would be to actively pursue more collaboration agreements with scientific institutes in the Region; for example, with universities that could assist in producing knowledge and at the same time would address the issue of lack of scientific personnel. Universities, having their own pools of funding, would not pose any additional financial burden to the process of EA implementation in the Region and they have the expertise and the experience to provide valuable assistance in accumulating additional scientific data for EA.

Another weakness, that of EA being complex and difficult to apply, may be approached together with the threat of different national agendas. It is very hard for one to address the threat of different national agendas adequately or to change the fact that EA is difficult to apply since at its very heart lies the marine ecosystem that is complex and unpredictable. However, one strategic action could be to increase the visibility of EA at the national level and especially to the broad public. Through events or workshops addressing different parts of the populations and different local groups, EA can be promoted and become less of a complex principle by actively educating the public, emphasizing their participation and support. The more visibility and recognition gains among the broad public, the more chances are for the official national attitudes to change towards it and be more actively involved in EA implementation if they have not been already.

4.3. Malawi Principles in the Baltic Sea Region

This section now considers how the implementation of the ecosystem approach in the Baltic Sea Region corresponds to the vision espoused in the Malawi principles (Table 2). This is based on the results from the SWOT analysis. Table 2 clearly illustrates that there are different degrees of implementation of the ecosystem approach in the Baltic Sea Region. There is management autonomy to choose actions and planning time frames, with management recognizing changes in line with the Malawi principles. However, there is still the sectoral approach, with silos operating, signaling the need for better collaboration and policy coherence at the local (e.g., municipalities), national (e.g., ministries), regional (EU level) and international levels. There is also not enough recognition and use of local knowledge. The local level is still not adequately represented in the decision-making process. One of the key areas for improvement is highlighted in principle 12, the need for better stakeholder engagement. Communication and capacity building of all stakeholders can lead to better implementation through an increased understanding of ecosystem interactions and hence the adoption of policies by stakeholders, leading to better science policy processes.

Table 2. Malawi principles in the Baltic Sea Region (Green-achieved; Red-not achieved; Grey-somewhat achieved).

Malawi Principle	Implementation in the Baltic Sea Region
Management objectives are a matter of societal choice.	This is in full implementation in the Baltic Sea Region as HELCOM gives contracting parties freedom in deciding how measures are met. For example, GEAR 24 meeting minutes state that "The Meeting concluded that national reporting, irrespective of which approach is used, needs to capture the scale and methodology used for the assessment results used".
Management should be decentralized to the lowest appropriate level.	GEAR meetings were attended mainly by national-level representatives. As such, this objective is not yet achieved in the Baltic Sea Region.
Ecosystem managers should consider the effects of their activities on adjacent and other ecosystems.	There is some evidence of this, but there is need for strengthening collaboration between sectors. GEAR 24 meeting minutes note that "The Meeting took note of the list of activities and pressures to be used across the HELCOM Working Groups for linking the activities and pressures to the proposed BSAP actions". HELCOM's work on linking activities and pressures is evidence of this principle in progress.
Recognizing potential gains from management, there is a need to understand the ecosystem in an economic context, considering, e.g., mitigating market distortions, aligning incentives to promote sustainable use and internalizing costs and benefits.	There is now recognition in the Baltic Sea community of the need to include socio-economic assessments into current targets. The 2021 GEAR 24 meeting minutes had an entire Annex 3 dedicated to the integration of social and economic analysis into an update of the Baltic Sea Action Plan.
A key feature of the ecosystem approach includes conservation of ecosystem structure and functioning.	This is acknowledged in the BSAP guiding principles but not achieved in practice as the Baltic Sea ecosystem is under pressure from stressors such as eutrophication.
Ecosystems must be managed within the limits to their functioning.	Due to the above, this principle is not met. The 'limit to functioning' is not fully understood for the Baltic Sea, as there is not sufficient data to determine what is the integrated limit to functioning (as opposed to the silo approach of, e.g., fisheries)
The ecosystem approach should be undertaken at the appropriate scale.	The holistic assessments compiled by HELCOM are done at different scales (such as the different bays of the Baltic Sea), so this is somewhat achieved. However, since there are not enough data, the appropriate scale is sometimes not achievable.
Recognizing the varying temporal scales and lag effects which characterize ecosystem processes, objectives for ecosystem management should be set for the long term.	The time frames for Baltic Sea action are generally long term, as HELCOM sets, e.g., 14-year targets for the first BSAP. It is generally recognized that the ecosystem of the Baltic Sea takes a long time to recover after management actions are taken.
Management must recognize that change is inevitable.	This is written into policy documents such as the BSAP, as it recognizes that monitoring should be continuous to cater for changes in the ecosystem.
The ecosystem approach should seek the appropriate balance between conservation and use of biodiversity.	This is not a principle achieved in practice in the Baltic Sea Region. Resources are used outside the limits of conservation in some areas, eg., in agriculturally-dependent economies
The ecosystem approach should consider all forms of relevant information, including scientific, indigenous and local knowledge, innovations and practices.	Whilst the Baltic Sea Region and HELCOM have been hailed as leaders in collecting and sharing information, there is still the need to incorporate local and indigenous knowledge into the decision-making process.
The ecosystem approach should involve all relevant sectors of society and scientific disciplines.	There is much room for improvement here. Whilst HELCOM allows observers in its meetings, they are not given decision-making powers. The practice of engaging the local public in environmental decision-making varies with each contracting party.

5. Conclusions

There is much literature on the ecosystem approach, but the novelty of this work is that it systematically studies the ecosystem approach using the strengths, weaknesses, opportunities and threats (SWOT) framework. This method not only highlights key aspects of the ecosystem approach but enables the development of strategies for further optimization of the implementation of the approach in the Baltic Sea Region and beyond.

This research presents the findings from the Baltic Sea Region, one of the largest water catchment regions in the world and a part of the European Union. It has shown that the ecosystem approach to management is a means by which to capture the intricacies of the relationship between humans and the Sea (the environment). Findings reveal that one of the advantages of the ecosystem approach that Baltic Sea experts cherish is its simplicity in communicating the importance of Baltic Sea protection. However, the caveat is that the language of the ecosystem approach can be complex and unclear, hence it can serve as the link between research and practical implementation. This research has recommended that the implementation of the ecosystem approach is most effective when using existing tools and incorporating into prevailing policies. On a wider scale, the ecosystem approach is not only useful in management decision making but also as a means of conveying the complexities of the relationship between humans and the environment. Resource allocation backed by strong political leadership can strengthen interdisciplinary research so that the ecosystem approach can be better integrated with sustainability policies and practices.

This study has utilized the SWOT framework to highlight key aspects of the ecosystem approach that can be further developed for better implementation in the Baltic Sea Region and beyond. This research can be seen as the platform from which other dialogue and further research emerges as outlined below:

- This study utilized a document analysis of key documents to identify themes relevant to the ecosystem approach in the Baltic Sea Region. Further research can employ a survey of key stakeholders in the Baltic Sea Region at all levels in the multi-level governance framework.
- As an extension to the point above, it would be interesting to carry out a survey that considers key demographics such as different groups in society and ages, to see how the perception of the ecosystem approach varies generationally.
- Given the interdisciplinary nature of the ecosystem approach, it would be interesting to convene stakeholders from different disciplines in focus groups and conduct a SWOT analysis in a workshop-style environment. This can feed into the survey results above and help to overcome any potential compiler bias.
- Studies on the implementation of the ecosystem approach at various levels in the Baltic Sea Region and comparisons with other areas of the world would provide valuable empirical data on strategies for success and implementation deficits.
- Another key area of study that would help in taking the ecosystem approach forward would be the development of new indicators that reflect the interdisciplinary focus of the approach. For example, governance indicators and stakeholder engagement indicators are some that would provide useful information.

Author Contributions: This article was conceptualized by S.J. and V.L. and outline drafted by S.J. and V.L. Literature review, methods and results and discussion written by both S.J. and V.L. Coding and data analysis done by both V.L. and S.J. Both authors have read and agreed to the published version of the manuscript.

Funding: This research received academy of Finland SeaHer funding, grant number 315715. The authors also wish to acknowledge support from the 'Sea' profiling area of Åbo Akademi University.

Institutional Review Board Statement: Not applicable.

Informed Consent Statement: Not applicable.

Data Availability Statement: All data is stored centrally at Åbo Akademi University.

Conflicts of Interest: The authors declare no conflict of interest.

References

1. Grumbine, R.E. What is ecosystem management? *Conserv. Biol.* **1994**, *8*, 27–38. [CrossRef]
2. Szaro, R.C.; Sexton, W.T.; Malone, C.R. The emergence of ecosystem management as a tool for meeting people's needs and sustaining ecosystems. *Landscape Urban Plan.* **1998**, *40*, 1–7. [CrossRef]

3. Kirkfeldt, T.A. Why choosing between ecosystem-based management, ecosystem-based approach and ecosystem approach makes a difference. *Mar. Policy* **2019**, *106*, 103541. [CrossRef]
4. Rodriguez, N.J.I. A comparative analysis of holistic marine management regimes and ecosystem approach in marine spatial planning in developed countries. *Ocean Coast. Manage.* **2017**, *137*, 185–197. [CrossRef]
5. Agardy, T.; Davis, J.; Sherwood, K. *Taking Steps toward Marine and Coastal Management—An Introductory Guide*; Series: UNEP Regional Seas Reports and Studies No. 189; UNEP: Nairobi, Kenya, 2011; pp. 1–68.
6. Engler, C. Beyond rhetoric: Navigating the conceptual tangle towards effective implementation of the ecosystem approach to oceans management. *Environ. Rev.* **2015**, *23*, 288–320. [CrossRef]
7. Yaffee, S.L. Three Faces of Ecosystem Management. *Conserv. Biol.* **1999**, *13*, 713–725. [CrossRef]
8. Arkema, K.K.; Abramson, S.C.; Dewsbury, B.M. Marine ecosystem-based management: From characterization to implementation. *Front. Ecol. Environ.* **2006**, *4*, 525–532. [CrossRef]
9. Söderström, S.; Kern, K. The Ecosystem Approach to Management in Marine Environmental Governance: Institutional interplay in the Baltic Sea Region. *Environ. Policy Gov.* **2017**, *27*, 619–631. [CrossRef]
10. Conference of the Parties (Convention on Biological Diversity). Report on the Workshop on the Ecosystem Approach. In Proceedings of the item 13 of the provisional agenda of Fourth meeting, Bratislava, Slovakia, 4–15 May 1998.
11. Conferences of the Parties (Convention on Biological Diversity). Decision V/6 'Ecosystem Approach'. In Proceedings of the Fifth Ordinary Meeting of the Conference of the Parties to the Convention on Biological Diversity, Nairobi, Kenya, 15–26 May 2000.
12. Convention on the Protection of the Marine Environment of the Baltic Sea Area (Helsinki Convention) and OSPAR Convention for the Protection of the Marine Environment of the North-East Atlantic. In Proceedings of the First Joint Ministerial Meeting of the Helsinki and OSPAR Commissions (JMM). Statement towards an Ecosystem Approach to the Management of Human Activities, "Towards an Ecosystem Approach to the Management of Human Activities", Bremen, Germany, 25–26 June 2003; Available online: https://www.ospar.org/site/assets/files/1232/jmm_annex05_ecosystem_approach_statement.pdf (accessed on 2 July 2021).
13. HELCOM Baltic Sea Action Plan. In Proceedings of the HELCOM Ministerial Meeting, Krakow, Poland, 15 November 2007; Available online: http://helcom.fi/baltic-sea-action-plan (accessed on 15 July 2021).
14. Hammer, M.; Balfors, B.; Mörtberg, U.; Petersson, M.; Quin, A. Governance of water resources in the phase of change: A case study of the implementation of the EU Water Framework Directive in Sweden. *Ambio* **2011**, *40*, 210–220. [CrossRef]
15. Directive 2008/56/EC of the European Parliament and of the Council of 17 June 2008 establishing a Framework for Community Action in the Field of Marine Environmental Policy (Marine Strategy Framework Directive). *Off. J. Eur. Union* **2008**, *164*, 1–22.
16. Directive 2014/89/EU of the European Parliament and of the Council of 23 July 2014 establishing a Framework for Maritime Spatial Planning. *Off. J. Eur. Union* **2014**, *257*, 1–11.
17. Doyen, L.; Thébaud, O.; Béné, C.; Martinet, V.; Gourguet, S.; Bertignac, M.; Fifas, S.; Blanchard, F. A stochastic viability approach to ecosystem-based fisheries management. *Ecol. Econ.* **2012**, *75*, 32–42. [CrossRef]
18. Garcia, S.M.; Rice, J.; Charles, A. Balanced harvesting in fisheries: A preliminary analysis of management implications. *ICES J. Mar. Sci.* **2016**, *73*, 1668–1678. [CrossRef]
19. Kolding, J.; Garcia, S.M.; Zhou, S.; Heino, M. Balanced harvest: Utopia, failure, or a functional strategy? *ICES J. Mar. Sci.* **2016**, *73*, 1616–1622. [CrossRef]
20. Cavanagh, R.D.; Hill, S.L.; Knowland, C.A.; Grant, S.M. Stakeholder perspectives on ecosystem-based management of the Antarctic krill fishery. *Mar. Policy* **2016**, *68*, 205–211. [CrossRef]
21. Coll, M.; Cury, P.; Azzurro, E.; Bariche, M.; Bayadas, G.; Bellido, J.M.; Chaboud, C.; Claudet, J.; El-Sayed, A.F.; Gascuel, D.; et al. Workshop Participants. The scientific strategy needed to promote a regional ecosystem-based approach to fisheries in the Mediterranean and Black seas. *Rev. Fish Biol. Fisher.* **2013**, *23*, 415–434. [CrossRef]
22. O'Boyle, R.; Sinclair, M.; Keizer, P.; Lee, K.; Ricard, D.; Yeats, P. Indicators for ecosystem-based management on the Scotian Shelf: Bridging the gap between theory and practice. *ICES J. Mar. Sci.* **2005**, *62*, 598–605. [CrossRef]
23. Rees, H.L.; Boyd, S.E.; Schratzberger, M.; Murray, L.A. Role of benthic indicators in regulating human activities at sea. *Environ. Sci. Policy* **2006**, *9*, 496–508. [CrossRef]
24. Rees, H.L.; Hyland, J.L.; Hylland, K.; Mercer Clarke, C.S.L.; Roff, J.C.; Ware, S. Environmental indicators: Utility in meeting regulatory needs. An overview. *ICES J. Mar. Sci.* **2008**, *65*, 1381–1386. [CrossRef]
25. Bayer, E.; Barnes, R.A.; Rees, H.L. The regulatory framework for marine dredging indicators and their operational efficiency within the UK: A possible model for other nations? *ICES J. Mar. Sci.* **2008**, *65*, 1402–1406. [CrossRef]
26. Curran, K.; Bundy, A.; Craig, M.; Hall, T.; Lawton, P.; Quigley, S. Recommendations for Science, Management, and an Ecosystem Approach in Fisheries and Oceans Canada, Maritimes Region. *DFO Can. Sci. Advis. Sec. Res.* **2012**, *Doc. 2012/061*. Available online: http://www.dfo-mpo.gc.ca/csas-sccs/Publications/ResDocs-DocRech/2012/2012_061-eng.html (accessed on 2 July 2021).
27. Stephenson, R.L.; Hobday, A.J.; Cvitanovic, C.; Alexander, K.A.; Begg, G.A.; Bustamante, R.H.; Dunstan, P.K.; Frusher, S.; Fudge, M.; Fulton, E.A.; et al. A practical framework for implementing and evaluating integrated management of marine activities. *Ocean Coast. Manage.* **2019**, *177*, 127–138. [CrossRef]
28. Leslie, H.; Sievanen, L.; Crawford, T.G.; Gruby, R.; Villanueva-Aznar, H.C.; Campbell, L.M. Learning from Ecosystem-Based Management in Practice. *Coast. Manage.* **2015**, *43*, 471–497. [CrossRef]
29. Gürel, E.; Tat, M. SWOT analysis: A theoretical review. *J. Int. Soc. Res.* **2017**, *10*, 994–1006. [CrossRef]

30. Minsky, L.; Aron, D. Are You doing the SWOT Analysis Backwards? Harvard Business Review. 2021. Available online: https://hbr.org/2021/02/are-you-doing-the-swot-analysis-backwards (accessed on 6 October 2020).
31. Houben, G.; Lenie, K.; Vanhoof, K. A knowledge-based SWOT-analysis system as an instrument for strategic planning in small and medium sized enterprises. *Decis. Support Syst.* **1999**, *26*, 125–135. [CrossRef]
32. CBD. *Secretariat of the Convention of Biological Diversity (CBD)*; The Ecosystem Approach CBD Guidelines; CBD: Montreal, QC, Canada, 2004.
33. Hsieh, H.-F.; Shannon, S.E. Three Approaches to Qualitative Content Analysis. *Qual. Health Res.* **2005**, *15*, 1277–1288. [CrossRef]
34. Williamson, K.; Given, L.M.; Scifleet, P. Qualitative data analysis. In *Research Methods, Information, Systems and Contexts*, 2nd ed.; Williamson, K., Johanson, G., Eds.; Elsevier Ltd.: Amsterdam, The Netherlands, 2018; pp. 453–476.
35. Maier, M.A. Content Analysis: Advantages and Disadvantages. In *The SAGE Encyclopedia of Communication Research Methods*; Allen, M., Ed.; SAGE Publications, Inc.: Newcastle upon Tyne, UK, 2017; pp. 240–242.
36. Bengtsson, M. How to plan and perform a qualitative study using content analysis. *Nurs. Open.* **2016**, *2*, 8–14. [CrossRef]
37. HELCOM Meeting Portal. Available online: https://portal.helcom.fi/Lists/MeetingInformation/GEAR%20meetings.aspx (accessed on 5 July 2021).
38. HELCOM GEAR Group on the Implementation of Ecosystem Approach. Available online: https://helcom.fi/helcom-at-work/groups/gear/ (accessed on 8 January 2021).
39. Minutes of the First Meeting of HELCOM Group for the Implementation of the Ecosystem Approach (HELCOM GEAR), Bonn, Germany. 2012. Available online: https://portal.helcom.fi/Archive/Shared%20Documents/GEAR%201-2012_Minutes%20GEAR1.pdf#search=%22GEAR%201%22 (accessed on 20 July 2021).
40. Outcome of the Sixth Meeting of HELCOM Group for the Implementation of the Ecosystem Approach (HELCOM GEAR), Berlin, Germany. 2014. Available online: https://portal.helcom.fi/Archive/archive2/GEAR%206-2014_Outcome%20of%20GEAR%206-2014.pdf#search=%22GEAR%206%22 (accessed on 25 July 2021).
41. Bayram, B.C.; Ucvncv, T. A Case Study: Assessing the Current Situation of Forest Products Industry in Taşköprü through SWOT Analysis and Analytic Hierarchy Process. *Kast. Univ. J. For. Faculty.* **2016**, *16*, 510–514. [CrossRef]
42. David, F.R.; Creek, S.A.; David, F.R. What is the Key to Effective SWOT Analysis, Including AQCD Factors. *SAM Adv. Manage.* **2019**, *84*, 25.
43. Benzaghta, M.A.; Elwalda, A.; Mousa, M.M.; Erkan, I.; Rahman, M. SWOT analysis applications: An integrative literature review. *J. Glob. Bus. Insights* **2021**, *6*, 54–72. [CrossRef]

 *sustainability*

Review

A Review of Key Sustainability Issues in Malaysian Palm Oil Industry

Lakshmy Naidu and Ravichandran Moorthy *

Research Centre for History, Politics and International Affairs, Faculty of Social Sciences and Humanities, Universiti Kebangsaan Malaysia, Bangi 43600, Selangor, Malaysia; p103087@siswa.ukm.edu.my
* Correspondence: drravi@ukm.edu.my

Abstract: The palm oil industry has contributed enormously to the economic growth of developing countries in the tropics, including Malaysia. Despite the industry being a development tool for emerging economies, the oil palm crop is inundated with allegations of its unsustainable plantation practices and viewed as environmentally detrimental and socially adverse. These negative perceptions are amplified through anti-palm oil campaigns and protectionist trade regulations in developed countries, particularly in the European Union (EU). This situation, if further exacerbated, could potentially affect the export of palm oil and the industry as a whole. As such, this article provides a critical review of the key sustainability issues faced by the Malaysian palm oil industry as the second biggest exporter of palm oil to the global market. The various insights and the interpretations of sustainability are contested according to the contexts and the interests of the countries involved. Hence, palm oil is constantly exposed to bias masked by non-tariff barriers from consumer countries to protect their domestically produced vegetable oils. This could constrain the commodity competitiveness in the international market. As issues on palm oil sustainability continue to evolve, policymakers at key stakeholder agencies need to devise strategies to manage global disruption in the palm oil trade.

Keywords: palm oil; oil palm; sustainability; trade; European Union; stakeholder agencies

Citation: Naidu, L.; Moorthy, R. A Review of Key Sustainability Issues in Malaysian Palm Oil Industry. *Sustainability* **2021**, *13*, 10839. https://doi.org/10.3390/su131910839

Academic Editors: Baojie He, Ayyoob Sharifi, Chi Feng and Jun Yang

Received: 27 August 2021
Accepted: 27 September 2021
Published: 29 September 2021

Publisher's Note: MDPI stays neutral with regard to jurisdictional claims in published maps and institutional affiliations.

1. Introduction

The rapid growth in the world population and the increase in demand for renewable energy, particularly towards mitigating greenhouse gas (GHG) emissions, have resulted in the rise in demand for oilseeds. This situation has contributed towards the expansion of oilseed cultivation, especially oil palm, in the tropical producing countries. As one of the major exporters of palm oil, Malaysia's oil palm plantation area reached 5.87 million hectares, which supplied 34.3% of the total palm oil trade and constituted 18.3% (17.37 million tonnes) of the global oils and fats industry last year [1]. With the anticipated increase in the global population to nine billion by 2050, the demand for oils and fats is expected to reach 35 million tonnes annually [2]. In addition to food, new uses of palm oil, such as for industrial applications and biodiesel, have further strengthened the demand to cultivate more oil palm.

However, as the total land area available is fixed it means that more crops have to be produced within the same area. In order to fully optimize land usage for oil palm crops, palm oil research has been focused on sustainable plantation and development. Despite various efforts made in research advancements and sustainable practices, oil palm cultivation has been exposed to increasing controversy and scrutiny over the years. The oilseed crop has been associated with environmental degradation, such as peatland conversion, tropical rainforest deforestation, loss of biodiversity, reduced carbon sink source, flood regulator, and the carbon footprint [3–7]. The industry has also been plagued with social issues, such as the violation of human rights, the rise in crime, the influx of foreign labor, conflicts over land rights, labor management, and the displacement of local and indigenous communities [8–10]. Moreover, concerns over the safety and the quality of palm oil contaminants, such as 3-monochloropropane-1,2-diol (3-MCPD) and glycidyl esters (GE) in the food sector have continued to be present over the past decade [11,12].

These negative perceptions have created concerns about the sustainability of the commodity, especially in the developed economies such as the European market. The European Union (EU) established certified and sustainable palm oil standards as part of its importation requirements, which started in 2015, for a completely sustainable palm oil supply chain in Europe by 2020 [13]. Furthermore, the European Parliament agreed that the main criteria for biodiesel are set for no deforestation and voted to ban biodiesels from palm oil beginning in 2021 and to completely phase them out by 2030 in adherence to the EU Renewable Energy Directive (RED) II [14,15]. In addition to the restriction of palm oil-based biodiesel, the European Commission (EC) implemented the maximum limits on 3-monochloropropane-1,2-diol esters (3-MCPDE) and GE in edible oils and fats in January 2021 and March 2018, respectively [16]. Apart from these interventions, non-governmental organizations (NGOs) and consumers advocate boycotting palm oil through palm oil-free campaigns, while private companies promote the use of certified and sustainably produced palm oil through labels on their products [17,18]. The imposed restrictions and the increase in public awareness could suppress the utilization of palm oil in the transportation and food sectors in Europe. It may jeopardize the market share of Malaysian palm oil in the European oils and fats sector if palm oil is continuously viewed as unsustainably produced.

Therefore, there is growing pressure on exporting countries such as Malaysia to meet the sustainable production and development standards in the palm oil supply chain in order to trade in the European market. The formation of the Roundtable of Sustainable Palm Oil (RSPO) in 2004 as a reliable global standard for certified and sustainable palm oil (CSPO) enabled the palm oil industry players to demonstrate commitment towards sustainable development [19]. In 2015, the Malaysian Sustainable Palm Oil (MSPO) was developed as a national standard to monitor the development of the sustainable palm oil sector, and compliance to the standard was made compulsory by 31 December 2019 [20]. These certification standards indicate the importance of responsible practices from the environmental, social, and safety perspectives. The principles and criteria in the standards require involvement from multi-stakeholder agencies to discuss sustainability issues from a policy formulation perspective in order to stay competitive internationally.

Hence, the purpose of this article is to provide a critical review of the numerous academic articles that have been written about sustainability issues in the palm oil industry. By elucidating the discourse on palm oil concerning key sustainability issues, this paper aims to bridge the gap between the issues of the misconceptions, the allegations, and the realities of palm oil. The critical review also intends to contribute intelligibility and coherently to the existing body of knowledge in the palm oil sector. This is important as policymakers could use the information at governmental agencies to develop effective sustainable palm oil strategies, and academicians and researchers could use it for further studies in the field.

The article is structured based on environmental, social, economic, and health sustainability themes. The sub-section on environmental sustainability examines the environmental challenges to Malaysia's palm oil industry and the actions taken by the Malaysian authorities to counter them. The social sustainability sub-section highlights how the palm oil industry has contributed to the social wellbeing of the local people and how the negative campaigning by international agencies is harming the industry and the people. In the economic sustainability sub-section, the article shows the palm oil industry's contribution to Malaysia's economy and the actions taken to reduce the carbon footprint and meet the high standards of the global market. In the health sustainability sub-section, issues such as the health problems resulting from the haze, forest fires, and slash-and-burn farming techniques are addressed. It also discusses issues related to nutritional value and the association with diseases. In the Section 4, the authors discuss the international dynamics that hamper the palm oil industry and the issues surrounding Sustainable Development

Goals (SDGs). The Section 5 summarizes the findings and provides recommendations to policymakers.

2. Materials and Methods

This review article investigates the sustainability issues of the palm oil industry in Malaysia. By using a macro-approach to examine the sustainability of the commodity, the article discusses the development of the palm oil industry at the international and national levels. The article consists of critically reviewed sources ranging from academic books, scholarly journals, proceeding papers, and reports from the leading researchers and academicians in oil palm studies. The data reviewed in this article was obtained from the international and national agencies involved in palm oil sustainability in order to present the most current developments. Among the data sources are the Amsterdam Declaration from the EU, the Report on the Status of Production Expansion from the European Commission, and background information from the RSPO, as well as oil palm statistics and development from the Malaysian Palm Oil Board (MPOB) and the Malaysian Palm Oil Council (MPOC). Collectively, the literature reviewed provides the insight to explore the complexity inherent in the palm oil sector with the aim of achieving sustainable development.

As a measure to discuss the latest progress in the industry, this article examines the literature published in the past two decades, between the years 2000 and 2021. The searching process was conducted in the Scopus database using five main keywords: palm oil, sustainability, trade, European Union, and stakeholder agencies. From the literature review, four main themes in palm oil sustainability are identified for review. The themes are based on the environmental, social, economic, and health perspectives. With the exclusion of health, the other themes are in conformance with the three pillars of sustainable development by the United Nations' SDGs.

Throughout the article, priority has been given to the findings and discussions from scholarly literature to ensure the reliability and validity of this review. However, news articles were reviewed to obtain information on the latest governmental policies and actions that have yet to be academically investigated. Careful consideration was taken to ensure the accuracy of these news articles by counterchecking with other related media releases.

3. Results

As a highly contested oilseed crop, palm oil dominates the discussion on sustainability compared to other vegetables in the oils and fats sector. The exchange of information on this commodity has been prevalent among palm oil-producing countries, palm oil-consuming countries, NGOs, traders, manufacturers, growers, processing and milling facilities, associations, consumers, and policymakers in the past decades. The findings of these discourses can potentially affect the supply chain and the trade in palm oil and its products as they may influence the policymaking processes made by stakeholder agencies. The complexity of palm oil poses a challenge to policymakers to increase the yield produced in each hectare of agricultural land to optimize its socio-economic benefits while reducing the detrimental impacts on the environment. Hence, policy formulation that underpins sustainable development of the palm oil industry needs to balance between environmental, social, and economic stances [21].

The following four sub-sections highlight the issues of palm oil sustainability based on the environmental, social, economic, and health perspectives.

3.1. Environmental Sustainability

The categorization of palm oil-based biodiesel as a renewable energy source has strengthened palm oil exports to Europe. Since 2000, the EU countries have increased their total use of palm oil by 63% and by the year 2017, 87% of palm oil imports were used to produce biodiesel [20,22]. As the third-biggest importer of palm oil after India and China, Malaysian palm oil exports to the EU increased by 9.5% from 1.91 million tonnes in 2018 to 2.09 million tonnes in 2019 [23,24]. The imported palm oil is used as a substitute for soybean oil as the main feedstock for biodiesel production [17]. Although the EU uses the term biofuel to describe the blending of renewable resources with fossil fuel, palm oil i exported as a feedstock to produce biodiesel due to its compatibility with diesel.

Amid countries expanding their oil palm plantations to meet growing market de mands, the EU implemented RED-2009/28/EC, which mandates a binding target of a 10% share of biofuel as a transport fuel by 2020 [25]. The inclusion of biofuel in the regiona bloc energy mix aims to reduce GHG emissions, improve energy security and protect dc mestic farmers' revenue. As stated in Article 17, biofuel products must be environmentall friendly and sustainable to decrease GHG emissions by at least 35% compared to fossi fuels [26]. Nevertheless, the biofuel policy as stipulated in RED has generated criticism from member states since its inception [27].

As a measure to increase the sustainability criteria to take into account member-state grievances, the resolution to eliminate palm oil in biofuel starting in the year 2021 wa made a part of the EU RED II renewed policy framework. Palm oil has been classified as a high Indirect Land Use Change (ILUC) risk biofuel, whereby an increase in the demand for palm oil will lead to the indirect conversion of forest into land for biodiesel feedstock te meet the raw materials required for the European market and hence will be phased out by 2030 [17]. Moreover, the risk in ILUC needs to be assuaged as the GHG savings are set a 65% and would impact the GHG balance of biofuels, if unaddressed [28]. Classifying palm oil for biofuel as unsustainable will not be factored into the EU renewable energy target a it is not considered green fuel. Therefore, it is evident that palm oil is specifically targetec due to its high ILUC risk.

The regulation serves the dual purpose of protecting the industry players of domesti cally produced vegetable oil by using regulatory barriers towards palm oil while adhering to the euro-centric notion of sustainable development. The plan " … to take measure to phase out the use of vegetable oil that drive deforestation, including palm oil, as component of biofuel, preferably by 2020", and the plan that " … recommends finding and promoting more sustainable alternatives for biofuel uses, such as European oil producec from domestically cultivated rape and sunflower seeds" reveal that palm oil is specificall mentioned in the resolution [29] (p. 9). The unwarranted attention has translated into discriminatory action that could impede the export of palm oil into the EU market and has tainted the reputation and acceptance of palm oil and its products internationall Moreover, there is contention regarding ILUC as defining the criteria for land conversion as it reflects the European perspective rather than an internationally accepted view on environmental issues [30]. The debate on sustainable palm oil is also perceived as an enforcement of Western environmental values on producing countries that are dominatec by Asian culture [31].

Recognizing that palm oil is a highly productive oilseed crop, replacing palm oil with other vegetable oils would create a need for more land for cultivation. Hence, it serve as an indicator that the potential replacement by other oils would not be suitable. Even as studies indicate that beef and soy imported into the EU created more deforestation a compared to that of the imported palm oil, the regulation emphasized palm oil only and referred to it as a forest-risk commodity [29,32]. The requirement for establishing a uniform and compulsory certification standard for imported palm oil, detailed product labels, such as palm oil-free and certified palm oil, and campaigns to enhance customer awareness and acceptance have distinguished palm oil from other vegetable oils. These actions can be deemed an unwarranted reinforcement of the misconception that palm oil is harmful to the environment. Notwithstanding the resolution of a shared global responsibility and the need for a global solution on palm oil issues, there exists reasonable doubt as to whether palm oil-producing countries have been engaged in the resolution itself [29].

However, studies on the impact of oil palm plantations on global natural resource indicate environmental consequences in the producing countries. These impacts have caused severe and widespread pressure on the environment as more landscapes are cleared to meet the growing global demand for affordable and versatile vegetable oil [3]. Among the adverse impacts on the ecological systems are deforestation and degradation of tropical forests, which cause destruction of biodiversity, soil erosion, water pollution, and deterioration of the water and air quality [33–35]. Moreover, using slash-and-burn techniques for forest and land clearing leads to forest fires and releases carbon dioxide into the atmosphere, which results in global warming and induces climate change [36,37]. These detrimental impacts on natural resources and ecosystems have positioned the palm oil industry badly, especially among EU member states, as palm oil products are alleged as being imported deforestation [38].

Nonetheless, the Malaysian oil palm plantations can be categorized as systematically organized as they have succeeded from the organized rubber plantations, which span more than a century and are mostly owned by the large plantation companies. Hence, several standard practices adopted to reduce environmental degradation were taken throughout the years. Some of the good agricultural practices are the usage of organic fertilizers to recycle nutrients through the placement of cut fronds, empty fruit bunches and chipped old palm trunks; the planting of cover crops; the construction of terraces and silt pits; the correct placement of cut fronds; the mulching with empty fruit bunches; the practice of precision agriculture; and the utilization of Integrated Pest Management (IPM) to reduce the use of pesticides by using livestock to control weeds and using predators to control pests [1,23,39]. Moreover, the compulsory zero-burning practice for all oil palm replanting returns organic material into the soil to increase fertility as well as minimizing GHG emissions and accidental forest fires [40]. Mature oil palm trees also serve as a green canopy and a carbon sink that captures carbon from the atmosphere [41,42].

In addition to extracting crude palm oil, the mill produces large quantities of organic waste in the form of Palm Oil Mill Effluent (POME), which is used as a potential source of renewable energy [43]. Methane captured from POME is used as biogas in mills to produce electricity and thus reduces the usage of fossil fuels and GHG emissions by 30% [44]. The biogas trapping in palm oil mills can reduce climate change impact, maximize available resources, and fulfil sustainability requirements while enhancing the image of palm oil as an environmentally friendly product. Moreover, POME can be used as an organic fertilizer as it restores soil fertility and provides nutrients to oil palm trees while reducing chemical fertilizer usage and cost. Although POME is a highly polluting waste and may have profound environmental implications if untreated, it has been converted to an economically useful and environmentally friendly by-product. By achieving zero-waste standards, the oil palm industry rebuts negative allegations of environmental impacts through cleaner production and greater sustainability.

Contrary to EU perceptions, the Malaysian palm oil industry is highly governed and regulated. There are currently more than 15 laws and regulations that the industry must conform with, such as the Land Conservation Act 1960; the Environmental Quality Act 1974; the Environmental Quality (Clean Air Regulations) 1978; the Pesticides Act 1974 (Pesticides Registration Rules); the Occupational Safety and Health Act (1977); the Protection of Wildlife Act 1972; and the National Park Act 1984 [45]. The industry also has to comply with the Hazard & Critical Control Points (HACCP) and the Environmental Impact Assessment (EIA) requirements in addition to the MPOB's Codes of Practices (CoP) certification scheme as a guide to retain food safety, quality, and sustainability throughout the palm oil supply chain [46].

Apart from domestic regulations, Malaysia is also a signatory to the Convention on Biological Diversity (CBD), the Charter of the Indigenous Tribal Peoples of the Tropical Forests, and the Cartagena Protocol on Biosafety. In 2006 the Malaysian Palm Oil Council (MPOC) launched the Malaysian Palm Oil Wildlife Conservation Fund (MPOWCF) to financially fund wildlife conservation activities and studies by academicians, government officials, and NGOs on the overall impact of the palm oil industry on wildlife, biodiversity and the environment [47]. The commitments towards national and international laws demonstrate Malaysia's responsibility and transparency in ensuring conservation of the environment and protecting wildlife impacted in the palm oil sector.

3.2. Social Sustainability

As the engine of growth for an emerging nation, the palm oil industry has contributed directly towards poverty eradication through employment opportunities. Land development schemes such as FELDA provided land to 112,635 families for oil palm plantations to create livelihoods and generate income [48]. The organized plantations spurred economic development and promoted social advancements, especially in the early years after independence. New job openings were created in the agricultural sector, which comprises the planting and harvesting of oil palm trees. These new downstream sectors include processing facilities and palm oil-related industries, research and innovation for high-quality planting materials, technology advancement, and new end-products. Further to employment opportunities at oil palm plantations, the palm oil industry has enabled job openings and positions in other spin-off auxiliary services, such as transportation, trading/brokering houses, bulking installations, financial services, and other related supporting industries. As the country ventures into the development of by-products in the palm oil sector, the utilization of biomass through new technology is expected to create an additional 66,000 new jobs and had generated MYR 30 billion by 2020 as stipulated in the National Biomass Strategy 2020 [49].

In spite of the benefits in alleviating the socio-economic status of the society, the industry has been inundated with social issues related to oil palm plantations. Social issues, such as the violation of human rights, the rise in crime, the influx of foreign labor, conflicts over land rights, labor management, and the displacement of local and indigenous communities are some of the pertinent concerns in the industry [8–10]. The conflicts over land rights and forest resources negatively impact the surrounding communities, plantation companies, and governments while working conditions unconducive to safety in the oil palm plantations further expose workers to health and safety risks and ergonomic injuries [50,51]. As development in the oil palm sector continues to progress, studies have found that the diminishing human interaction with nature could lead to a sense of disaffection towards the environment [52].

The recent ban on palm oil import by multinational consumer goods manufacturers in the United States (US) due to allegations of forced labor has also garnered the attention of the industry. Companies such as Nestle, General Mills and Unilever's restriction on the purchase of palm oil was made to exclude the commodity from their global supply chain. Major oil palm growers, such as FGV Holding and Sime Darby Plantation, were accused of exploitative practices of workers, with the latter company being recognized as the world's largest producer of CSPO by RSPO [53]. The US Customs and Border Protection claimed to possess evidence of sexual and physical violence, retaining salary, and restrictions on movement among workers in the plantation. However, there are uncertainties as to whether these contentions were systematically and academically researched to ensure the validity and reliability of the findings.

Moreover, palm oil mills are regulated by the authorities through laws that protect workers' welfare, such as the Occupational Safety and Health Act (OSHA) 1994; the Employment Act 1990; the Passport Act 1966; the Factories and Machinery Act (FMA) 1989; and the Children and Young Persons (Employment) Act 1966 [46]. The implementation of these acts through regulations and policies demonstrates social sustainability in the areas of human rights and labor rights as well as corporate governance. Hence, there exist reasonable doubts that the allegations hurled against the industry are substantiated.

3.3. Economic Sustainability

Many countries, especially those rich in natural resources, are dependent on the agricultural sector as an important mainstay of the economy. Agrarian sectors have an important role in food security and employment in many nations and have historical significance. Economic development studies frequently argue that agricultural productivity is usually followed by rapid industrialization and thus has become a source of long-term sustainable economic growth [54–56]. However, other scholars contend that the natural resource-rich countries grow at a slower rate than the resource-poor countries, and the countries tend to exhibit an inferior institutional quality. This phenomenon is known as the 'resource curse, the staple trap, or the paradox of plenty' [57]. The revenues generated by the commodity sector may have a positive impact on manufacturing. However, this dependence on natural resources may translate into low economic diversification and greater deindustrialization. Nonetheless, studies suggest that while agricultural exports contribute to the shrinkage of the manufacturing sector, the agricultural sector with strong links with the domestic economy may increase the prospects of industrialization [57,58]. This is true with the case of Malaysia; the palm oil industry has spurred downstream manufacturing activities.

Strong economic growth and trade liberalization in the international economy over the past few decades have led to an increase in the export of palm oil and its related products. The total export of oil palm and palm oil products in 2020 was recorded at 26.59 million tonnes with a revenue of MYR 73.25 billion compared to MYR 67.55 billion in 2019 [1]. Table 1 provides data on the export volume and value of oil palm and its products for 2019 and 2020. Based on the table, this review supports the idea that the income generated from the agricultural commodity exports relates to the production sector output [57]. As an integral agricultural sector, the industry mitigates the food security risks in many countries [54,55]. Moreover, economic modernization in this sector induced technological improvement that reduced dependence on the unskilled workforce [58].

Table 1. Export of oil palm and palm oil products for 2019 and 2020.

	Volume (Million Tonnes)			Value (MYR Million)		
	2020	2019	Difference (%)	2020	2019	Difference (%)
Palm oil	16.22	17.43	(7.0)	45,656	39,128	16.7
Palm kernel oil	1.14	1.01	13.3	4151	3306	25.6
Palm-based oleochemicals	4.41	5.46	(19.4)	16,415	18,121	(9.4)
Biodiesel	0.33	0.72	(53.7)	1194	1994	(40.1)
Other palm-based products	1.96	1.84	6.8	4542	3951	15.0
Palm kernel cake	2.53	2.58	(2.1)	1295	1045	23.9
Total	26.59	29.04	(8.5)	73,253	67,546	8.4

Source: MPOB (2021) [1].

The export competitiveness of palm oil is attributed to several factors. Oil palm is a highly productive oil crop as it produces up a yield up to ten times higher than other oilseed crops, such as soybean, sunflower and rapeseed, and has a productive life span of over 20 years [23,59]. As such, oil palm plantations record the highest land productivity. With the growth in population and the surge in renewable energy, palm oil can address those demands by increasing the yield from the existing plantation areas. Oil palm growers can maximize the Oil Extraction Rate (OER) through breeding and cloning highly efficient plants. In addition, the innovation and development of scientific research in planting materials can increase palm oil productivity while ensuring sustainable practices.

By virtue of its high productivity, the cost competitiveness of palm oil makes it the most affordable vegetable oil in price-sensitive markets in emerging countries, such as India and China. As a cost-effective solution, palm oil is a suitable replacement for animal fats. The versatility of the crop is evident as it has been used as an ingredient in the food and non-food industries due to its strong stability and ease of conversion [20,60,61]. In addition, the biomass from the palm mesocarp fiber, palm kernel shell, and empty fruit bunches is a source of a renewable energy which is used to generate heat in plantation mills as a cost-saving measure, while palm oil-based biodiesel has been identified as an alternative fuel due to its price competitiveness [62–64]. The utilization of palm oil in various sectors is attributed to technological innovations, quality enhancements, and scientific advancement.

In addition to reducing the carbon footprint, POME can potentially securitize energy supply while generating additional revenue from the sale of surplus energy. Using POME as an alternative energy source not only saves on the costs of mill operation and waste treatment but mill operators can earn up to MYR 3.8 million per year through the generation of electricity [65]. The ability to generate income from POME has changed the perception of the by-product from waste to be treated to a resource that could generate earnings. Unused electricity can be connected to the national grid under the Feed in Tariff (FIT) scheme introduced in the National Renewable Energy Act 2011. This supports the Fifth Fuel Policy target of achieving 5% of national grid-connected electricity generated from renewable sources [66].

The stringent observations of quality and regulatory control and the capacity of domestic processing facilities are instrumental for the industry to meet the high standards of the global market. These factors have enabled Malaysia to consolidate its position as the second leading producer of palm oil in the international trade of oils and fats. The development brought about by palm oil has led the commodity crop to be referred to as the development engine [30] and the driver of development [67].

However, the rise of palm oil from the 1960s onwards has created disruptions in the global market dominated by other vegetable oils, notably the soybean oil and the rapeseed oil industries. Hence, countries, especially member states in the EU, sought to protect their domestic oil industries by introducing tariff barriers and disguising non-tariff barriers such as safety, societal and environmental concerns. Furthermore, negative information was disseminated through the anti-palm oil campaigns as part of the politics to defend locally produced vegetable oil in order to win back its market share. The backlash against palm oil is attributed to the competitiveness of palm oil in comparison with soybean oil and rapeseed oil as feedstock for biodiesel production. These barriers are discriminatory as they inhibit producing countries from exporting palm oil on the global market. The restriction could limit the use of palm oil-based biodiesel in importing countries and hence contradicts free trade practice principles regulated by the World Trade Organization (WTO).

Therefore, this review posits that the protectionism policy has evolved from traditional trade barriers such as import tariffs, quota restrictions, and import bans to contemporary issues such as health, workers' protection, and the environment. These current concerns are multifaceted and are interlinked with international trade. Moreover, preventing the conversion of forest land to cultivate commodity crops would hamper developing countries such as Malaysia from using palm oil as a development tool. Denying the growth of developing economies by enforcing natural-resource conservation might be seen as prejudice because established nations in the West have abused their environment for economic advancement for millennia.

3.4. Health Sustainability

Constant clearing of land for oil palm plantation and expansion using the slash-and-burn technique has led to accidental fires. The fires resulted in the recurring haze that has increased the occurrences of ill health among the people in Southeast Asia [68]. In spite of national laws and the plantation companies' practice of restricting the use of fire to clear forests, transboundary haze continues to be an annual air pollution problem that affects people's livelihood and health. Moreover, oil palm trees which are planted on peatlands are highly susceptible to the fires that lead to the transboundary haze. The transboundary haze issue is intricate, and politically motivated interventions are ineffective as the system is safeguarded by patron-client relationships [69].

Another health-related issue concerning palm oil relates to its perceived nutritional value and association with diseases. One prominent incident regarding the health standard of palm oil occurred in the 1980s in the US [70]. National campaigns were undertaken to convince the public that tropical oils such as palm oil and coconut oil in food products contributed to coronary heart disease due to their higher saturated fat. Interest groups successfully compelled food manufacturers to replace tropical oils with hydrogenated soybean oil. The substitution was successful initially as "almost all of the nearly 2 billion pounds of tropical oils removed from annual use in the US food supply in the late 1980s was replaced, pound for pound, by hydrogenated oil … " [71] (p. 236). However, it backfired as the partially hydrogenated oil in food products produces trans fatty acid, which is harmful to cardiovascular health as it increases total cholesterol. This has led to a surge in palm oil exports to the US as the solid-fat component makes it an ideal substitute to eliminate trans-fatty acid in hydrogenated vegetable oils.

In spite of the assertion that the consumption of a high amount of saturated fats from palm oil is linked with cardiovascular diseases, studies revealed that there is no concrete evidence associating palm oil intake with the risk of heart disease or the negative impact on children's health [72,73]. Furthermore, the link between the dietary intake of palm oil and obesity, in comparison with other vegetable oils, is not supported by adequate evidence [74–76]. There is also no scientific proof to validate the association between palm oil intake and the incidence of cancer [77]. In fact, the nutraceuticals and phytonutrients from palm oil reduce cholesterol and diabetic levels in adults [78,79]. Research on the effects of palm oil tocotrienol as an antioxidant in the prevention and treatment of bone-related illness was conducted [80]. In addition, the anti-inflammatory effect of oil palm phenolics may prevent the formation of neurodegenerative diseases such as Alzheimer's and Parkinson's [81,82]. The findings by these scholars have strengthened the neutral and positive effects of palm oil intake in the context of obesity and non-communicable diseases in humans.

Thus, this review postulates that the nutritional content of palm oil has contributed significantly to improving the health of the global population. Existing literature and studies conducted on the health aspects of palm oil indicate that there are potential benefits that require further scientific investigation. As palm oil is essential in developing good human health, this review suggests the need for more research to gather concrete evidence to support the industry and invalidate adverse claims regarding palm oil safety and quality. The most pertinent question is the relationship between the type and quantity of oils and fats with health, while taking into consideration other nutrients in the dietary intake as well as lifestyle habits.

4. Discussion

Based on the deliberation of sustainability issues in this paper, this review contends that there is a shift in the context of international attacks on the palm oil industry. The alleged health hazards of palm oil consumption, as claimed in the 1980s, have evolved to that of the abuse of worker human rights and the detrimental impact on the environment in current times. The reposition of bombardment towards palm oil indicates the tendency of importers and consumers in developed economies to continuously develop negative perceptions towards commodities that are in direct competition with their domestically produced vegetable oils. Despite the nature of oil palm as a perennial crop with a competitive edge, the continuous and damaging accusations on the oilseed crop can be surmised as unsubstantiated and politically motivated. Hence, this review premises that palm oil is constantly exposed to bias masked by non-tariff barriers on its cultivation and industrial application in the varying contexts of usage and consumer countries. Therefore, further research is required to mitigate the impact of the evolution these discriminatory measures. The findings of the research would be beneficial in protecting the interests of palm oil-producing countries such as Malaysia.

Through analysis of the literature, it is clear that environmental, social, economic, and health aspects form the basis of the sustainable development of the palm oil industry. These aspects are diverse and are intertwined with one another. Supporting the sustainable production of palm oil is a constructive approach compared to limiting palm oil development and shunning oil palm plantations. Due to its global impact, the palm oil sector supports multiple SDGs, such as the SDG 2 on ending hunger, SDG 3 by promoting good health, SDG 7 through the development of renewable energy and SDG 12 on sustainable consumption and production patterns. As such, the increase in accountable and responsible practices in the palm oil industry should be strategically and consistently encouraged. This would strengthen the industry's ability to address multi-dimensional challenges while remaining sustainable in order to enhance its international reputation.

This review also establishes that the discourse on sustainability provides various insights. The interpretation of sustainability is contested according to the views and contexts of the stakeholders. There exist patterns of differing perspectives in the palm oil industry where palm oil-producing countries focus on the economic benefits while palm oil-consuming countries stress social, environmental, and safety concerns. Furthermore, although sustainability standards have been asserted regarding palm oil, other vegetable oils have been subjected to fewer requirements for compliance. This has led to the perception that the regulation on sustainability has been targeted at disqualifying palm oil, especially as the feedstock for biodiesel production. The regulation can be interpreted as a form of trade protectionism with a strong political clout that specifically targets palm oil. As such, issues on palm oil sustainability need to be addressed effectively through comprehensive policies to ensure its competitiveness and acceptance in the international market.

5. Conclusions

It is evident that the oil palm tree is a highly productive and versatile crop that produces competitively priced oil. The broad functionality of the oil seed crop, which is achieved through the chemical modification process, has widened the application of palm oil in diverse industries. As such, the production and trade of palm oil are of great concern due to their socio-economic significance in national development. Hence, the sustainability issues need to be addressed effectively for the industry to remain resilient in the future.

Based on the discussion, concerns about palm oil sustainability, such as commodity driven deforestation, which results in a loss of biodiversity and induces climate change continue to dominate the discourse on palm oil cultivation. The increase in the crime rate, the high employment of foreign labor, and the violation of indigenous community land rights are some of the social issues that further place the entire industry in a bad light. On the economic front, the commodity has faced immense scrutiny from the international community despite its price competitiveness and high productivity rate. Emerging issues concerning palm oil quality and nutritional value have been extensively researched compared to other vegetable oils by scientists around the world. These challenges that confront the palm oil industry are multi-dimensional and involve numerous stakeholders at the national and international levels. The way forward to enhance the image and increase the competitiveness of Malaysian palm oil in the global market is through compliance with sustainability certification. The adherence will assist in positioning Malaysia as the preferred global supplier of CSPO.

This review shows that the palm oil industry constantly faces pressure and challenges in conforming to sustainable production and development. This warrants further research using empirical data as issues on sustainability continue to evolve in the palm oil supply chain. Other emerging factors, such as advancement in the research, innovation, and development of new products and technologies and the requirements of importing countries can influence the criteria and principles of sustainability imposed on palm oil. Although this article provides an overall review of the palm oil sector, the study utilizes sources from the Scopus database only. Nevertheless, this research area on palm oil sustainability can provide valuable preliminary insights to policymakers in developing strategies to manage disruption in the global palm oil trade. These strategies serve to safeguard the country's economic interest in the export market and protect the environment and ensure societal wellbeing.

Author Contributions: Conceptualization, L.N.; methodology, L.N.; resources, L.N.; writing—original draft preparation, L.N.; writing—review and editing, R.M.; supervision, R.M.; project administration, R.M.; funding acquisition, R.M. All authors have read and agreed to the published version of the manuscript.

Funding: The APC was funded by Geran Kursi Endowmen MPOB-UKM (EP-2019-047).

Institutional Review Board Statement: Not applicable.

Informed Consent Statement: Not applicable.

Data Availability Statement: Not applicable.

Conflicts of Interest: The authors declare no conflict of interest.

References

1. Parveez, G.K.A.; Tarmizi, A.H.A.; Sundram, S.; Soh, K.L.; Ong-Abdullah, M.; Palam, K.D.P.; Salleh, K.M.; Ishak, S.M.; Idris, Z. Oil palm economic performance in Malaysia and R&D progress in 2020. *J. Oil Palm Res.* **2021**, *33*, 181–214. [CrossRef]
2. Malaysian Palm Oil Council. Putting on the Brakes—Palm Oil Today. Available online: http://palmoiltoday.net/putting-on-thebrakes/ (accessed on 6 August 2021).
3. Austin, K.G.; Mosnier, A.; Pirker, J.; McCallum, I.; Fritz, S.; Kasitbhatla, P.S. Shifting patterns of oil palm driven deforestation in Indonesia and implications for zero-deforestation commitment. *Land Use Policy* **2017**, *69*, 41–48. [CrossRef]
4. Meijaard, E.; Brooks, T.M.; Carlson, K.M.; Slade, E.M.; Garcia-Ulloa, J.; Gaveau, D.L.A.; Lee, J.S.H.; Santika, T.; Juffe-Bignoli, D.; Struebig, M.J.; et al. The environmental impacts of palm oil in context. *Nat. Plants* **2020**, *6*, 1418–1426. [CrossRef]
5. Kiew, F.; Hirata, R.; Hirano, T.; Xhuan, W.G.; Aries, E.B.; Kemudang, K.; Wenceslaus, J.; San, L.K.; Melling, L. Carbon dioxide balance of an oil palm plantation established on tropical peat. *Agric. For. Meteorol.* **2020**, *295*, 108189. [CrossRef]
6. Lupascu, M.; Varkkey, H.; Tortajada, C. Is flooding considered a threat in the degraded tropical peatlands? *Sci. Total Environ.* **2020**, *723*, 137988. [CrossRef] [PubMed]
7. Addressing the Impact of Large-Scale Oil Palm Plantations on Orangutan Conservation in Borneo a Spatial, Legal and Political Economy Analysis. Available online: https://pubs.iied.org/sites/default/files/pdfs/migrate/12605IIED.pdf (accessed on 1 August 2021).
8. Alam, A.S.A.F.; Er, A.C.; Begum, H. Malaysian oil palm industry: Prospect and problem. *J. Food Agric. Environ.* **2015**, *13*, 143–148.
9. Kaniapan, S.; Hassan, S.; Ya, H.; Patma Nesan, K.; Azeem, M. The utilisation of palm oil and oil palm residues and the related challenges as a sustainable alternative in biofuel, bioenergy, and transportation sector: A review. *Sustainability* **2021**, *13*, 3110. [CrossRef]
10. Kaur, A. Plantation systems, labour regimes and the state in Malaysia, 1900–2012. *J. Agrar. Chang.* **2014**, *14*, 190–213. [CrossRef]
11. Sulin, S.N.; Mokhtar, M.N.; Mohammed, M.A.P.; Baharuddin, A.S. Review on palm oil contaminants related to 3-monochloropropane-1,2-diol (3- MCPD) and glycidyl esters (GE). *Food Res.* **2020**, *4*, 11–18. [CrossRef]
12. Gao, B.; Jin, M.; Zheng, W.; Zhang, Y.; Yu, L.Y. Current progresses on monochloropropane diol esters in 2018-2019 and their future research trends. *J. Agric. Food Chem.* **2020**, *68*, 12984–12992. [CrossRef]
13. European Union. The Amsterdam Declaration in Support of a Fully Sustainable Palm Oil Supply Chain by 2020. Available online: http://www.euandgvc.nl/documents/publications/2015/december/7/declarations-palm-oil (accessed on 1 August 2021).
14. Kannan, P.; Mansor, N.H.; Rahman, N.K.; Tan, S.P.; Mazlan, S.M. A review on the Malaysian sustainable palm oil certification process among independent oil palm smallholders. *J. Oil Palm Res.* **2021**, *33*, 171–180. [CrossRef]
15. European Commission. Report on the Status of Production Expansion of Relevant Food and Feed Crops Worldwide [COM(2019)142 final]. Available online: https://eur-lex.europa.eu/legal-content/en/TXT/?uri=CELEX%3A52019DC0142 (accessed on 1 August 2021).
16. Beekman, J.K.; Popol, S.; Granvogl, M.; MacMahon, S. Occurrence of 3-monochloropropane-1,2-diol (3-MCPD) esters and glycidyl esters in infant formulas from Germany. *Food Addit. Contam. Part A* **2021**, 1–16. [CrossRef] [PubMed]
17. Oosterveer, P. Sustainability of palm oil and its acceptance in the EU. *J. Oil Palm Res.* **2020**, *32*, 365–376. [CrossRef]
18. Murphy, D.J. The future of oil palm as a major global crop: Opportunities and challenges. *J. Oil Palm Res.* **2014**, *26*, 1–24.
19. Roundtable on Sustainable Palm Oil. RSPO Background. Available online: https://rspo.org/about (accessed on 29 July 2021).
20. Abdul Majid, N.; Ramli, Z.; Md Sum, S.; Awang, A.H. Sustainable palm oil certification scheme frameworks and impacts: A systematic literature review. *Sustainability* **2021**, *13*, 3263. [CrossRef]
21. Hansmann, R.; Mieg, H.A.; Frischkencht, P. Principal sustainability components: Empirical analysis of synergies between the three pillars of sustainability. *Int. J. Sustain. Dev. World Ecol.* **2012**, *19*, 451–459. [CrossRef]
22. Bentivoglio, D.; Bucci, G.; Finco, A. Revisiting the palm oil boom in Europe as a source of renewable energy: Evidence from time series analysis. *Qual.-Access Success* **2018**, *19*, 59–66.
23. Parveez, G.K.A.; Hishamuddin, E.; Soh, K.L.; Ong-Abdullah, M.; Salleh, K.M.; Bidin, M.N.I.Z.; Sundram, S.; Hasan, Z.A.; Idris, Z. Oil palm economic performance in Malaysia and R&D progress in 2019. *J. Oil Palm Res.* **2020**, *32*, 159–190. [CrossRef]
24. Malaysian Palm Oil Board. *Malaysian Oil Palm Statistics 2019*, 39th ed.; MPOB: Bangi, Malaysia, 2020; p. 289.

25. European Commission. *Green Paper—Towards a European Strategy for the Security of Energy Supply*; COM_2000_776; EC: Brussels, Belgium, 2000; p. 90.
26. El Qudsi, M.I.; Kusumawardhana, I.; Kyrychenko, V. The garuda strikes back: Indonesian economic diplomacy to tackle European Union protectionism on crude palm oil. *J. Int. Stud. Energy Aff.* **2020**, *1*, 110–135. [CrossRef]
27. Bürgin, A. National binding renewable energy targets for 2020, but not for 2030 anymore: Why the European Commission developed from a supporter to a brakeman. *J. Eur. Public Policy* **2015**, *22*, 690–707. [CrossRef]
28. Brinkman, M.L.J.; Hilst, F.V.D.; Faaji, A.P.C.; Wicke, B. Low-ILUC-risk rapeseed biodiesel: Potential and indirect GHG emission effect in Eastern Romania. *J. Biofuels* **2018**, *1292*, 171–186. [CrossRef]
29. Hinkes, C. Adding (bio)fuel to the fire: Discourses on palm oil sustainability in the context of European policy development. *Environ. Dev. Sustain.* **2019**, *22*, 7661–7682. [CrossRef]
30. Kurniaty, T. Indonesia environmental diplomacy in President Joko Widodo's era (2014-2019) of the issue rejection Indonesia's CPO by European Union. *Sociae Polites Maj. Ilm. Sos. Polit.* **2020**, *21*, 74–95. [CrossRef]
31. Pye, O. Commodifying sustainability: Development, nature and politics in the palm oil industry. *World Dev.* **2019**, *121*, 218–228. [CrossRef]
32. European Commission. The Impact of EU Consumption on Deforestation. Available online: http://ec.europa.eu/environment/forests/pdf/1.%20Report%20analysis%20of%20impact.pdf (accessed on 1 August 2021).
33. Fitzherbert, E.B.; Struebig, M.J.; Morel, A.; Danielsen, F.; Brühl, C.A.; Donald, P.F. How will oil palm expansion affect biodiversity? *Trends Ecol. Evol.* **2008**, *23*, 538–545. [CrossRef] [PubMed]
34. Jiwan, N. The political ecology of the Indonesian palm oil industry: A critical analysis. In *The Palm Oil Controversy in Southeast Asia: A Transnational Perspective*, 1st ed.; Pye, O., Bhattacharya, J., Eds.; ISEAS Publishing: Singapore, 2013; Volume 1, pp. 48–75.
35. Vijay, V.; Pimm, S.L.; Jenkins, C.N.; Smith, S.J. The impacts of oil palm on recent deforestation and biodiversity loss. *PLoS ONE* **2016**, *11*, e0159668. [CrossRef] [PubMed]
36. Miettinen, J.; Shi, C.; Liew, S.C. Land cover distribution in the peatlands of Peninsular Malaysia, Sumatra and Borneo in 2015 with changes since 1990. *Glob. Ecol. Conserv.* **2016**, *6*, 67–78. [CrossRef]
37. Varkkey, H. Patronage politics, plantation fires and transboundary haze. *Environ. Hazards* **2013**, *12*, 200–217. [CrossRef]
38. Pehnelt, G.; Vietze, C. Recalculating GHG emissions saving of palm oil biodiesel. *Environ. Dev. Sustain.* **2013**, *15*, 429–479. [CrossRef]
39. Turner, P.D.; Gillbanks, R.A. *Oil Palm Cultivation and Management*, 2nd ed.; The Incorporated Society of Planters: Kuala Lumpur, Malaysia, 2013; pp. 154–196.
40. Noor, M. Zero burning techniques in oil palm cultivation: An economic perspective. *Oil Palm Ind. Econ. J.* **2003**, *3*, 16–24.
41. Yarak, K.; Witayangkurn, A.; Kritiyutanont, K.; Arunplod, C.; Shibasaki, R. Oil palm tree detection and health classification on high-resolution imagery using deep learning. *Agriculture* **2021**, *11*, 183. [CrossRef]
42. Henson, I.E. The carbon cost of palm oil production in Malaysia. *Planter* **2008**, *84*, 445–464.
43. Harsono, S.S.; Grundmann, P.; Soebronto, S. Anaerobic treatment of palm oil mill effluents: Potential contribution to net energy yield and reduction of greenhouse gas emissions from biodiesel production. *J. Clean. Prod.* **2014**, *64*, 619–627. [CrossRef]
44. Kaewmai, R.; Aran, H.; Musikavong, C. Greenhouse gas emissions of palm oil mills in Thailand. *Int. J. Greenh. Gas Control* **2012**, *11*, 141–151. [CrossRef]
45. Malaysian Palm Oil Council. *The Impossible Task The Incredible Jouney*, 1st ed.; MPOC: Petaling Jaya, Malaysia, 2017; p. 144.
46. Shukri, N.F.M.; Kuntom, A.; Jamaluddin, M.F.; Mohamed, M.N.; Menon, N.R. Code of good milling practice in enhancing sustainable palm oil production. *J. Oil Palm Res.* **2020**, *32*, 688–695. [CrossRef]
47. Malaysian Palm Oil Wildlife Conservation Fund (MPOWCF). Available online: http://mpoc.org.my/malaysian-palm-oil-wildlife-conservation-fund-mpowcf/ (accessed on 4 August 2021).
48. Parid, M.M.; Miyamoto, M.; Nor Aini, Z.; Lim, H.F.; Michinakq, T. Eradicating Extreme Poverty through Land Development Strategy. In Proceedings of the Workshop on REDD+ Research Project in Peninsular Malaysia, Kuala Lumpur, Malaysia, 4 February 2013; pp. 81–91.
49. Umar, M.S.; Jennings, P.; Urmee, T. Strengthening the palm oil biomass: Renewable energy industry in Malaysia. *Renew. Energy* **2013**, *60*, 107–115. [CrossRef]
50. Dhiaulhaq, A.; Gritten, D.; De Bruyn, T.; Yasmi, Y.; Zazali, A.; Silalahi, M. Transforming conflict in plantations through mediation: Lessons and experiences from Sumatera, Indonesia. *For. Policy Econ.* **2014**, *41*, 22–30. [CrossRef]
51. Nawi, N.S.M.; Deros, B.M.; Rahman, M.N.A.; Sukadarin, E.H.; Nordin, N. Malaysian oil palm workers are in pain: Hazards identification and ergonomics related problems. *Malays. J. Public Health Med.* **2016**, *16*, 50–57.
52. Soga, M.; Gaston, K.J. Extinction of experience: The loss of human–nature interactions. *Front. Ecol. Environ.* **2016**, *14*, 94–101. [CrossRef]
53. US Palm Oil Ban Baffles Industry Watchdogs. Available online: https://bepi.mpob.gov.my/news/detail.php?id=31613 (accessed on 6 August 2021).
54. Niftiyev, I. The Role of Agriculture and Agrarian Sectors in the Azerbaijan Economy: Main Trends and Dynamics Since 1991. Available online: https://ssrn.com/abstract=3681896 (accessed on 24 September 2021).
55. Niftiyev, I.; Czech, K. Dutch disease perspective on vegetable exports in the Azerbaijan economy. *J. Appl. Econ. Sci.* **2020**, *70*, 813–827.

56. Timmer, C.P. The agricultural transformation. In *Handbook of Development Economics*, 1st ed.; Chenery, H., Srinivasan, T.N., Eds.; Elsevier: Amsterdam, The Netherlands, 1988; Volume 1, pp. 275–331.

57. Sadik-Zada, E.R.; Loewenstein, W.; Hasanli, Y. Commodity revenues, agricultural sector and the magnitude of deindustrialization: A novel multisector perspective. *Economies* **2019**, *7*, 113. [CrossRef]

58. Sadik-Zada, E.R. Natural resources, technological progress, and economic modernization. *Rev. Dev. Econ.* **2021**, *25*, 381–404. [CrossRef]

59. Khatun, R.; Reza, M.I.H.; Moniruzzaman, M.; Yaakob, Z. Sustainable oil palm industry: The possibilities. *Renew. Sust. Energ. Rev.* **2017**, *76*, 608–619. [CrossRef]

60. Kushairi, A.; Soh, K.L.; Azman, I.; Hishamuddin, E.; Ong-Abdullah, M.; Bidin, M.N.I.Z.; Ghazali, R.; Sundram, S.; Parveez, G.K.A. Oil palm economic performance in Malaysia and R&D progress in 2017. *J. Oil Palm Res.* **2018**, *31*, 163–195. [CrossRef]

61. Aguiar, L.K.; Martinez, D.C.; Caleman, S.M.Q. Consumer awareness of palm oil as an ingredient in food and non-food products. *J. Food Prod. Mark.* **2017**, *25*, 875–895. [CrossRef]

62. Kuss, V.V.; Kuss, A.V.; da Rosa, R.G.; Aranda, D.A.; Cruz, Y.R. Potential of biodiesel production from palm oil at Brazilian amazon. *Renew. Sust. Energ. Rev.* **2015**, *50*, 1013–1020. [CrossRef]

63. Mekhilef, S.; Siga, S.; Saidur, R. A review on palm oil biodiesel as a source of renewable fuel. *Renew. Sust. Energ. Rev.* **2011**, *15*, 1937–1949. [CrossRef]

64. Sumathi, S.; Chai, S.P.; Mohamed, A.R. Utilization of oil palm as a source of renewable energy in Malaysia. *Renew. Sust. Energ. Rev.* **2008**, *12*, 2404–2421. [CrossRef]

65. Chin, M.J.; Poh, P.E.; Tey, B.T.; Chan, E.S.; Chin, K.L. Biogas from palm oil mill effluent (POME): Opportunities and challenges from Malaysia's perspective. *Renew. Sust. Energ. Rev.* **2013**, *26*, 717–726. [CrossRef]

66. Foong, S.Z.Y.; Chong, M.F.; Ng, D.K.S. Strategies to promote biogas generation and utilisation from palm oil mill effluent. *Process Integr. Optim. Sustain.* **2021**, *5*, 175–191. [CrossRef]

67. Rival, A.; Levang, P. *Palms of Controversies: Oil Palm and Development Challenge*, 1st ed.; Center for International Forestry Research: Bogor Barat, Indonesia, 2014; p. 21.

68. Padfield, R.; Drew, S.; Syayuti, K.; Page, S.; Evers, S.; Campos-Arceiz, A.; Kangayatkarasu, N.; Sayok, A.; Hansen, S.; Schouten, G.; et al. Landscape in transition: An analysis of sustainable policy initiatives and emerging corporate commitments in the palm oil industry. *Landsc. Res.* **2016**, *41*, 744–756. [CrossRef]

69. Varkkey, H. *The Haze Problem in Southeast Asia: Palm Oil and Patronage*, 1st ed.; Routledge: London, UK, 2016; p. 123.

70. Yakob, S. Government, business and lobbyists: The politics of palm oil in US-Malaysia relations. *Int. Hist. Rev.* **2018**, *41*, 909–930. [CrossRef]

71. Teicholz, N. *The Big Fat Surprise: Why Butter, Meat and Cheese Belong in a Healthy Diet*, 1st ed.; Simon & Schuster: New York, NY, USA, 2014; p. 236.

72. Ismail, S.R.; Maarof, S.K.; Ali, S.S.; Ali, A. Systematic review of palm oil consumption and the risk of cardiovascular disease. *PLoS ONE* **2018**, *13*, e0193533. [CrossRef] [PubMed]

73. Di Genova, L.; Cerquiglini, L.; Penta, L.; Biscarini, A.; Esposito, S. Pediatric age palm oil consumption. *Int. J. Environ. Res. Public Health* **2018**, *15*, 651. [CrossRef] [PubMed]

74. Sun, G.; Xia, H.; Yang, Y.; Ma, S.; Zhou, H.; Shu, G.; Wang, S.; Yang, X.; Tang, H.; Wang, F.; et al. Effects of palm olein and olive oil on serum lipids in a Chinese population: A randomized, double-blind, cross-over trial. *Asia Pac. J. Clin. Nutr.* **2018**, *27*, 572–580. [CrossRef] [PubMed]

75. Teh, S.S.; Mah, S.H.; Gouk, S.W.; Voon, P.T.; Ong, A.S.H.; Choo, Y.M. Effects of palm olein-olive oil blends on fat deposition in diet-induced obese mice. *J. Food Nutr. Res.* **2018**, *6*, 39–48. [CrossRef]

76. Lee, S.T.; Voon, P.T.; Ng, T.K.W.; Esa, N.; Lee, V.K.M.; Abu Saad, H.; Loh, S.P. Effects of palm based high-oleic blended cooking oil diet on selected biomarkers of inflammation and obesity compared to extra virgin olive oil diet in overweight Malaysian adults. *J. Oil Palm Res.* **2018**, *30*, 289–298. [CrossRef]

77. Gesteiro, E.; Guijarro, L.; Sanchez-Muniz, F.J.; Vidal-Carou, M.D.C.; Troncoso, A.; Venanci, L.; Jimeno, V.; Quilez, J.; Anadon, A.; Gonzalez-Gross, M. Palm oil on the edge. *Nutrients* **2019**, *11*, 2008. [CrossRef]

78. Loganathan, R.; Vethakkan, S.R.; Radhakrishnan, A.K.; Razak, G.A.; Kim-Tiu, T. Red palm olein supplementation on cytokines, endothelial function and lipid profile in centrally overweight individuals: A randomised controlled trial. *Eur. J. Clin. Nutr.* **2019**, *73*, 609–616. [CrossRef]

79. Tan, S.M.Q.; Chiew, Y.; Ahmad, B.; Kadir, K. Tocotrienol-rich vitamin E from palm oil (tocovid) and its effects in diabetes and diabetic nephropathy: A pilot phase ii clinical trial. *Nutrients* **2018**, *10*, 1315. [CrossRef]

80. Radzi, N.F.M.; Ismail, N.A.; Alias, E. Tocotrienols regulate bone loss through suppression on osteoclast differentiation and activity: A systematic review. *Curr. Drug Targets* **2018**, *19*, 1095–1107. [CrossRef] [PubMed]

81. Weinberg, R.P.; Koledova, V.V.; Schneider, K.; Sambandan, T.G.; Grayson, A.; Zeidman, G.; Artamonova, A.; Sambanthamurthi, R.; Fairus, S.; Sinskey, A.J.; et al. Palm fruit bioactives modulate human astrocyte activity in vitro altering the cytokine secretome reducing levels of TNFα, RANTES and IP-10. *Sci. Rep.* **2018**, *8*, 1–19. [CrossRef] [PubMed]

82. Weinberg, R.P.; Koledova, V.V.; Shin, H.; Park, J.H.; Tan, Y.A.; Sinskey, A.J.; Sambanthamurthi, R.; Rha, C. Oil palm phenolics inhibit the in vitro aggregation of β-amyloid peptide into oligomeric complexes. *Int. J. Alzheimer's Dis.* **2018**, *2018*, 1–12. [CrossRef] [PubMed]

Review

What Nudge Techniques Work for Food Waste Behaviour Change at the Consumer Level? A Systematic Review

Hannah Barker [1,*], Peter J. Shaw [1], Beth Richards [2], Zoe Clegg [2] and Dianna Smith [1]

1   Geography and Environmental Science, University of Southampton, Southampton SO17 1BJ, UK; p.j.shaw@soton.ac.uk (P.J.S.); d.m.smith@soton.ac.uk (D.S.)
2   Hampshire County Council, Winchester SO23 8UJ, UK; beth.richards@hants.gov.uk (B.R.); zoe.clegg@hants.gov.uk (Z.C.)
*   Correspondence: h.r.barnes@soton.ac.uk

Abstract: In European countries over 40% of food loss and waste occurs at the retail and consumer stages; this situation cannot be sustained and remediation is urgently needed; opportunities for change must be created. "Nudge" techniques have been shown to be effective in changing behaviour in areas related to food consumption (e.g., healthy diet), but the effectiveness of interventions using nudge techniques to change food waste behaviours remains unclear, despite a growing body of research. The aim of this review is to elucidate means to change household food waste behaviour using nudge approaches and identify priority needs for further research. Four databases, grey literature and reference lists were searched systematically to identify relevant research on nudges to change food waste behaviours. This search identified sixteen peer-reviewed research articles and two grey literature reports that were critically appraised using a critical appraisal checklist framework for descriptive/case series. Four studies deemed reliable show interventions using nudges of social norms, reminders or social norms with disclosure were effective in changing food waste behaviours at the household level, while disclosure alone, i.e., revealing environmental costs of food waste, was not. This review, unique in the application of a critical appraisal, suggests there is reliable information on the effectiveness of nudge for food waste recycling interventions when incorporating nudges of social norms, reminders or disclosure alongside use of social norms. If food waste recycling behaviour is considered an upstream measure to raise consumers' consciousness on the amount of food waste they produce, this may have a positive impact on food waste reduction and therefore has important policy implications for food waste behaviour change at the household level.

Keywords: food waste; behaviour change; consumer; household; nudge

Citation: Barker, H.; Shaw, P.J.; Richards, B.; Clegg, Z.; Smith, D. What Nudge Techniques Work for Food Waste Behaviour Change at the Consumer Level? A Systematic Review. *Sustainability* 2021, 13, 11099. https://doi.org/10.3390/su131911099

Academic Editors: Jun Yang, Ayyoob Sharifi, Baojie He and Zhi Feng

Received: 4 September 2021
Accepted: 30 September 2021
Published: 8 October 2021

Publisher's Note: MDPI stays neutral with regard to jurisdictional claims in published maps and institutional affiliations.

1. Introduction

Sustainability aims to protect the natural environment specifically human and environmental health, while compelling innovation so as not to compromise lifestyle for future generations [1]. Approaches to sustainability intend to maintain the delicate ecosystems of earth in balance, usually through encouraging renewable fuel sources, protecting physical environments and decreasing carbon emissions. At present, poor sustainability is a key concern affecting the global food system [2]. This situation is a significant problem globally for societies and governments. Food processing and production create environmental problems along the entire food supply chain [3,4] with direct effects on environmental resources through use of fertile soils, fresh water, energy, fertilisers and release of carbon emissions in the production and transport of food [5–9]. Globally, approximately a third of total food produced is wasted or lost [7] along the food supply chain [10]. In European countries 21–33% of food is lost during agricultural production, 21–25% during manufacturing, storage, processing and distribution, and over 40% at retail and consumer stage [11–13]. In the UK, more than £19 billion worth of food is lost or wasted annually [14]. A focus on tackling this complex problem at consumer level is an essential part of the multifarious puzzle. The targets set under the United Nations' Sustainable Development Goals call for halving per capita food waste at the consumer level by 2030 [15].

In order to shift from the current situation of high household food waste to a more sustainable future for food waste, behaviour change will be necessary. General information can affect the motivation for and ability to change behaviour [16,17], for example information awareness campaigns [18,19] as frequently individuals lack awareness of environmental sustainability issues relating to food [20]. However, to create change at appropriate scale and speed, additional approaches to awareness campaigns would be required [21]. Research shows that although information is valuable, when offered alone this is not where the key to motivating change lies [19]. Opportunities for change must be created. Constructing opportunities to change household food waste behaviour can be simple, e.g., making preferred choices more accessible. Examples include positioning food waste caddies in households in easy reach to support recycling of food waste, offering household food deliveries containing optimal food amounts to avoid surplus, or encouraging food portioning tools that help to avoid over-portioning and thus reduce plate waste.

Psychologists and neuroscientists have developed a description of brain function based on two systems, system 1—processes that are automatic, unconscious and fast and system 2—reflective, controlled, slow and effortful [22–24]. This dual process is a theoretical basis for nudge theory, with nudge proposing that system 1, automatic decisions, can be systematically triggered to change behaviours and improve outcomes going with the flow of human nature [22,25].

Nudging [26] behaviours in this way, has considerable merit in this context. Nudging was developed from ideas advanced by Daniel Kahneman [27]. It is challenging to offer a universal definition of the term nudge as understandings of nudge can vary broadly [22]. Economist Thaler and legal scholar Sunstein convey the concept of "nudging," defined as "any aspect of the choice architecture that alters people's behaviour in a predictable way without forbidding any options or significantly changing their economic incentives" [28]. There is a growing interest in nudging as despite usually incurring low cost, they can deliver results and be highly effective all the while negating unpopular rule setting [29]. Subsequently, Bornemann and Smeddinck [30] identify five criticisms of nudge: conceptual, normative, functional, empiric and practical [31]. Conceptual criticisms question the reach of nudge and boundary between nudges and other behavioural influences, normative criticisms express concern over potential manipulation of moral concepts relevant to freedom, independence and objective information [31]. Functionality concerns the effectiveness of the nudge approach, while empiric broaches the efficiency of the method and long-term success [31]. Moreover, practical issues concerning knowledge on the decision context reflect resource demand and cost of implementation of nudge approaches [32]. These five concerns mainly centre around hard-to-avoid hidden automatic defaults, which is inconsistent with the definition offered by Thaler and Sunstein [28], thus most normative, functional and empiric criticisms may be overcome with judicious planning and implementation [31]. If resources are available then practical issues may be overcome; this is the same as with countless other interventions. Conceptual concerns are reflective of the wide-reaching applications and understandings possible to the nudge approach and will always provoke discussion due to individual perspectives of nudge [31].

In the food domain, nudging has been applied largely in response to the obesity epidemic [33] and, to a lesser extent to encourage environmental initiatives affected by food consumption, e.g., reducing red meat consumption [34]. For information on nudge interventions more broadly related to the food supply chain a systematic review shows that there is evidence to support "green nudging" as effective in leveraging more sustainable practices for farmers and consumers [35]. There remains minimal application of nudges in the context of household food waste behaviours, and those published tend to focus on eating-out options [36–38]. However, research on food waste behaviours has expanded in recent years, leading to a requirement for a systematic review to appraise critically the body of research evidence and to understand what works to change household food waste behaviours and what are the priority needs for further research. A recent review developed a systematic map of existing research on behaviourally informed interventions targeting changes in consumer food waste and consumption behaviour [39]. However, a feature missing was 'a critical appraisal of each individual study ... (as) ... this is not a common standard for systematic maps' [39]. Previous reviews on food waste in households have not included a quality assessment of the studies included and have focused on policy actions, interventions for food waste reduction, food waste drivers, causal mechanisms for food waste behaviour, comparison of food waste amounts or avenues for future research. This review addresses the gap in critical assessment.

Frequently applied theoretical bases for behaviour change interventions include the transtheoretical model, social cognitive theory and the theory of planned behaviour [40]. This present review acknowledges that often these theoretical frameworks can work in parallel with nudges [41] and these theories explain why various nudges may be effective. Self-monitoring and other self-regulatory techniques (goal-setting, prompting, self-monitoring, providing feedback on performance, goal review) are consistently reported as effective behaviour change tools [42,43]. Some of these elements also constitute nudges per se. From a policy perspective, nudge interventions have advantages. Firstly, relative affordability, and secondly, ease of implementation and scope for adaptation in different contexts [28,29,44]. As shown [35,44–46] nudges can be considered not as a replacement to firmer environmental and food policies, but rather as a complement.

Ten optimal nudges have been identified with examples to define the scope of nudge in this review as shown in Table 1. They are in line with definitions described by originators of the nudge concept [29], these definitions are also used specifically in the context of food waste in a peer-reviewed primary data study [31].

Table 1. Identification of Nudges.

A. Default rules, e.g., automatic enrolment in programs such as external meal planning and fee-based strategically portioned food ingredient delivery
B. Simplification, e.g., reducing barriers of target behaviour
C. Use of social norms, e.g., Regular exchange about personal experiences on the reduction in food waste with friends and neighbours
D. Increase in ease and convenience, e.g., making low-waste food options visible
E. Disclosure, e.g., revealing environmental costs associated with food waste
F. Warnings, graphic, or otherwise, e.g., Pictures that demonstrate how food waste damages the environment
G. Pre-commitment strategies, e.g., A challenge on household food waste reduction with a friend
H. Reminders, e.g., Tips on shopping planning via email
I. Eliciting implementation intentions, e.g., asking "do you plan to reduce food waste?"
J. Informing people of the nature and consequences of their own past choices, e.g., Feedback on financial costs of an individual's food waste

The aim of this review is to determine what nudge techniques work for food waste behaviour change through comprehensive literature search, review, critical appraisal and discussion of relevant papers.

2. Materials and Methods

To identify the peer-reviewed literature on this topic, four databases were searched: Scopus, IBSS, Web of Science and Psych Info in March 2021. The search terms used were (Nudg* OR "Architect* OR Choice Architect* OR "Behavioural insights") AND ("Food Waste" OR "Food Loss") AND (Consumer* OR domest* OR Household*). Studies were identified on the basis of inclusion (Table 2) and exclusion criteria and then assessed as full text articles. Figure 1 shows the search process as conducted. For the initial screening of titles and abstracts the free website Rayyan, developed by Mourad Ouzzani, Hossam Hammady and Ahmed Elmagarmid, was used to sort and organise the literature; articles were included or excluded on the basis of the title and abstract fitting the inclusion criteria (Table 2) and exclusion criteria, i.e., not inclusion criteria and no review study designs. Next a file was created on Elsevier Mendeley reference manager for full text PDFs identified. The inclusion and exclusion criteria (Table 2) were applied again to these texts on full reading.

Table 2. Inclusion criteria to select relevant papers.

Subject Inclusion Criteria	
Dates	2011 to March 2021. Rationale for 2011 cut off is that 2010 was the year the Nudge Unit was established in the UK government cabinet office [26].
Subject intervention	Any intervention or exploratory study that investigates interventions using nudges to change household food waste behaviours. (Food waste: the definition of food waste is taken as authors own definition and use of term 'food waste'. Rationale for this approach is due to heterogeneity of definitions of food waste in the literature [6])
Setting/sample	Household
Published and peer reviewed	Europe
Language	English
Study design	Qualitative and quantitative studies

Reference lists were searched in all papers identified for full text articles. The grey literature search was carried out in UK-only institutions due to limitations of resource and English language inclusion. In this review eight major UK supermarket websites (Tesco, Sainsbury's, ASDA, ALDI, LIDL, Morrisons, Waitrose and Coop) and the UK's Waste and Resources Action Programme (WRAP) webpages were searched.

When all papers were collated and those meeting the criteria selected, a critical appraisal of the studies was completed. Previous systematic reviews on food waste have not included a critical appraisal [39,41,47–49].

The quality appraisal is therefore a novel contribution and focused on aspects likely to affect the validity of the results including design, the methods of observation, adequate reporting, statistical analysis, sample sizes and allocation. A framework based on the Joanna Briggs Institute Critical Appraisal checklist for descriptive/case series [50] was applied. No meta-analysis was carried out as the identified studies were heterogeneous in type of design and results. A process of assessment was carried out to determine quality studies, following formalised rules detailed in Table 3:

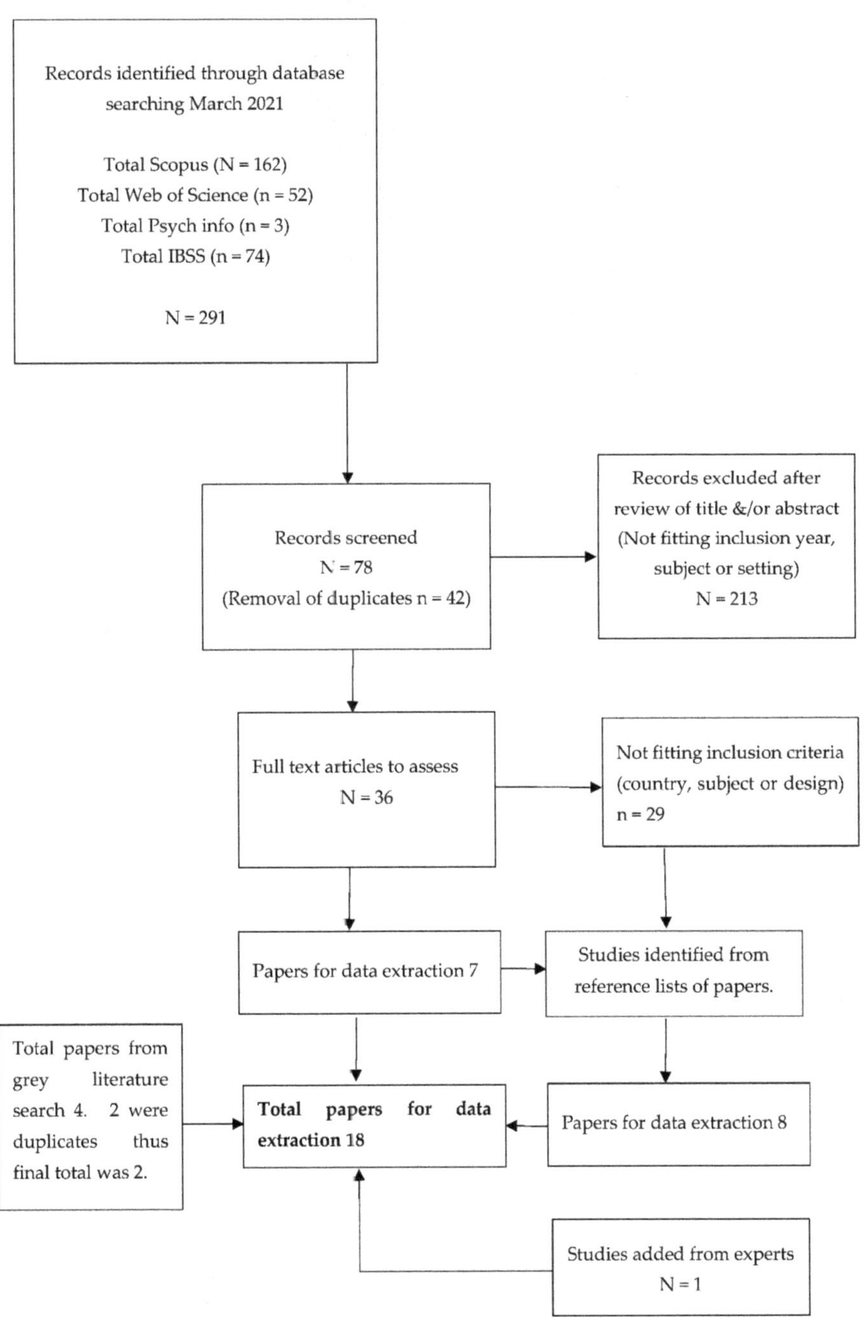

Figure 1. Search strategy and results to select relevant papers.

Table 3. Inclusion and exclusion criteria for critical appraisal of relevant papers.

Subject	Inclusion and Exclusion Criteria
Population	Include studies recruiting from a specific geographical area, social media or supermarket customer base. Exclude studies recruiting using personal contacts.
Population	Include studies that are representative demographically of the population.Include studies that represent demographics of a residential area (i.e., local authority) of a town/city even if not representative of the whole population.
Intervention	Include studies with a detailed description of methods
Comparison	Include studies with a control group
Outcome	Exclude all self-reported measures, i.e., self-reported surveys or qualitative interviews/focus groups
Outcome	Include studies with a clear description of statistical analysis and measure of precision, i.e., confidence interval, standard deviation or p value.

3. Results

Following the search strategy described above, a total of 291 potential articles were identified (Figure 1). The initial screening identified 78 research papers, of which 42 were duplicates (duplicated two, three or four times over the four databases), leaving 36 for full paper search. On reading the 36 papers fully the database search produced papers for consideration. Reference list searching of the 36 papers for full paper search identified another 8 papers for data extraction. One extra paper was identified by an expert. Grey literature studies on food waste were discovered for three UK supermarkets: Tesco, Sainsburys and ASDA. The ASDA study was already captured in two journal publications. Thus, the grey literature search brought 2 extra studies for data extraction. The total number of papers identified for data extraction, from the database search, reference list search, expert advice and grey literature search was 18. Supplementary material Table S1 shows the critical appraisal applied to the eighteen selected studies. Table 4 shows the key results of papers deemed higher quality from the review and illustrates good practice to be replicated. Supplementary material Tables S2–S4 show the summarised results of lower quality papers in the review. Table S1 shows changes to food waste behaviours or perceptions of food waste behaviours in relation to food waste behaviour interventions using nudge techniques. Six studies were published between 2011 and 2016, and twelve studies were published from 2017 to 2020, illustrating the increase in research on food waste in recent years [49]. Ten of the studies were UK based, three were from Sweden, two from Germany, two from The Netherlands and one from Denmark.

3.1. Nudge Interventions

Nudge interventions and associated research were wide–ranging. Three studies did not run an intervention and instead discussed consumer perception(s) of food waste behaviour interventions [31,51,52]. Eight studies used more than one key intervention [53–60]. Most frequently interventions used written information. Five studies used written information interventions incorporating a variety of nudges ranging from disclosure, e.g., environmental impact of food waste from an average household, to individualized consequences to the environment or financial impacts, to reminders, to descriptions of food waste behaviours of other people in the same community, i.e., social norms [53,54,57,60,61]. Four studies with supermarket awareness campaigns all incorporated social norm and reminder nudges [55,56,58,59]. One campaign also included pre-commitment strategies e.g., making pledges on food waste behaviour [59] and another included a number of tools to increase the convenience of behaviour change e.g. food bag clips [56]. Another intervention included, in addition to nudge, economic marketing techniques, i.e., incentives and positive communication [55]. Three studies used a food waste sorting bin, arguably a visual prompt or reminder nudge for food waste behaviour [53,62,63]. Other interventions included: sticker prompts for a food waste bin, i.e., a visual reminder nudge [60]; social recipes whereby participants shared ingredients to make recipes together to reduce waste incorporating social norm nudges [57]; verbal information with environmental disclosure [54]; written social comparison feedback on food waste behaviours of nearby streets, i.e., social norms nudging [64]; written reminders and recommendations to change food waste behaviours, i.e., including nudges of pre-commitment strategies, reminders and setting implementation intentions [65]; a measuring tool for portioning of rice and pasta, i.e., increase in ease of convenience [66]; and environmental impact feedback on food waste habits, i.e., a nudge informing people of their individual consequences [57].

Table 4. Results following application of inclusion and exclusion criteria for critical appraisal of relevant papers.

Study Population	Intervention and Nudge Approach	Comparison	Outcome Measure and Methods	Results	Results Overall
Shaw et al. 2018 [60] UK N = 60 Purposive Sample; attempt at representative sample; allocation: geographical area	Intervention included households receiving a leaflet using nudge E: disclosure, either emphasizing financial impacts or environmental impacts of avoidable food waste in order to encourage avoidable food waste reduction.	Control and 2 Treatment Groups	Grams/household/week Pre- and Post-Intervention Only study in this table that differentiates between avoidable and unavoidable food waste and that breaks down food waste by food type.	No statistically significant difference in the weekly total weight of avoidable food waste before and after the intervention. Statistically significant? No	No change
Linder et al. 2018 [61] Sweden N = 474 Convenience sample; clear detail on representative sample compared to population; allocation: geographical area	Intervention: Information leaflet and recycling station. Control group received no information leaflet and recycling station. Information leaflet used C: social norms, encouraging participants to 'Join your neighbours'; attitudes of residents described as considering FW recycling as very important. E: disclosure: vivid and tangible info on benefits of recycling FW to biofuel. Recycling station includes nudge D: increase in ease and convenience	Control and Treatment Group	Kilograms of food waste/per sorting station/2 weeks Pre- and Post-Intervention Reported how missing data was managed.	Food waste Pre-intervention Control: 37.67 (29.76) Treatment: 57.31 (55.67) Difference 18.64 Food waste Post-intervention Control: 27.81 (13.67) Treatment: 59.77 (25.04) Difference 31.96 Statistically significant? Yes	Positive
Nomura et al. 2011 [64] UK N = 9082 RCT; Representative and random; allocation: geographical area	Households in the treatment group were sent two postcards that provided feedback on how their street performed on food waste recycling compared with the average for their neighbourhood (nudge was C: use of social norms)	Control and Treatment Group	Effect size (Regression) Pre- and Post-Intervention	Positive effect 2.8% Statistically significant? Yes	Positive

Table 4. *Cont.*

Shearer et al. 2017 [67] UK N = 64,284 RCT; Representative and random; allocation geographical area	Intervention included stickers, affixed to the lids of refuse bins, to encourage the separate collection of household food waste for recycling. Nudge was E: reminder, i.e., a visual-prompt as a reminder to engage in a behaviour.	Control and Treatment Group	Mean tonnage of food waste/collection round/week Pre- and Post-Intervention Reported how missing data was managed.	Control: No change. Baseline: 1.24 (SD 0.36) and Experimental: 1.24 (SD 0.36). The difference was −0.0091%. Treatment: mean weight of food waste collected increased by 20% from 1.23 (SD = 0.35) to 1.49 (SD = 0.37) tonnes. Statistically significant? Yes	Positive

3.2. Study Design and Samples

Of the eighteen studies selected (Table S1), fifteen were quantitative and three were qualitative. The most common sampling strategy used was convenience sampling (thirteen studies); however, purposive sampling [63], ad hoc sampling [31], random sampling [64,67] and unclear or unstated methods were also used [53,56]. Sixteen studies included participants who were self-selected. Total sample sizes of all eighteen studies ranged from 15 to 64,284, 9 studies used individuals [31,51,52,57–59,62,65,66] and 8 studies used households as the sampling unit [53–55,60,61,63,64,67] and 1 study was unclear [56]. In one study the sample sizes were unclear [56]. Five studies had sample sizes of under 100 [55,57,60,62,63] Five studies had sample sizes between 100 and 500 [31,52,53,61,65]. Seven studies had total sample sizes of over 500 [51,54,58,59,64,66,67].

Regarding the allocation of the interventions, nine studies used geographical area [53,54,56,60,61,63–66]. Once the geographical area was selected, two studies used random sampling methods [64,67]. Four studies used convenience sampling [54,61,62,65] Two studies used purposive sampling [60,63] and two studies were unclear on sampling methods [53,56]. Five studies recruited from a supermarket customer base [51,55,58,59,66] One study [66] included random selection of customer base—all others used convenience sampling. Two studies allocated their sampling strategy from specific locations, i.e., fair or shop [31,66] both studies used convenience sampling and ad hoc sampling, respectively. Two studies recruited from social media [52,57], both used convenience sampling. Two studies recruited from personal contacts [57,62], both used convenience sampling.

The studies largely did not have samples representative of the country's population in which they took place. Eight studies included no statement on how representative the sample was [52,55–59,62,63]. Six studies included a detailed description on demographics indicating how representative the sample was in relation to the local town or area [31,53,54,60,61,65]. Three were representative [51,64,67]. Three studies had randomised samples [64,66,67]. Two studies were random and representative [64,67].

3.3. Methods of Assessment

The majority of studies were clear and transparent concerning their methods of assessment and inherent limitations. In Bernstad [53], which measured food waste weight at multiple time points with comparisons taken over 10 weeks before and 33 weeks after campaigns A and B, the method description of how many households per intervention were included in the food waste weight measurement (written information or bin equipment) was ambiguous. Three studies had two methods of measurement clearly indicated [54,55,57]. Eleven studies used questionnaire methods [31,51,52,54–59,65,66] Eight studies used weights of food waste as measurement to varying degrees of accuracy [53–57,60,61,67]. Of these studies only one or two of five recommended methods for optimal physical measurement of food waste were used [68]. One study used observations of food caddy placement [64] and three studies used qualitative responses [52,62,63]. Nine studies relied solely on self-reported data [31,51,52,58,59,62,63,65,66] and nine used an objective measure [53–57,60,61,64,67].

The majority of studies did not specify whether the food waste measured was inedible, or edible-avoidable-food waste; a crucial oversight given the importance of these definitions in practically informing solutions to the issue of food waste. Seven studies did specify the food waste measured was either edible food waste or avoidable food waste [55,57–60,65,66]. Five studies referred to the type of food that was wasted [57–60,66]. Hubbard and Tesco [55] used a photo diary to measure food waste however these findings were not reported in the study summary accessed through grey literature searches.

In all fifteen quantitative studies drop-out rates from interventions were not indicated. It is unclear how easy interventions were for participants to complete. Four studies did report on missing data, i.e., missed bin collections [61,67] and missed responses on online surveys [58,59].

3.4. Reliability and Precision

For the majority of studies, whether or not the assessment was reliable was unclear. It is established from van Herpen and van der Lans [69] and van Herpen et al. [70] that self-reported measures of food waste behaviour change via questionnaire are not reliably accurate as a measure of food waste unless used purely for comparative methods to assess differences between households and ideally within a specified recent timeframe, i.e., the last week. Hence, all studies that used this method of measurement via questionnaire or interview or focus group were classified as "unclear" regarding reliability. The results of the self-reported studies that used quantitative methods are summarised (Table S2–S4). These studies all had relatively small sample sizes of approximately 0–500 and did tend to report positively with regard to the effect of intervention on food waste reduction, however it may be that the positive results from these smaller studies were subject to publication bias.

There were nine studies that used objective measures [53–57,60,61,64,67]. Sainsburys et al. was not considered reliable as sample sizes, methods of recruitment and analysis were consistently unclear [56]. In Bernstad [53] the methods description of how many households per intervention were included in the food waste weight measurement (written information or bin equipment) was ambiguous, hence reliability was classified as unclear, although this study had merit in weighing food waste objectively at multiple time points before and after the intervention. The Hubbard and Tesco study [55] was classified unreliable and imprecise as the report did not clearly describe the statistical analysis for the results and there were no confidence intervals, standard deviations or p values indicated. There was a summary comparison of the average waste from the first week compared to the final week and no other data available. The Lim et al. study [57] was marked as unreliable because the sample size was only fifteen and unrepresentative (all university students between the ages of 20–28). Furthermore, the study did not take travel into account for logistics involved with the social recipe intervention.

Bernstad et al. [54] in Sweden split their sample into two intervention groups. One group consisted of 420 households and included an intervention using nudges of disclosure in written and oral information and nudges of increased ease and convenience, i.e., being given food waste recycling bags. The second group consisted of 210 households and the intervention included written information using disclosure nudges. Food waste weights were recorded at multiple time points over 24 months. P values were included along with clear details of statistical analysis. The results showed overall that there was no change in either group for food waste recycling. This study also included a clear description of the population and discussed how representative the sample was in comparison with the population average of the City of Malmo. However, the study did not include a control group in the design hence it was excluded on the application of quality rules (Table 3). Despite this exclusion, it was noted that there was no change in either group as both received nudge interventions, however without a control group it is difficult to know whether this would have been replicated or different with other households in the same community without any intervention.

Four studies were considered reliable assessments on application of the quality rule applied (Table 3). Their results were summarised in Table 4 [60,61,64,67]. All four studies addressed food waste recycling or food waste reduction, their interventions were simple and well-articulated, sample sizes adequate or large and precision of results calculated. Three of these studies were based in the UK [60,64,67] and one in Sweden [61].

Nomura et al. [64] in a UK study incorporated a randomised control trial design with two groups, treatment and control, of 5009 and 4073, respectively. The intervention used social norms nudges by applying social feedback on local recycling rates. Regression analysis, standard errors and level of significance (p values) were calculated and showed a statistically significant positive effect of the intervention on household food waste recycling. Mode of measurement was one of observation of recycling food bin to indicate participation with food waste recycling. There were, however, no weight measures or compositional measures of the food waste. Whether the food waste was properly separated or what amount of food waste was to be recycled was not therefore specified.

Shearer et al. [67] in their UK study included a randomised control trial design with treatment and control group of 33,716 and 30,568 participants, respectively. The treatment group received a visual prompt nudge reminder as a sticker on their food waste caddy. Weights of food waste were measured for both groups at multiple time points pre and post intervention and standard deviation and p values calculated, with statistically significant changes in food waste recycling observed.

Linder et al. [61] in Sweden sent an information leaflet with nudges of social norms and disclosure for food waste recycling to the treatment group. Both treatment and control groups were subject to food waste recycling stations near their homes, i.e., arguably a nudge of increase in ease and convenience. Treatment and control groups (264 and 210 respectively) had food waste weighed pre- and post-intervention. Standard deviation was indicated and level of statistical significance (p values) calculated. A positive and statistically significant change in food waste recycling was noted in the treatment group.

The study by Shaw et al. [60] in the UK comprised a sample size of 60, including 3 groups (n = 20 for leaflet using the nudge disclosure for environmental impact, n = 20 for leaflet using the nudge disclosure for economic impact and n = 20 for control). Food waste was measured via compositional analysis and weight. It was the only study, of the four studies (Table 4), that differentiated between avoidable and unavoidable food waste and that separated food waste by food type. A standard error was included in the results. The results showed a lack of differences between the three groups which negated the need to fully conduct statistical analyses usually involved with a before-after-control-impact experimental design.

4. Discussion

This systematic review aimed to gather and appraise the evidence around interventions using nudge for food waste behaviour change. The results contribute to this field of research by identifying the most effective nudge interventions for altering food waste behaviour in households in Europe, providing insights for future policy formation and nudge applications.

There were four studies that were determined to be of higher quality that showed reliable results with three nudges used: use of social norms, reminders and disclosure. The use of social norms and reminders were both shown to have positive influence on change in food waste behaviours [61,64,67]. Disclosure was shown to have a positive influence when incorporated in an intervention for food waste recycling [61]. However, disclosure showed no change for an intervention to reduce food waste [60]. Despite these interventions all using objective measures, optimal methods for physical measurement of food waste as outlined by Elimelech et al. (2018) [68] were not used in any of the studies indicating that although the results have rigorous elements there is room to increase rigor in the methods used to obtain a more robust result.

The outcomes of the present study provide some insight into the application of nudges in changing food waste behaviour, particularly in relation to food waste recycling. If food waste recycling is considered an upstream nudge (visual reminder) that increases awareness of food waste for the consumer the outcome could arguably be a reduction in food waste in households. There exist implications to local government and individuals, and for the practical application of the findings.

4.1. Explanation

Other reviews support the use of social norms as being one of the most influential elements affecting sustainable consumer behaviour [41,47]. It is well documented that consumer behaviour in relation to food is affected by a wide range of personal, social and environmental factors, i.e., personal beliefs, attitudes, knowledge and genetics; social interaction with friends, family and community and the environment—shops, schools, work place, facilities, the economy and technology [31]. The theory of planned behaviour explains this phenomenon as it indicates that attitudes, social norms and perceived behavioural control influence intentions which predict behaviour [40]. Despite good intentions the value-action gap is well documented and it is broadly understood that behavioural nudges may help to bridge this gap. Social cognitive theory also explains why the use of social norms in nudging for food waste behaviour change is effective as it suggests that a focus on observing and learning from others has influence on positive and negative reinforcement of behaviour [40]. This also suggests that social norms should be used with care as social norms have the potential to reinforce negative behaviour [71].

Shearer et al. [67] (Table 4) showed food waste behaviour was changed by the use of nudge reminders. This outcome was supported by other studies [53,55–59,62,65,66] that were not considered for the purpose of this review as there were no objective measures of change in food waste. However, the methods were clear and the sample sizes adequate, thus it was useful to understand the perspectives of consumers towards nudges for food waste reduction to explain the findings (Tables S1 and S4). Aschemann-Witzel [51] used a Likert scale of 1–7 (with 1 being least agreeable and 7 most agreeable) for four demographically different sample groups, the combined total N = 826. The fourth most accepted nudge interventions by all four groups out of thirteen nudges was: "I would like to avoid that food goes bad while stored at my home with the help of very easy tricks and tips" [51]. Von Kameke and Fischer [31] used a 1–5 Likert scale (1 = great support; 5 = no support at all) N = 101. Participants were recruited by ad hoc sampling outside one organic and one discount food store in the City of Lüneburg in Germany. In contrast to Aschemann-Witzel [51] one of the nudges that received the least support was: tips on shopping planning via mail/post (median = 4.49, standard deviation = 1.09); though it was better received online (median = 3.49, standard deviation = 1.69) but support was still lacking [31]. It appears perceptions on nudge reminders are divided but certainly for some groups of people it is perceived as effective in changing behaviour.

This pattern can be explained by the transtheoretical model of behaviour change which splits receptivity to behaviour change into stages: precontemplation, contemplation, preparation, action, maintenance and termination [72,73]. This model outlines that an intervention may be successful—or not—depending on the stage in which an individual is at the time. If an individual is at the action stage a reminder may be well received and effective, but, if an individual is at the precontemplation stage, they may not be interested. Equally some individuals may have more pressures from their environment and social background that may influence their response [47,48]. There would also be a difference between individuals which may be explained by the self-determination theory which references motivations and aspects required for lasting change. The theory suggests motivation 'exists in the individual and is driven by interest or enjoyment of the task itself'. The individual must believe the behaviour is enjoyable or compatible with their 'sense of self', values and life goals [74]. This is also compatible with the SHIFT (Social influence, Habit formation, Individual self, Feelings and cognition, and Tangibility) framework of sustainable behaviour change whereby there is focus on the individual self, having powerful influences on consumer behaviour, i.e., positivity of self, self-interest, self-efficacy [71].

Two of the studies in Table 4 used disclosure. Linder et al. [61] used disclosure and social norms and did show change in food waste behaviours while Shaw et al. [60] only used disclosure as a nudge and showed no change. This difference may be due to the type of food behaviour change the intervention aimed to disrupt, the former pertained to food waste recycling, while the latter pertained to food waste reduction behaviours. Alternatively, the dual use of social norms and disclosure may be more compelling than disclosure alone. One reason for this outcome may be that the use of social norms lends a positivity to the intervention that offsets the negativity often associated with disclosure. Disclosure may reveal environmental costs associated with food waste. This may not be effective due to the problem of abstractness, information on climate change can be fear provoking and vague with overwhelmingly large-scale consequences making individual acts feel inconsequential which may lead to green fatigue or demotivate as a result of information overload [71].

In one study individuals' perceptions of nudges of warnings, i.e., pictures that demonstrate the extent of the food waste amounts were collected. Overall the rating offered was 2.91 by 101 participants, the scale ranged from 1 ("great support") to 5 ("no support at all"). This was the only mention of the nudge warning within the review. It is unclear why this has not been used more frequently and whether it is a nudge that could be effective. The WRAP "love food hate waste" campaign commenced this line of engagement in social media campaigns [75]. In other areas, i.e., cigarette smoking, the impact of pictures of tobacco health warnings is shown to have an effect [76]. As food waste connects to a lesser degree immediately with the individual it may be that this approach is less effective due to its relative abstractness. Highlighting minimisation of food waste as a way of boosting nutrition and saving money may be more immediately beneficial to the individual and therefore a useful angle to exploit for mutual benefit of changing food waste behaviours and improving health and food security. Other nudge techniques only used once in the studies reviewed include: pre-commitment strategies, eliciting implementation intentions and simplification. It may be that pre-commitment strategies and eliciting implementation intentions are less used as these nudges require that participants already wish to change their food waste behaviours which may not be the case at all. Simplification by removing barriers is little used as it may be that barriers to food waste behaviour are frequently aspects outside of the consumer's control.

Further explanation for why social norms and reminders can be effective relate to tangibility, that is bringing sustainable behaviour to the personal human level. Often green actions can seem vague, distant from the self or abstract [71,77], only for realisation in the future or not feasible in the face of daily challenges [71]. Changes slowly emerge and uncertainty surrounds problems, solutions and outcomes. Social norms and simple reminders prompting actions at the individual and social level are tangible and key to individuals paying attention and taking part [71]. There is much long-term thinking associated with sustainable behaviours regarding cost to current pleasure to promote a sustainable result in the future. This poses problems as people are often hesitant to sacrifice their own benefit [71]. Yet, carrying out actions with others that help others can offer a positive feeling occasionally described the 'warm glow' effect [78], focusing on these kinds of benefits to the self in the present may increase sustainable behaviours [71]. Framing social norms or reminders as nudges for food waste reduction in this way, e.g., 'reducing food waste will benefit your children's future' may improve their effectiveness.

4.2. Implications for Policy

Nudges of social norms and reminders could be useful policy actions for changing food waste behaviours, particularly because they are inexpensive and adaptable approaches

Such approaches should not replace stricter policy measures for food waste reduction at the household level, but as a complement [35,44]. This discussion will consider usage of these identified nudges for food waste behaviour change in local government contexts.

Overwhelmingly the sub context of policy around food waste is one of inefficient legislation. Filippini et al. [79] state that generally policies underlining production, social equity and governance closely connected with social and economic dimensions of food production and consumption are prioritised and food waste actions are poorly represented. The cost of food is clearly prohibitive as in 2018 the UN FAO data showed approximately 2.2 million people in the UK were severely food insecure [80]. If nudges facilitate better use of purchased food, that could also help to reduce food waste, we may positively impact on food security and diet quality (as often fresh produce is most wasted) in households [81]. Policy is often made without full understanding of the benefits and costs to society [82]. Policy measures of fields connected to food waste such as food (in)security, food safety and low cost of waste disposal may take priority [83]. It is worth acknowledging this and considering how best to frame a policy on food waste so it has more backing and traction. Framing actions for food waste within a food security policy may serve both goals as the two, although distinct, are deeply connected along the food supply chain [11–13].

Governments and stakeholders are keen to find ways to effectively improve healthier food behaviours to encourage improvement in public health [45]. The WHO recently asked for retail environments to encourage consumers in this way [84]. Interventions using nudges have gained increased attention in the international policy debate, particularly in the food context they have been applied to promote healthier food patterns of consumption such as increased fruits and vegetables [44,85–88]. Food waste is a point of intersection between these key issues. As nudges have been shown useful to both these issues, an intervention using nudge to encourage consumers towards plant-based diets could also incorporate nudges towards food waste behaviour change. This is especially key as it has been commonly shown that an increase in fresh fruit and vegetable consumption can lead to an increase in food waste [89]. A suggestion on how this could work would be to nudge consumers to buy fruit and vegetables in forms such as canned or frozen—items that are less often wasted compared with fresh produce but offer nutritional gains [90]. Alternatively, nudging storage of apples in the fridge rather than a fruit bowl would increase their shelf life. Discussion about the approach of linking healthy nutrition and food waste awareness is often neglected in the discussion around food waste. It is an approach that may benefit health and environmental outcomes for local government.

4.3. Limitations and Priorities for Future Research

Overall there is no assessment of study quality and robustness in previous reviews of food waste behaviour interventions. This review adds to the literature by indicating the paucity of quality primary studies using interventions with nudge for food waste behaviour change. This review indicates there is some information on the benefit of nudges (namely use of social norms, reminders or disclosure alongside use of social norms) for food waste recycling interventions, which as an upstream measure may have a positive impact on food waste reduction. However, there is currently limited information on the benefit of nudge for food waste reduction interventions.

There was lack of distinction, in the included studies, between whether or not food waste was edible or inedible, which is key information when considering the effectiveness of edible food waste reduction interventions. It is also key information for food waste recycling interventions as it is helpful to understand whether the increase in food waste recycling is due to edible or inedible food waste as this gives an indication of how to target food waste reduction interventions. Another key limitation included understanding the duration of effectiveness of nudge interventions as studies rarely evaluate long term outcomes [91]; some research articles state nudge may only have short term effects [92]. Thus, methods to attempt to measure the longer-term effect of interventions using nudge should be incorporated in future studies, studies in other areas, health not food consumption, have achieved this [93] and could thus help to inform this methodology.

Regarding limitations to the current review, qualitative studies hold strength to uncover subject matter and anomalies to add to the body of research, however, they are not a reliable method to uncover whether or not nudge interventions are effective for food waste behaviour change in a generalisable sample. Thus, despite robust qualitative method from studies reviewed in this paper, we have not deemed them reliable for the purpose of this review and research question to hand [52,62,63]. Some studies included in the present study carried out mixed methods and not all parts of the study were relevant to this review [31,46,51,56,62,66]. In these instances, the parts of the studies that did adhere to the inclusion criteria were included.

In the future we need more food waste behaviour studies that use nudge interventions and measure changes in food waste before and after the intervention using either physical weight measurements using robust methods [68], or photo diary studies using appropriate methods [94] to identify edible and inedible food waste, and capture more data on food groups wasted. There is also a need for studies to use representative sample and control groups when testing the effectiveness of a nudge intervention to change food waste behaviours as well as precision in statistical analysis. There are different outcome measures and effect sizes in almost every paper included in the review; future research could work to overcome these challenges which a more standardised approach so that a synthesis of results could be undertaken with meta-analysis. Future research could also assess effectiveness of nudges to change food waste behaviours in different demographic to find out whether there are differences in the kinds of approaches that work depending on demographics. Another area for future research would be to explore the effectiveness of interventions incorporating social norms and reminder nudges to change food waste recycling behaviours and public health nutrition given that both are shown to be effective in each area and population.

5. Conclusions

In conclusion there is no assessment of study quality in previous reviews of food waste behaviour interventions, thus this review indicates a lack of quality primary studies using interventions with nudge for food waste behaviour change. This review suggests there is reliable information on the effectiveness of nudge for food waste recycling interventions when incorporating nudges of social norms, reminders or disclosure alongside use of social norms. If food waste recycling behaviour is considered an upstream measure to raise consumer consciousness on the topic of food waste this may have a positive impact on food waste reduction. This review illustrates the limited information on the effectiveness of nudge for food waste reduction interventions. Behaviour change models and frameworks indicate nudges work when they are tangible, relevant and beneficial to the individual and their lifestyle. Nudges are inherently flexible and adaptable which lends them to policy implementation in different contexts. Incorporating policy on food waste within policy for food security and public health nutrition may maximise impact.

Supplementary Materials: The following are available online at https://www.mdpi.com/article 10.3390/su131911099/s1, Table S1: Critical Appraisal; Table S2: Food waste weight/household or individual/timeframe reported for food waste reduction; Table S3: Percentages reported for food waste reduction; Table S4: Likert scales measuring nudge intervention on food waste behaviour change.

Author Contributions: Conceptualization, H.B.; methodology, H.B., D.S. and P.J.S.; writing—original draft preparation, H.B.; writing—review and editing, H.B., D.S., P.J.S., B.R. and Z.C.; supervision D.S., P.J.S., B.R. and Z.C.; project administration, H.B.; funding acquisition, H.B. and D.S. All authors have read and agreed to the published version of the manuscript.

Funding: The research for this article was funded by the Economic and Social Research Council South Coast Doctoral Training Partnership (Grant Number ES/P000673/1).

Institutional Review Board Statement: Not applicable.

Informed Consent Statement: Not applicable.

Data Availability Statement: Supporting data are available in the Supplementary Materials.

Conflicts of Interest: The authors declare no conflict of interest.

References

1. United Nations. *Our Common Future (The Brundtland Report): World Commission on Environment and Development*; United Nations General Assembly: New York, NY, USA, 1987.
2. Swinburn, B.A.; Kraak, V.I.; Allender, S.; Atkins, V.J.; Baker, P.I.; Bogard, J.R.; Brinsden, H.; Calvillo, A.; De Schutter, O.; Devarajan, R.; et al. The Global Syndemic of Obesity, Undernutrition, and Climate Change: The Lancet Commission report. *Lancet* **2019**, *393*, 791–846. [CrossRef]
3. Garcia-Herrero, I.; Hoehn, D.; Margallo, M.; Laso, J.; Bala, A.; Batlle-Bayer, L.; Fullana, P.; Vazquez-Rowe, I.; Gonzalez, M.J.; Durá, M.J.; et al. On the estimation of potential food waste reduction to support sustainable production and consumption policies. *Food Policy* **2018**, *80*, 24–38. [CrossRef]
4. Springmann, M.; Wiebe, K.; Mason-D'Croz, D.; Sulser, T.B.; Rayner, M.; Scarborough, P. Health and nutritional aspects of sustainable diet strategies and their association with environmental impacts: A global modelling analysis with country-level detail. *Lancet Planet. Heal.* **2018**, *2*, e451–e461. [CrossRef]
5. Food and Agriculture Organization of the United Nations (FAO). *Food Wastage Footprint: Impacts on Natural Resources-Summary Report*; FAO: Rome, Italy, 2013.
6. Garske, B.; Heyl, K.; Ekardt, F.; Weber, L.M.; Gradzka, W. Challenges of food waste governance: An assessment of European legislation on food waste and recommendations for improvement by economic instruments. *Land* **2020**, *9*, 231. [CrossRef]
7. Östergren, K.; Gustavsson, J.; Bas-Brouwers, H.; Timmermans, T.; Hansen, O.-J.; Møller, H.; Anderson, G.; O'Connor, C.; Soethoudt, H.; Quested, T.; et al. *Fusions Definitional Framework for Food Waste-Full Report*; The Swedish Institute for Food and Biotechnology: Gothenburg, Sweden, 2014.
8. Scherhaufer, S.; Moates, G.; Hartikainen, H.; Waldron, K.; Obersteiner, G. Environmental impacts of food waste in Europe. *Waste Manag.* **2018**, *77*, 98–113. [CrossRef] [PubMed]
9. Szulecka, J.; Strøm-Andersen, N.; Scordato, L.; Skrivervik, E. Multi-level governance of food waste. Comparing Norway, Denmark and Sweden. In *Book from Waste to Value. Valorisation Pathways for Organic Waste Streams in Circular Bioeconomics*; Klitkou, A., Fevolden, A.M., Capasso, M.E., Eds.; Routledge: Oxon, UK, 2019; pp. 253–271.
10. Parfitt, J.; Barthel, M.; MacNaughton, S. Food waste within food supply chains: Quantification and potential for change to 2050. *Philos. Trans. R. Soc. B Biol. Sci.* **2010**, *365*, 3065–3081. [CrossRef] [PubMed]
11. Flanagan, K.; Robertson, K.; Hanson, C. *Reducing Food Loss and Waste: Setting a Global Action Agenda*; WRI: Washington, DC, USA, 2019.
12. Gustavsson, J.; Cederberg, C.; Sonesson, U. *Global Food Losses and Food Waste*; Swedish Institute for Food and Biotechnology (SIK): Gothenburg, Sweden, 2011.
13. Zeinstra, G.; van der Haar, S.; van Bergen, G. *Drivers, Barriers and Interventions for Food Waste Behaviour Change: A Food System Approach*; Wageningen Food & Biobased Research: Wageningen, The Netherlands, 2020; p. 36.
14. WRAP. Food Surplus and Waste in the UK Key Facts (Updated June 2021). Available online: https://wrap.org.uk/resources/report/food-surplus-and-waste-uk-key-facts (accessed on 24 September 2021).
15. Pérez-Escamilla, R. Food security and the 2015–2030 sustainable development goals: From human to planetary health. *Curr. Dev. Nutr.* **2017**, *1*, e000513. [CrossRef]
16. Parry, A.; James, K.; LeRoux, S. *Strategies to Achieve Economic and Environmental Gains by Reducing Food Waste*; Waste and Resources Action Programme (WRAP): Banbury, UK, 2015.
17. Van Trijp, H. *Encouraging Sustainable Behaviour: Psychology and the Environment*; Taylor & Francis: New York, NY, USA, 2014.
18. Halloran, A.; Clement, J.; Kornum, N.; Bucatariu, C.; Magid, J. Addressing food waste reduction in Denmark. *Food Policy* **2014**, *49*, 294–301. [CrossRef]
19. Langen, N.; Göbel, C.; Waskow, F. The effectiveness of advice and actions in reducing food waste. *Proc. Inst. Civ. Eng. Waste Resour. Manag.* **2015**, *168*, 72–86. [CrossRef]
20. Maciejewski, G. Consumers towards Sustainable Food Consumption. *Mark. Sci. Res. Organ.* **2020**, *36*, 19–30. [CrossRef]
21. Aschemann-Witzel, J.; De Hooge, I.E.; Rohm, H.; Normann, A.; Bossle, M.B.; Grønhøj, A.; Oostindjer, M. Key characteristics and success factors of supply chain initiatives tackling consumer-related food waste—A multiple case study. *J. Clean. Prod.* **2017**, *155*, 33–45. [CrossRef]
22. Vlaev, I.; King, D.; Dolan, P.; Darzi, A. The Theory and Practice of Nudging: Changing Health Behaviors. *Public Adm. Rev.* **2016**, *76*, 550–561. [CrossRef]
23. Evans, J.S.B. Dual-Processing Accounts of Reasoning, Judgement, and Social Cognition. *Annu. Rev. Psychol.* **2008**, *59*, 255–278. [CrossRef]
24. Strack, F.; Roland, D. Reflective and Impulsive Determinants of Social Behavior. *Personal. Soc. Psychol. Rev.* **2004**, *8*, 220–247. [CrossRef]

25. Marteau, T.M.; Hollands, G.J.; Fletcher, P.C. Changing Human Behavior to Prevent Disease: The Importance of Targeting Automatic Processes. *Science* **2012**, *337*, 1492–1495. [CrossRef] [PubMed]

26. Rutter, J. Nudge Unit [Internet]. Institute for Government. 2020. Available online: https://www.instituteforgovernment.org.uk/explainers/nudge-unit (accessed on 24 September 2021).

27. Kahneman, D. *Thinking, Fast and Slow*; Farrar, Straus, and Giroux: New York, NY, USA, 2011.

28. Thaler, R.; Sunstein, C. *Nudge: Improving Decisions about Health, Wealth, and Happiness*; Yale University Press: New Haven, CT, USA, 2008.

29. Sunstein, C.R. Nudging: A Very Short Guide. *J. Consum. Policy* **2014**, *37*, 583–588. [CrossRef]

30. Bornemann, B.; Smeddinck, U. Anst€oßiges Anstoßen? Kritische Beo-bachtungen zur Nudging: Diskussion im deutschen Kontext. *Z. Parlam.* **2016**, *47*, 437–459.

31. Von Kameke, C.; Fischer, D. Preventing household food waste via nudging: An exploration of consumer perceptions. *J. Clean. Prod.* **2018**, *184*, 32–40. [CrossRef]

32. Barton, A.; Grüne-Yanoff, T. From libertarian paternalism to nudging and beyond. *Rev. Philos. Psychol.* **2015**, *6*, 341–359. [CrossRef]

33. Broers, V.J.; De Breucker, C.; Van den Broucke, S.; Luminet, O. A systematic review and meta-analysis of the effectiveness of nudging to increase fruit and vegetable choice. *Eur. J. Public Health* **2017**, *27*, 912–920. [CrossRef] [PubMed]

34. Lehner, M.; Mont, O.; Heiskanen, E. Nudging—A promising tool for sustainable consumption behaviour? *J. Clean. Prod.* **2016**, *134*, 166–177.

35. Ferrari, L.; Cavaliere, A.; De Marchi, E.; Banterle, A. Can nudging improve the environmental impact of food supply chain? A systematic review. *Trends. Food Sci. Technol.* **2019**, *91*, 184–192. [CrossRef]

36. Kallbekken, S.; Sælen, H. Nudging hotel guests to reduce food waste as a win-win environmental measure. *Econ. Lett.* **2013**, *119*, 325–327. [CrossRef]

37. Papargyropoulou, E.; Wright, N.; Lozano, R.; Steinberger, J.; Padfield, R.; Ujang, Z. Conceptual framework for the study of food waste generation and prevention in the hospitality sector. *Waste Manag.* **2016**, *49*, 326–336. [CrossRef] [PubMed]

38. Filimonau, V.; Matute, J.; Kubal-Czerwińska, M.; Krzesiwo, K.; Mika, M. The determinants of consumer engagement in restaurant food waste mitigation in Poland: An exploratory study. *J. Clean. Prod.* **2020**, *247*, 119105. [CrossRef]

39. Reisch, L.A.; Sunstein, C.R.; Andor, M.A.; Doebbe, F.C.; Meier, J.; Haddaway, N.R. Mitigating climate change via food consumption and food waste: A systematic map of behavioral interventions. *J. Clean. Prod.* **2021**, *279*, 123717. [CrossRef]

40. Michie, S.; West, R.; Campbell, R.; Brown, J.; Gainforth, H. *ABC of Behaviour Change Theories*; Silverback Publishing: Sutton, UK, 2014.

41. Reynolds, C.; Goucher, L.; Quested, T.; Bromley, S.; Gillick, S.; Wells, V.K.; Evans, D.; Koh, L.; Kanyama, A.C.; Katzeff, C.; et al. Review: Consumption-stage food waste reduction interventions—What works and how to design better interventions. *Food Policy* **2019**, *83*, 7–27. [CrossRef]

42. Greaves, C.J.; Sheppard, K.E.; Abraham, C.; Hardeman, W.; Roden, M.; Evans, P.H.; Schwarz, P. Systematic review of reviews of intervention components associated with increased effectiveness in dietary and physical activity interventions. *BMC Public Health* **2011**, *11*, 1–12. [CrossRef] [PubMed]

43. Michie, S.; Abraham, C.; Whittington, C.; McAteer, J.; Gupta, S. Effective Techniques in Healthy Eating and Physical Activity Interventions: A Meta-Regression. *Heal. Psychol.* **2009**, *28*, 690–701. [CrossRef]

44. Liu, P.J.; Wisdom, J.; Roberto, C.A.; Liu, L.J.; Ubel, P.A. Using behavioral economics to design more effective food policies to address obesity. *Appl. Econ. Perspect. Policy* **2014**, *36*, 6–24. [CrossRef]

45. Sunstein, C.R.; Reisch, L.A. Automatically green: Behavioral economics and environmental protection. *Harv. Env. Law Rev.* **2014**, *38*, 127–158. [CrossRef]

46. Vecchio, R.; Cavallo, C. Increasing healthy food choices through nudges: A systematic review. *Food Qual. Prefer.* **2019**, *78*, 103714. [CrossRef]

47. Schanes, K.; Dobernig, K.; Gözet, B. Food waste matters-A systematic review of household food waste practices and their policy implications. *J. Clean. Prod.* **2018**, *182*, 978–991. [CrossRef]

48. Boulet, M.; Hoek, A.C.; Raven, R. Towards a multi-level framework of household food waste and consumer behaviour: Untangling spaghetti soup. *Appetite* **2020**, *156*, 104856. [CrossRef]

49. Porpino, G. Household food waste behavior: Avenues for future research. *J. Assoc. Consum. Res.* **2016**, *1*, 41–51. [CrossRef]

50. Munn, Z.; Moola, S.; Riitano, D.; Lisy, K. The development of a critical appraisal tool for use in systematic reviews addressing questions of prevalence. *Int. J. Health Policy Manag.* **2014**, *3*, 123–128. [CrossRef]

51. Aschemann-Witzel, J. Helping You to Waste Less? Consumer Acceptance of Food Marketing Offers Targeted to Food-Related Lifestyle Segments of Consumers. *J. Food Prod. Mark.* **2018**, *24*, 522–538.

52. Wakefield, A.; Axon, S. I'm a bit of a waster: Identifying the enablers of, and barriers to, sustainable food waste practices. *J. Clean. Prod.* **2020**, *275*, 122803. [CrossRef]

53. Bernstad, A. Household food waste separation behavior and the importance of convenience. *Waste Manag.* **2014**, *34*, 1317–1323. [CrossRef] [PubMed]

54. Bernstad, A.; La Cour Jansen, J.; Aspegren, A. Door-stepping as a strategy for improved food waste recycling behaviour-Evaluation of a full-scale experiment. *Resour. Conserv. Recycl.* **2013**, *73*, 94–103. [CrossRef]

5. Hubbub and Tesco. Tesco No Time for Waste Challenge. 2020. Available online: https://issuu.com/hubbubuk/docs/tesco_impact_report_oct_2020__4_ (accessed on 24 September 2021).
6. Sainsburys. Inspiring Food Waste Behaviour Change: Year one Results and Analysis. 2017. Available online: https://www.about.sainsburys.co.uk/~{}/media/Files/S/Sainsburys/documents/making-a-difference/Copy%20of%20WLSM2606.pdf (accessed on 24 September 2021).
7. Lim, V.; Funk, M.; Marcenaro, L.; Regazzoni, C.; Rauterberg, M. Designing for action: An evaluation of Social Recipes in reducing food waste. *Int. J. Hum. Comput. Stud.* **2017**, *100*, 18–32. [CrossRef]
8. Young, W.; Russell, S.V.; Robinson, C.A.; Barkemeyer, R. Can social media be a tool for reducing consumers' food waste? A behaviour change experiment by a UK retailer. *Resour. Conserv. Recycl.* **2017**, *117*, 195–203.
9. Young, C.W.; Russell, S.V.; Robinson, C.A. Sustainable Retailing—Influencing Consumer Behaviour on Food Waste. *Bus. Strat. Env.* **2018**, *15*, 1–15. [CrossRef]
10. Shaw, P.J.; Smith, M.M.; Williams, I.D. On the prevention of avoidable food waste from domestic households. *Recycling* **2018**, *3*, 24. [CrossRef]
11. Linder, N.; Lindahl, T.; Borgström, S. Using Behavioural Insights to Promote Food Waste Recycling in Urban Households-Evidence From a Longitudinal Field Experiment. *Front. Psychol.* **2018**, *9*, 352. [CrossRef]
12. Comber, R.; Thieme, A. Designing beyond habit: Opening space for improved recycling and food waste behaviors through processes of persuasion, social influence and aversive affect. *Pers. Ubiquitous Comput.* **2013**, *17*, 1197–1210. [CrossRef]
13. Metcalfe, A.; Riley, M.; Barr, S.; Tudor, T.; Robinson, G.; Guilbert, S. Food waste bins: Bridging infrastructures and practices. *Sociol. Rev.* **2012**, *60*, 135–155.
14. Nomura, H.; John, P.; Cotterill, S. The Use of Feedback to Enhance Environmental Outcomes: A Randomized Controlled Trial of a Food Waste Scheme. *Local Environ.* **2011**, *16*, 637–653. [CrossRef]
15. Schmidt, K. Explaining and promoting household food waste-prevention by an environmental psychological based intervention study. *Resour. Conserv. Recycl.* **2016**, *111*, 53–66. [CrossRef]
16. Van Dooren, C.; Mensink, F.; Eversteijn, K.; Schrijnen, M. Development and Evaluation of the Eetmaatje Measuring Cup for Rice and Pasta as an Intervention to Reduce Food Waste. *Front. Nutr.* **2020**, *6*, 197. [CrossRef] [PubMed]
17. Shearer, L.; Gatersleben, B.; Morse, S.; Smyth, M.; Hunt, S. A problem unstuck? Evaluating the effectiveness of sticker prompts for encouraging household food waste recycling behaviour. *Waste Manag.* **2017**, *60*, 164–172. [PubMed]
18. Elimelech, E.; Ayalon, O.; Ert, E. What gets measured gets managed: A new method of measuring household food waste. *Waste Manag.* **2018**, *76*, 68–81. [CrossRef] [PubMed]
19. Van Herpen, E.; van der Lans, I. A picture says it all? The validity of photograph coding to assess household food waste. *Food Qual. Prefer.* **2019**, *75*, 71–77.
20. Van Herpen, E.; van der Lans, I.A.; Holthuysen, N.; Nijenhuis-de Vries, M.; Quested, T.E. Comparing wasted apples and oranges: An assessment of methods to measure household food waste. *Waste Manag.* **2019**, *88*, 71–84. [CrossRef] [PubMed]
21. White, K.; Habib, R.; Hardisty, D.J. How to SHIFT consumer behaviors to be more sustainable: A literature review and guiding framework. *J. Mark.* **2019**, *83*, 22–49. [CrossRef]
22. Prochaska, J.O.; DiClemente, C.C. The Transtheoretical Approach. In *Handbook of Psychotherapy Integration Oxford Series in Clinical Psychology*, 2nd ed.; Norcross, J.C., Goldfried, M.R., Eds.; Oxford University Press: Oxford, UK, 2005; pp. 147–171.
23. Prochaska, J.O.; Velicer, W.F. The Transtheoretical Model of Health Behavior Change. *Am. J. Heal. Promot.* **1997**, *12*, 38–48. [CrossRef]
24. Gillison, F.B.; Rouse, P.; Standage, M.; Sebire, S.J.; Ryan, R.M. A meta-analysis of techniques to promote motivation for health behaviour change from a self-determination theory perspective. *Health Psychol. Rev.* **2019**, *13*, 110–130. [CrossRef]
25. Love Food Hate Waste. Available online: https://www.lovefoodhatewaste.com/?_ga=2.66326894.1117890828.1632425891-897301705.1632425891 (accessed on 9 September 2021).
26. Fong, G.T.; Hammond, D.; Hitchman, S.C. The impact of pictures on the effectiveness of tobacco warnings. *Bull. World Health Organ.* **2009**, *87*, 640–643. [CrossRef]
27. Reczek, R.W.; Trudel, R.; White, K. Focusing on the forest or the trees: How abstract versus concrete construal level predicts responses to eco-friendly products. *J. Environ. Psychol.* **2018**, *57*, 87–98. [CrossRef]
28. Giebelhausen, M.; Chun, H.E.H.; Cronin, J.J.; Hult, G.T.M. Adjusting the warm-glow thermostat: How incentivizing participation in voluntary green programs moderates their impact on service satisfaction. *J. Mark.* **2016**, *80*, 56–71. [CrossRef]
29. Filippini, R.; Mazzocchi, C.; Corsi, S. The contribution of Urban Food Policies toward food security in developing and developed countries: A network analysis approach. *Sustain. Cities. Soc.* **2019**, *47*, 101506. [CrossRef]
30. FAO; IFAD; UNICEF; WFP; WHO. The State of Food Security and Nutrition in the World. Available online: http://www.fao.org/publications/sofi/2021/en/ (accessed on 29 September 2021).
31. Rutten, M.M. What economic theory tells us about the impacts of reducing food losses and/or waste: Implications for research, policy and practice. *Agric. Food Secur.* **2013**, *2*, 1–13. [CrossRef]
32. Canali, M.; Amani, P.; Aramyan, L.; Gheoldus, M.; Moates, G.; Ostergren, K.; Silvennoinen, K.; Waldron, K.; Vittuari, M. Food Waste Drivers in Europe, from Identification to Possible Interventions. *Sustainability* **2017**, *9*, 37. [CrossRef]
33. Benson, C.; Daniell, W.; Otten, J. A qualitative study of United States food waste programs and activities at the state and local level. *J. Hunger. Environ. Nutr.* **2018**, *13*, 553–572. [CrossRef]

84. Betty, A.L. Using financial incentives to increase fruit and vegetable consumption in the UK. *Nutr. Bull.* **2013**, *38*, 414–420. [CrossRef]
85. Carroll, K.A.; Samek, A.; Zepeda, L. Food bundling as a health nudge: Investigating consumer fruit and vegetable selection using behavioral economics. *Appetite* **2018**, *121*, 237–248. [CrossRef] [PubMed]
86. Hollands, G.J.; Cartwright, E.; Pilling, M.; Pechey, R.; Vasiljevic, M.; Jebb, S.A.; Marteau, T.M. Impact of reducing portion sizes in worksite cafeterias: A stepped wedge randomised controlled pilot trial. *Int. J. Behav. Nutr. Phys. Act.* **2018**, *15*, 1–14. [CrossRef] [PubMed]
87. Stämpfli, A.E.; Stöckli, S.; Brunner, T.A. A nudge in a healthier direction: How environmental cues help restrained eaters pursue their weight-control goal. *Appetite* **2017**, *110*, 94–102. [CrossRef]
88. Wilson, A.L.; Buckley, E.; Buckley, J.D.; Bogomolova, S. Nudging healthier food and beverage choices through salience and priming. Evidence from a systematic review. *Food Qual. Prefer.* **2016**, *51*, 47–64. [CrossRef]
89. Cooper, C.; Booth, A.; Varley-Campbell, J.; Britten, N.; Garside, R. Defining the process to literature searching in systematic reviews: A literature review of guidance and supporting studies. *BMC Med. Res. Methodol.* **2018**, *18*, 1–14. [CrossRef]
90. Janssen, A.M.; Nijenhuis-de Vries, M.A.; Boer, E.P.J.; Kremer, S. Fresh, frozen, or ambient food equivalents and their impact on food waste generation in Dutch households. *Waste Manag.* **2017**, *67*, 298–307. [CrossRef]
91. Bucher, T.; Collins, C.; Rollo, M.E.; McCaffrey, T.A.; De Vlieger, N.; Van Der Bend, D.; Truby, H.; Perez-Cueto, F.J. Nudging consumers towards healthier choices: A systematic review of positional influences on food choice. *Br. J. Nutr.* **2016**, *115*, 2252–2263. [CrossRef] [PubMed]
92. Allcott, H.; Rogers, T. The short-run and long-run effects of behavioral interventions: Experimental evidence from energy conservation. *Am. Econ. Rev.* **2014**, *104*, 3003–3037. [CrossRef]
93. Venema, T.A.; Kroese, F.M.; De Ridder, D.T. I'm still standing: A longitudinal study on the effect of a default nudge. *Psychol. Health* **2018**, *33*, 669–681. [CrossRef]
94. Quested, T.E.; Palmer, G.; Moreno, L.C.; McDermott, C.; Schumacher, K. Comparing diaries and waste compositional analysis for measuring food waste in the home. *J. Clean. Prod.* **2020**, *262*, 121263. [CrossRef]

sustainability

Article

When Harmful Tax Expenditure Prevails over Environmental Tax: An Assessment on the 2014 Mexican Fiscal Reform

Sugey de Jesús López Pérez * and Xavier Vence *

Department of Applied Economics, ICEDE, University of Santiago de Compostela, 15701 Galicia, Spain
* Correspondence: sugeydejesus.lopez@usc.es (S.d.J.L.P.); xavier.vence@usc.es (X.V.)

Abstract: This article examines the role of environmental taxation in mitigating environmental problems and contributing to sustainability in Mexico. It focuses on environmental tax revenues and tax expenditures since the 2014 Public Financial Reform (PFR), according to pro- or anti-environmental orientation. The research carried out combines the study of the regulation of the selected tax instruments, their classification and the empirical analysis of the tax revenues and tax expenditures associated with the different taxes over the periods of validity of the taxes and benefits studied, using the databases of the CIAT and the Mexican SHCP. A critical analysis addresses the weak environmental function of environment-related taxes (IEPS, ISAN . . .), as well as the late implementation and reduced impact of the carbon and pesticide taxes introduced in 2014. The evolution of tax incentives and expenditure is thoroughly examined by examining both environmental measures, which have evolved positively but within a very reduced level, and the most prevalent tax expenditure measures, with harmful impacts to the environment. Based on the results obtained, long-term structural changes in the Mexican tax system are suggested. As for the short to medium term, profound changes in tax expenditure are proposed to eliminate of those tax benefits harmful to the environment, introduce of tax benefits for circular activities (e.g., repairing, reusing and remanufacturing) and broaden the carbon tax base and rates. The conclusions include recommendations for moving towards a systemic green tax reform that assists the transformation towards a sustainable economy.

Keywords: environmental taxation; tax benefits; tax expenditure; carbon tax; harmful tax expenditure

JEL Classification: E62; H22; H23; H24; H25

Citation: López Pérez, S.d.J.; Vence, X. When Harmful Tax Expenditure Prevails over Environmental Tax: An Assessment on the 2014 Mexican Fiscal Reform. *Sustainability* **2021**, *13*, 11269. https://doi.org/10.3390/su132011269

Academic Editors: Baojie He, Ayyoob Sharifi, Chi Feng and Jun Yang

Received: 8 September 2021
Accepted: 9 October 2021
Published: 13 October 2021

Publisher's Note: MDPI stays neutral with regard to jurisdictional claims in published maps and institutional affiliations.

1. Introduction

Environmental problems are increasingly prominent and currently figure among the most important and urgent items on the global agenda, and consequently in Mexico also [1–7]. In fact, in successive reports by the Intergovernmental Panel on Climate Change [5] scientific teams warn that both greenhouse gas (GHG) emissions and climate change have been accelerating at a strong pace in recent years. An exhaustive analysis by Rockström et al. [2] of the nine main planetary boundaries, indicates that we have already exceeded three of those boundaries (rate of biodiversity loss, nitrogen cycle and climate change) and are approaching the point of no return on several others (phosphorus cycle, ocean acidification . . .). Meanwhile, studies on the evolution of the circularity gap estimate that the circular economy will account for merely 8.6% of the global economy in 2019 [8].

The market does not generate the appropriate corrective mechanisms, nor do policies subordinated to market imperatives [9]. In fact, a recent study by the International Monetary Fund clearly warns that "markets alone cannot provide sufficient mitigation. Market failures, unaddressed and exacerbated by government failures, prevent an adequate market response to the challenge of climate change mitigation" [7]. The problem lies in the essential characteristics of the current economic system. The very dynamic capitalist production model also turned out to be extremely intensive in the use of raw materials, energy and oil, without internalizing the environmental costs derived from them [3,10–14].

While developed countries are certainly at the origin of most environmental problems, all countries have seen increases in their own environmental issues and their contribution to global problems. Within this dynamic, Mexico has also become a country with serious environmental challenges [4,15]. In fact, Mexico is in 12th place in CO_2 emissions in the world (significantly lower in per capita terms), and emissions of CO_2 have increased by almost 74.5% since 1990, although moderating in pace in recent years [15]. Environmental policy in Mexico has gone through different transformations that began in the 1980s and generally correspond to global trends and the successive international agreements signed by Mexico (Rio Agreement, Kyoto Protocol, Paris Agreement . . .). The aim has been to combat environmental pollution through direct regulation, soft technological and ecological measures, voluntary instruments (environmental certifications), and some fiscal measures (mainly, subsidies and benefits) [16–18]. A new step was taken with the 2014 Public Finance Reform (PFR), when environmental taxes on fossil fuels (CO_2) and pesticides were set up for the first time [19].

This paper seeks to contribute to the existing literature on environmental taxation in two directions: on one hand, by analyzing the situation and the changes introduced in a large country with an intermediate level of development; on the other hand, it tries to balance the efforts for the introduction of new environmental taxes (carbon tax and pesticides) and those related to the environment against the measures focused on tax benefits on the major taxes of the tax system (VAT, corporate, income, etc.). The aim is to examine particularly the latter in order to test the hypothesis that these tax benefits have an enormously greater weight in the revenue of public finances than environmental and environment-related taxes, with the aggravating factor, moreover, that the former are for the most part tax benefits that are harmful to the environment.

This article provides an analysis of the more recent changes. Following the introduction in Section 1, we review the theoretical and empirical literature on environmental taxation in Section 2, highlighting the advantages and limitations of these environmental policy instruments. Section 3 gives information on the databases used for the empirical study on Mexico. In Section 4, we analyse the implementation of environmental taxation in Mexico, particularly from 2014 onwards, giving attention to tax revenues and tax expenditure. In Section 5, we summarize the main results, and Section 6 offers some conclusions and recommendations.

2. Literature Review on Environmental Fiscal Policy

Environmental fiscal policy is directly aiming at creating economic incentives to promote positive environmental behaviour by the various economic actors (producers, dealers, retailers, consumers, public institutions, finance . . .). It is transmitted to economic agents through tax instruments, public expenditure and fiscal or tax expenditure, including subsidies, reductions and benefits [7,20–22]. Tax expenditure differs from public spending in that the former involves the renunciation of public revenue that is made operative through the existence of incentives or benefits that reduce the direct or indirect tax burden of certain taxpayers in relation to a reference tax system or 'benchmark' [23].

Most of the literature analysed on fiscal policy and environmental problems starts with the so-called Pigouvian taxes [3,24–33]. To Pigou and the environmental fiscal literature, taxation by the public sector is the best way of internalizing, through prices, the social cost of negative externalities that were not reflected in the market price. This idea was firstly developed in the theoretical framework of welfare economics [24], which, though often overlooked, poses analytical conditions that are difficult problems to solve in real implementation and practice. A precise calculation of externality would be required to establish an optimal tax, one equivalent to the social cost and added private benefit. The idea has been transferred to the debate on environmental policies in a more general and pragmatic formula: the polluter pays principle. One way of making this happen is through environmental taxation [26,32].

Environmental taxes are defined in the literature as compulsory tax payments at a fixed or variable rate, which must be paid by polluting agents based on facts related to the pollution caused by their production or consumption. These facts set the tax base for using the mechanism of environmental taxation to address pollution problems and social costs. While there is no single definition, the environmental tax is expected to re-orient production patterns and alter consumption patterns, regardless of the destination of the revenue obtained [3,21,32,34–37]. Environmental taxes are economic instruments to the extent that they affect the cost and price of goods, and by that means influence their consumption. In other words, the environmental tax works as a price altering agent to discourage uses and consumption that pollute the environment and encourages innovations in a more environmentally friendly direction. Environment tax instruments can create market incentives to develop and invest in emission-reduction technologies, to encourage behavioural changes in consumption and production and to achieve least-cost solutions. Moreover, according to the double dividend argument, the tax instruments can generate revenue that could be used to finance environmental expenditures or to mitigate adverse impacts on the diverse social groups [27,28,36,38].

Although the environmental taxes are preferable to pollution market [3,29,38–44], taxes also have some critical flaws. Current environmental taxation focuses on penalizing pollution through price, but does not prevent it. The implementation of the Pigouvian approach resulted in the first generation of environmental taxes, leading to set a tax for each specific environmental problem (different chemical pollutions, energy, NO_x, CO_2 ...). Such an approach generated a multitude of scope-limited proposals that eventually proved ineffective in addressing growing, complex and interrelated environmental problems [5,6,28,45,46].

Though taxation is increasingly recognized as an important environmental policy instrument, real progress in implementing environmental fiscal policies have been modest in Mexico and around the world [6,20,28,47,48]. Indeed, the numerous environmental taxes that have been implemented in recent decades—probably more than a hundred—have not met their environmental objectives [5], and their use is currently limited. According to Groothuis [45], "over the past 15 years, environmental tax as a share of GDP has declined in 52 out of 79 countries in the Organization for Economic Co-operation and Development (OECD) database. In addition to relatively low green tax levels, global fossil fuel subsidies amounted to \$373 billion in 2015."

The real evolution of the most serious environmental problems, such as climate change or urban and ocean pollution, show the insufficiency or failure of the measures that have been adopted in the last three decades. This applies to emissions market mechanisms especially, but also to CO2 taxes insofar as they have been implemented.

A recent paper issued by Best et al. [49] ran a cross-country empirical study on the efficacy of carbon pricing, showing that "a negative association between carbon pricing and the subsequent CO_2 emissions growth rate, with a one euro increase in the effective carbon price rate per tonne of CO_2 emissions being associated with a 0.3 percentage point reduction in the annual rate of emissions growth." At the same time, the most recent scientific reports unequivocally indicate that CO_2 and other GHG emissions have increased and even accelerated in the years since the 1997 Kyoto Protocol, and despite the launching of the EU Emissions Trading Scheme (EU ETS) in Europe and other countries, with only a brief and apparent respite caused by the Great Recession at the end of the last decade [5]. This disappointing result could suggest that carbon taxes (and other market solutions) run the risk of failing if the taxable events are narrow and the tax rates very low. Through the influence of companies and interest groups, carbon prices end up being set so low that they are ineffective in curbing emissions [13,46]. "The problem is in the economy: if the tax is very moderate, it fails to remove enough fossil fuel to help the climate; but if it is high enough to actually reduce it, then business and consumers resist the tax—because without some safety cushion for business and consumers, the whole problem falls on them and they rationally resist—to save profits and jobs" [46].

As previously stated, the environmental tax policy includes other relevant instruments particularly public and tax expenditures. Tax benefits (or tax expenditure) materialize into government fiscal waivers [50–53] which can be granted to economic agents in very diverse ways (incentives, tax relief, deductions, accelerated depreciation, etc.). From an economic point of view, justification for these tax benefits may be linked to industrial and trade policies favouring certain economic activities or supporting some consumption patterns, a part of a social or redistributive policy [52,54]. The usage of tax benefits as an incentive—a carrot—to promote environmental objectives is well established, in fact, is one of the first instruments of environmental tax policy [52,55,56].

Tax expenditure is an opaque and little-studied subject, instead of existing estimates placing tax expenditure at between 14–24% of total revenue in most countries and, in some cases (e.g., US, UK) that proportion exceeds 30%; as a proportion of GDP, the available estimations go from 3.7% in the Latin American countries to 8% in the US. Mexico also makes extensive use of tax expenditure measures, reaching up to 23.8% of total revenues and 3.15% of GDP in 2019 [23,57]. The question here concerns whether these instruments have been used in environmental policy to correct the negative environmental impacts caused by the various economic sectors, or the other way around. The literature on this issue is scarce, is mostly focused on discussing which tax expenditure instruments have an environmental purpose and, where appropriate, is concerned with assessing their effectiveness, either theoretically or empirically. However, the above-mentioned literature does not take stock of tax expenditures that serve other purposes and have a harmful impact on the environment.

This article attempts to fill this gap and provides empirical evidence of the evolution of the environment tax policy in Mexico in recent years, balancing the evolution of taxes and the benefits, including both environmental-friendly benefits and harmful ones.

3. Databases

The database used for this empirical study includes official sources such as the Inter American Center of Tax Administrations (CIAT) [58,59] for revenues (1990 to 2018) and tax expenditure (2014 to 2018), updated with 2019 information from the SHCP and the SAT (Tax Administration Service-Government of Mexico). In addition, the statistics corresponding to the IEPS on fossil fuels (CO_2) (2016 and 2017) were adapted according to the SHCP methodological note on carbon [60]. We focused on environmental fiscal instruments (taxes, incentives and tax benefits). An overview of environment-related taxes (1990–2019) is also developed, including information on six different taxes collected at the federal level (Figure 1).

The tax expenditure statistics (2014–2018) were compiled from CIAT data using dynamic tables. The incentives and benefits were classified by type (exemptions, reduced rates, depreciation, deferral, deductions, etc.). To identify the categories in the working database, we applied the following process: the filtering of the data group of environmentally friendly and environmentally harmful tax measures was done based on the description of the regulatory source (tax law for each tax instrument and Mexican federal income law) and according to the tax measure included in the tax expenditure database for each indicator or tax (VAT, ISR, ISAN and IEPS). Thus, the categories (a) environmentally friendly (including all measures with clear environmental characteristics and/or renewable components) and (b) environmentally harmful (fiscal measures that stimulate non-environmental investment opposed to the purposes of environmental policy and incentives that generate stimuli for consumption of products such diesel and fossil fuels) were obtained. The classification of the category by type of tax expenditure was made according to [23,62,63].

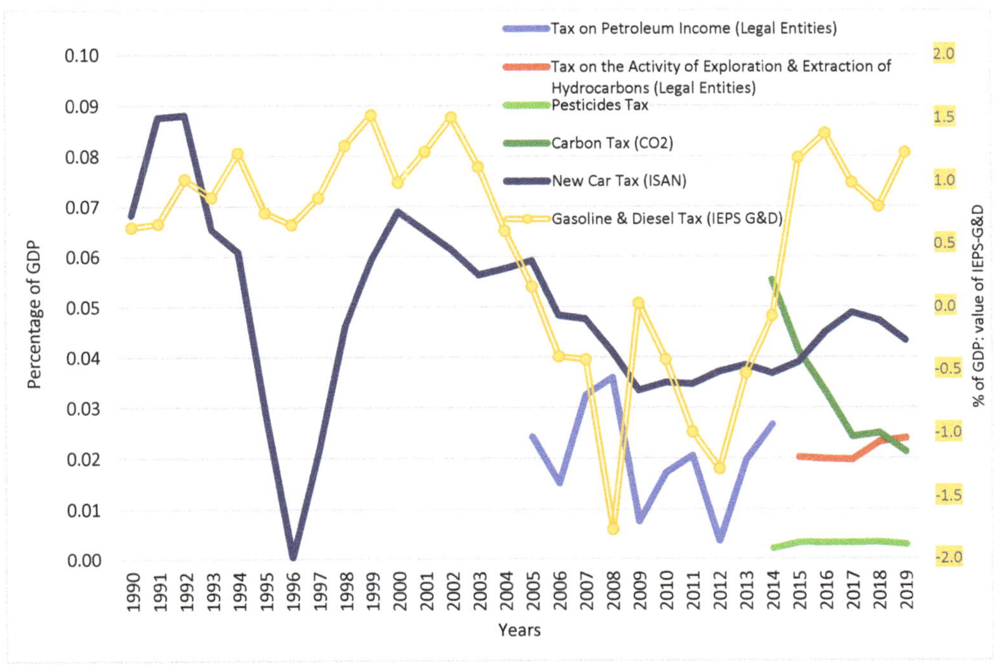

Figure 1. Evolution of environment-related taxes in Mexico, 1990–2019 (IEPS-G&D right scale) (Units: Percentage of GDP). Source: prepared by the authors, based on [58,61].

4. Implementation of Environmental Taxation in Mexico

4.1. Environmental Tax and Environmental-Related Tax in Mexico

Environmental fiscal policy took its very first steps in the form of tax expenditure (stimulus, incentives and tax benefits) in the early 1980s, as a second-order instrument to reinforce the environmental objectives being implemented through national development programs. In 1981, a government decree established the Fiscal Incentives for the Promotion of the Preventive Activity of Environmental Pollution [64]. In addition to direct regulation measures and voluntary instruments (ecological certifications), fiscal instruments were introduced to provide incentives for the immediate depreciation of investments in equipment for controlling and preventing pollution, along with tariff reductions on imported industrial equipment of this type. Some taxes and fiscal instruments concerning the use of energy resources or specific consumption, which are sometimes considered environmentally-related taxes (e.g., IEPS-G&D, ISAN), have been around for decades. However, environmental taxes that were explicitly designed as such appeared in 2014, resulting from the country's commitments to international agreements on climate change. The 2014 Public Finance Reform (PFR) established nine major objectives for improving the state's income capacity, promoting equity and reducing tax evasion and abuse by reducing existing fiscal incentives and expenditures and creating conditions for the liberalization and privatization of the energy and hydrocarbon sector. The environmental goals behind the measures included "fighting obesity and protecting the environment: fiscal provisions are established to discourage the consumption of goods that are harmful to health and the environment." In particular, two new green taxes were introduced: a tax on fossil fuels, or a carbon tax (quota per CO_2 content of various energy products), and a tax on pesticides (according to the level of toxicity), categorised as IEPS-Other consumption [16,19,65].

A conceptual and terminological clarification is needed before entering into data analysis. The OECD and the European Union have agreed to distinguish environmental taxes from so-called environment-related taxes, as a broader category that includes taxes "whose basis is a physical unit (or a proxy of a physical unit) of something that has a specific and proven harmful impact on the environment [66], in accordance with the tax base, regardless of whether the tax is intended to change behaviour or is imposed for another reason." In practice, these two categories are used interchangeably, which undoubtedly causes some confusion. Arlinghaus and van Dender [16] distinguish four subsets of environment-related taxes: energy taxes, transport taxes, pollution taxes and resource taxes. For its part, the CEPAL [20] distinguishes three conventional categories for classifying all taxes as environmentally related according to the tax base under consideration: (i) energy taxes, which include taxes on products related to energy generation with polluting effects, such as fossil fuels and electricity, in addition to those used in transport, such as gasoline and diesel; (ii) taxes on transport, which includes the full range of taxes on motor vehicles and other motorised means of transport by virtue of their marketing (domestic or imported), ownership (recurrent taxes), registration and circulation permits or road use, and (iii) other taxes on pollution and the use of natural resources. The latter include many of the least developed, least used instruments, regionally and internationally, along with taxes on gaseous substances, water extraction and disposal, extraction of natural resources and pesticides and fertilizers, among others.

According to this definition, the following environment-related taxes exist in Mexico: gasoline and diesel tax (IEPS G&D), tax on petroleum income (Legal Entities), tax on the activity of exploration and extraction of hydrocarbons (Legal Entities), new car tax (ISAN) and of course, the carbon tax (CO_2) and pesticides tax (Figure 1). However, a more detailed analysis reveals that some of them (the first four) are not environmental taxes, since they were designed for increasing public revenues or to adjust the price of certain resources to scarcity perspectives and not to reduce pollution or other harmful effects. Their taxable fact and taxable base do not include explicit incentives to modify the agent's economic behaviour. Moreover, some benefits applied to these taxes are clearly anti-environmental. Consequently, they do not really work as environmental taxes.

In this section, we will review the entire family of taxes in force in Mexico, starting with those which are environment-related (IEPS-G&D, ISAN, Petroleum and Hydrocarbons), and then the environmental ones (IEPS-Pesticides and CO_2) (See Figure 1).

The IEPS G&D was introduced in 1980. Despite being a tax on a highly polluting products—with CO_2 and other GHG emissions and effects on urban pollution and human health—its specific design clearly does not present environmental tax features and the evolution of revenues does not fit with the environmental policy objectives. The IEPS rate would be calculated according to the following formula as presented by Hernández and Antón (2014): IEPS rate = $(\alpha_{i,j} PVP - C - F - PP)/PP$, where α_i = 0.9091 when VAT = 11% and α_j = 0.8696 when VAT = 16%. PVP is the public price of gasoline, F is the freight and transport cost, and C is the commission to distributors, PP is the final producer price (PP = PS + AC + CT + CM, where PP is the final producer price, PS is the spot reference price (average US Gulf Coast gasoline price), AC is the quality adjustment, CT is the transport cost and CM is the handling cost). The saw shape of the curve over the period analysed—including a decade of negative revenues—is conditioned by the international crude oil reference prices, since the IEPS-G&D was designed with this complex formula that seeks to adjust the differences between international and national prices but does not really seek to reduce consumption [67].

There are also a number of subsidies that are implicit in the consumption of these products and clearly encourage pollution [16,48]. These strong subsidies explain the negative fiscal revenues from 2005 to 2014 (see Figure 1). Therefore, the reduced price does not provide sufficient signals for a reorientation of the consumption and production patterns that would lead to reductions in the pollution generated by these activities. In fact, the main specific changes of the 2014 Energy and Tax Reform were the gradual elimination of diesel subsidies in the transport sector [68]. These changes partially explain the evolution of the fiscal revenue curve, which takes positive values after 2014, reaching a maximum of 1.4% of GDP in 2016 (right axis), then falls in 2017–2018 and rises again in 2019 up to 1.2%.

This analysis of the characteristics of the IEPS-G&D and its revenue trend, including a decade of negative results, lead us to question its classification as an environmental or environment-related tax by the OECD, CEPAL or the World Bank. In fact, it does not really even qualify as an actual tax, as its special design makes it a pseudo-tax [57]. This reconsideration would imply that the implementation of environmental taxation in Mexico is still in a very incipient state. If this tax is excluded from the group of environmental taxes, an evaluation of the degree of compliance with international environmental commitments would reveal a much more critical state than that indicated by the OECD report on the carbon pricing gap (CPG) [69]. In fact, the estimated CPG for Mexico indicates that 69% of the parameters established for the 2030 scenario have not been met. Obviously, if we exclude this tax, which brings in the highest revenues of all those considered "environmental" by that organization, then the carbon pricing gap for that year would be much higher than the current OECD estimate [69].

ISAN is a tax on the purchase of new cars and applies a progressive rate to the purchase of luxury cars. Revenue from this tax reached 0.04% of GDP in 2019. As a tax that implies purchasing power, economic growth informs the evolution of fiscal revenues (falling down during the tequila crisis of 1994–1996 and the financial crisis or Great Recession of 2008). This instrument mainly taxes luxury goods, but does not incorporate the achievement of environmental objectives.

The tax on Petrol Rent and the Exploration and Extraction of Hydrocarbons is a unique case in Mexico, a nationalized producer of crude oil and hydrocarbons that is organized through the public company Pemex. This tax is not related to conventional income tax (since Pemex is exempt from it) and revenues are linked to Pemex control mechanisms and operational management [70]. Environmental motivation in the resource appropriation policy still very marginal, given that Mexico has prioritized a policy of economic growth that has always been supported in this sector in times of boom [71].

In any case, all environmental and environment-related taxes show a very limited revenue-raising capacity. Recent CEPAL [20] and OECD et al., [72] reports show that revenues from environmental taxes in the 23 Latin America and the Caribbean countries averaged 1.1% of GDP in 2018 (and 1.3% since 2006), which is considerably lower than the also-modest 2.3% average for OECD nations as a whole for the same year. Mexico is well below the OECD average, slightly below the Latin American Countries (LAC) average, and behind countries such as Costa Rica, Chile, Uruguay or Argentina. Even so, energy taxation is the most prominent component in the Mexican case, given the minimal presence of the other three tax bases (pollution, resources and transport).

IEPS-Other consumption includes a conglomerate of specific taxes applied to different products that are intended to provide a price disincentive to harmful consumption. With PFR 2014, two environmental taxes were created in this group: the tax on fossil fuels, or carbon tax (CO_2) and the tax on pesticides. The first was determined following the IPCC Guidelines for National GHG Inventories, which is part of the strategy to reduce CO_2 and other greenhouse gas emissions. The second, the tax on pesticides, was determined using rates ranging from 0 to 9 percent, according to the level of acute toxicity from exposure to these products, and is subject to the parameters established by the Official Mexican Standard NOM-232-SSA1-2009 for pesticides, "which establishes the requirements for the packaging, packing and labelling of technical grade products and for agricultural, forestry, livestock, gardening, urban, industrial and domestic use" [73]. This proposal will induce the replacement of more toxic agrochemicals with others that are less harmful.

The environmental taxes on CO_2 and pesticides were established with two central objectives: (1) to reduce pollution levels (of GHG emissions and harmful products, respectively), and (2) to increase tax revenues (designated to the general budget). Here we will look at the extent to which these objectives have been met. The carbon tax (CO_2) is levied on a quota linked to the amount of CO_2 in each fossil fuel and based on exchange quotes for the main carbon markets. The petrol tax, for example, was set in 2014 at 10.3 cents peso per liter, diesel at 12.6, fuel oil at 13.4 and coal at 27.5 per ton. This tax amount will remain

constant in real terms, as it is adjusted annually by the variation in the National Consume
Price Index (INPC in Mexico). However, natural gas and turbosine were exempted from
the tax at the express request of the private sector, which effectively undermines emission
reductions in large sectors of the economy. The CO_2 tax is also not applied when oil i
used for manufacturing, e.g., for the production of plastics, rather than combustion [74]. In
fact, Mexico shows a low effective carbon rate, taxing above EUR 30 per tonne of CO_2 fo
only the 30% of emissions from energy use (mainly from road transport sector), with mos
of emissions (68%) from industry, electricity, and the residential and commercial sector
remaining unpriced [68].

Environmental taxes are barely significant in collecting revenues and even show de
creasing performance (Table 1). Revenues from these taxes represented a meagre 0.024% o
GDP in 2019 (0.003% from pesticides and 0.021% from the CO_2 tax). Even more relevant i
the continuous decrease in the capacity of CO_2 tax to create resources over time (Figure 2
Revenue in 2019 amounted to 53.3% of what was obtained in its first year. Meanwhile
the pesticide tax has been basically irrelevant as far as revenue is concerned. Arlinghaus
et al. [16] observed that the effectiveness of pesticide taxes in reducing the use of harmfu
products is hard to establish due to a relatively dynamic market, the wide variety of prod
ucts, storage behaviour prior to tax increases, large seasonal and geographical variations in
the intensity and frequency of treatments, and the effects of competing regulations.

Table 1. Annual revenues from environmental taxes (CO_2 and pesticides) in Mexico, 2014–2019. Source: Prepared by the authors, based on [58,61,75].

Periods/ Environmental Taxes	(In Millions of Pesos)		(In % Total Tax Income of Country)		(In % of GDP)		GDP (% Annual Growth)
	Pesticides	CO_2	Pesticides	CO_2	Pesticides	CO_2	
2014	358.61	9670.35	0.016	0.425	0.002	0.055	2.85
2015	606.93	7648.51	0.021	0.267	0.003	0.041	3.59
2016	647.24	6657.74	0.020	0.205	0.003	0.033	2.63
2017	705.24	5325.17	0.021	0.156	0.003	0.024	2.11
2018	775.06	5883.55	0.021	0.160	0.003	0.025	2.19
2019	687.00	5153.20	0.018	0.136	0.003	0.021	−0.055

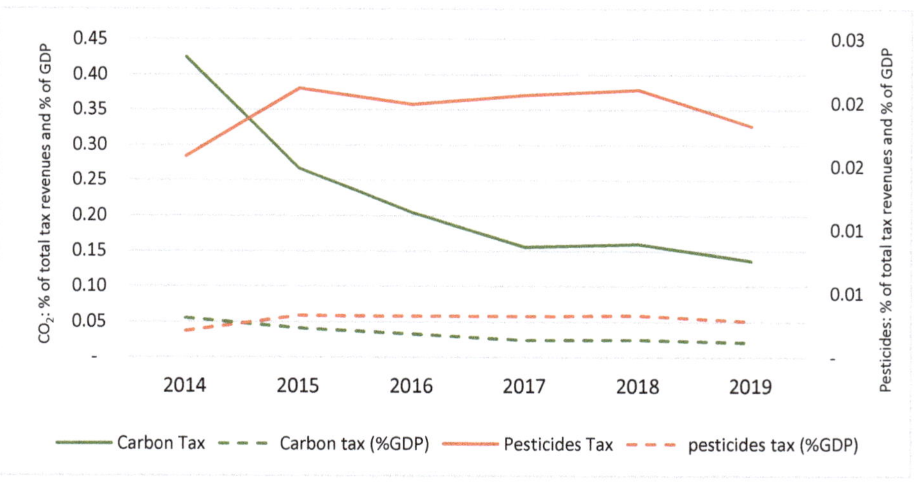

Figure 2. Trend of environmental taxes in relation to total revenues and GDP (left-CO_2 and right-Pesticides) in Mexico, 2014–2019. Units: percentage; source: prepared by the authors, based on [58,61].

Regarding the actual evolution of the emissions burden, long-term data series show
that CO_2 emissions in Mexico have increased by almost 74.5% since 1990, albeit moderating

in recent years [15]. It is interesting to observe the parallel evolution between the drop in CO_2 tax revenue both in absolute terms and as a proportion of GDP and the drop in the GDP growth rate over this period. Indeed, there is a very close statistical correlation of $R^2 = 0.446$ between the two variables, which suggests several hypotheses (Figure 3). First of all, the design of the CO_2 tax in Mexico applies to a reduced tax base (a limited number of activities) and has a very low price per ton of CO_2. Second, its evolution seems to be directly dependent on the growth rate, since growth deceleration is most strongly reflected in CO_2 emissions and in tax revenues. This also reveals how fossil fuel consumption is extremely sensitive to the country's economic progress. Thirdly, the evidence suggests that the implementation of a carbon tax with this specific design has had no significant effect on the evolution of emissions or economic activity.

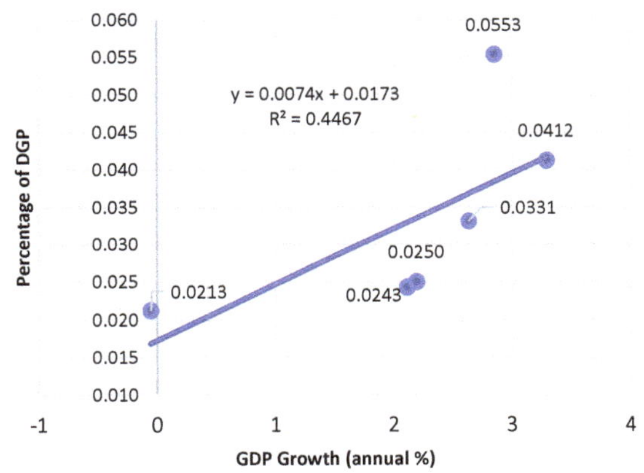

Figure 3. CO_2 tax revenue (% GDP) and annual GDP growth (right axis) in Mexico, 2014–2019. Source: prepared by the authors, based on [58,75].

These hypotheses suggest some nuances relative to the empirical literature on the topic. The meta-regression study conducted by Galindo et al. [76] on the potential effects of a CO_2 tax in Latin American countries shows that the impact of implementing a CO_2 tax on GDP will depend on the structural conditions of each country. For Mexico, all the estimated long-term results show negative impacts [48]. However, these results seem to be associated with being evaluated as an OECD member country and applying the OECD coefficients (the same happens for Chile). Alatorre et al. [77] conducted a study to determine the effects of the CO_2 tax on economic, social and technology transfer variables through a recursive system of equations. That author suggests that for LAC, the effect of the tax on these three variables depends largely on the non-linear relationship between the increase in relative energy prices and per capita GDP. Thus, in a context of low prices, economic and social policy would have negative effects, while a context of high prices would result in greater environmental benefits. In order to reduce the negative effects associated with the tax, both studies [76,77] suggest a combination of mitigating policies (e.g., tax reductions on the labour factor). They also highlight the importance of applying a CO_2 tax in order to encourage the implementation of new technologies (energy efficiency), the development of less-energy-intensive sectors and the generation of jobs that improve environmental conditions.

To sum up, the introduction of the carbon tax in 2014 made Mexico the 27th world jurisdiction to implement a CO_2 tax in accordance with the Kyoto Protocol and the Paris Agreement, both signed by Mexico. Mexico was the pioneer in the LAC region, followed by Chile (in 2017), Colombia (in 2017) and Argentina (in 2019) [78]. However, the price established per type of fossil fuel was too marginal to influence the behaviour of economic agents (the minimum threshold is suggested around EUR 30 per ton of CO_2, including all activities, and preferably above 100 in the rich countries). After 6 years of implementing both environmental taxes, the initial objectives are very far from being met. The taxes have been ineffective in both reducing pollution levels (GHG emissions and harmful products and increasing tax revenue beyond a symbolic level which; it is actually declining. So, the empirical results show that environmental taxation in Mexico has not achieved any of the objectives that were established in the 2014 General Economic Policy Criteria [19] and still has a very wide gap to reduce in terms of pollution levels.

However, this problem is not exclusive to Mexico; it is general and global, as emphasized by different global reports. "After a quarter century of academic debate and experimentation, a gap persists with respect to the 'carbon price change' needed to trigger rapid changes (. . .) The Mexico performance is relatively poorer as a result of the low-price set for each Tm of CO_2 emissions and the reduced range of sectors affected" [5].

Due to the environment taxation is failing in promoting a real change towards sustainability, we are dealing with other fiscal instruments such as tax expenditures in order to assess real performance.

4.2. Environmental-Frienly and Environmental-Harmful Tax Expenditure in Mexico

The many existing tax benefits can be classified from different perspectives [52,55]. Here we will focus on three main categories: the objectives pursued, their environmental effects (positive or negative) and all other objectives, be they social, cultural, employment competitiveness, etc. The last categories were grouped into a single block and isolated them from the analysis, though they are more abundant in Mexico and in many other countries [23].

We are focusing here on incentives and benefits that affect consumption, investments that favour the environment or reduce environmental impacts, whether explicitly established or not. Meanwhile, we will also examine benefits granted for consumption and investments favouring activities harmful for the environment. By doing so for the period 2014–2019, we are providing an assessment of the real environmental content of the 2014 Public Financial Reform.

The benefits and incentives related to environmental protection apply, above all, to the investment cost (deductions and accelerated depreciation) for investments in fixed assets of electric vehicles, vehicles powered by rechargeable electric batteries, electric bicycles and machinery and equipment for generating energy from renewable sources. Together, these accounted for only 0.07% of tax revenue and 0.01% of GDP in 2018 (Table 2 and Figure 4). In examining the magnitude of the incentives and benefits at the level of each tax for the last available year, it can be seen that the environmental cut was granted mainly through the ISR (0.31% of the revenue from this tax) and, to a lesser extent, the ISAN (which represents 0.45% of revenue). Most of these were reductions for qualifying investments (0.05% of total revenue) and other deductions (0.02% of total revenue). For income tax purposes, the incentive consists of a 100% deduction of the investment in clean technology and the depreciation of renewable assets. This reduces the taxable base for determining profit, allowing companies to declare high investment costs, which could even be reflected as losses and thus be tax-free. Direct benefits are also granted through an ISAN exemption (it is not included in the payment) for the purchase of ecological cars.

On the opposite side is the group of fiscal measures promoting productive activities and consumption practices with clearly unsustainable components that contradict the objectives of environmental policy in Mexico [79–81]. These environmentally harmful benefits include incentives for diesel and fuel consumption, which generate CO2 emissions, or benefits for resource extractive industries and the consumption of materials (Table 2). This relatively more substantial group accounted for 1.66% of total tax revenue and 0.26% of GDP in 2018. Most represent IEPS benefits (5.02% of this tax) from the consumption of diesel, fossil fuels and gasoline in the agricultural and transportation sectors, corporate income tax (3.14% of this tax) from reductions on investments and other deductions, and ISAN exemptions that are applied to the purchase of high-priced vehicles (1.48% of this tax).

Similarly, the reduced VAT rate (1.41% of this tax) related to drinking water supply services is understood as a social-economic policy, but has been categorized as anti-environmental, as it is equivalent to tax-free use of a natural resource.

Table 2. Tax revenues and environmental-friendly and environmental-harmful tax expenditure in Mexico, 2018. Source: prepared by the authors, based on [59].

Taxes	Revenues as % of GDP	Tax Expenditure as % GDP	Benefits as % of Total Revenues	Environmental Benefits as % of Revenue	Environmental Harmful Benefits as % of Revenue
Corporate Income Tax	3.7	0.6	3.9	0.07	0.69
Personal income tax	3.4	0.9	5.8	0.0	0.0
VAT	3.7	1.5	9.5	0.0	0.35
Special tax on production and services (IEPS)	1.7	0.2	1.5	0.0	0.61
Other	3.1	0.0	0.0	0.0	0.0
Total	**15.6**	**3.24**	**20.70**	**0.07**	**1.66**

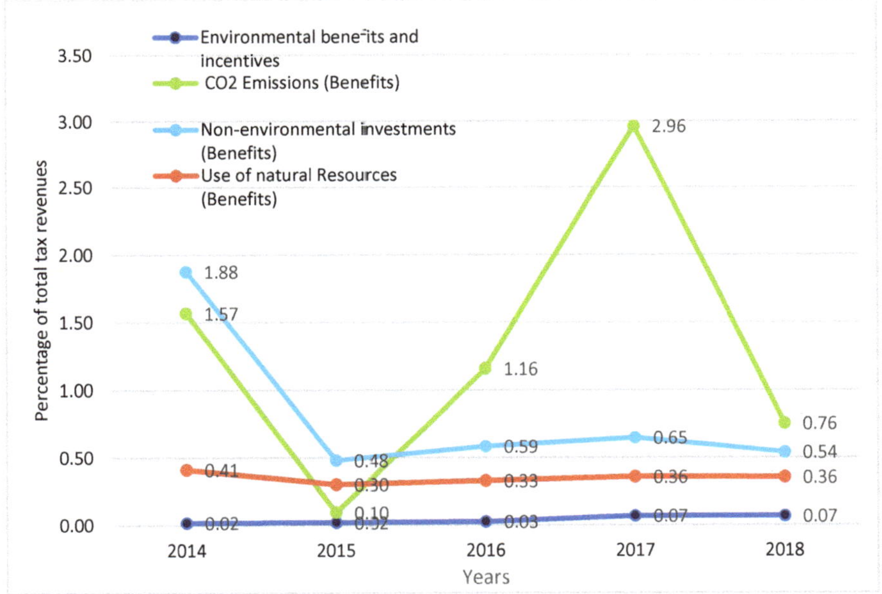

Figure 4. Environmental-friendly and environmental-harmful tax benefits as a percentage of total revenue in Mexico, 2014–2018. Units: percentage of total tax revenues; source: authors, based on [59].

Incentives for the consumption of new products, such as investment allowances or investment exemptions, especially when such benefits are not conditional on the use of the most environmentally friendly technologies, could encourage consumption practices that are undesirable from an environmental point of view. Such incentives encourage the consumption (and waste) of new materials, components and inputs from non-renewable resources. Accordingly, they are contradictory to the objectives of the circular economy, which seeks to extend the useful life of goods that can be repaired or maintained as long as possible. Although the investment incentives have declined significantly since the 2014 Reform in terms of GDP (from 0.24% in 2014 to 0.08% in 2018), more decisive cuts are desirable. Furthermore, even when the 2014 Reform introduced the tax on fossil fuels (CO tax), the benefits of this tax to the transportation and agricultural sectors remained in place at least until 2018 (reaching a peak in 2017: 2.96% of total revenue and 0.46% of GDP only starting to fall after that year from the definitive withdrawal of diesel benefits for the transport sector.

In short, evolution in the period since the reform (Figure 4) shows a slow, partial, but progressive reduction of incentives and tax benefits that have a clear anti-environmental bias. Though slimmer than before, they continue to be very broad and carry much more weight than environmental incentives and benefits. The latter have quadrupled since the reform (with a very striking jump in 2017 when deductions increased for electric vehicles and bicycles, and clean technologies), but the absolute and relative levels of these incentives are still extremely low (0.01% of total revenue).

As we have already pointed out, most of the incentives and benefits granted for other purposes (social, culture, foreign trade, etc.) are difficult to classify using environmental criteria. These tax expenditures represent 18.97% of total tax revenues and 2.97% of GDP

To summarize, the reality of tax expenditure in Mexico still disappointing. Tax benefits and subsidies have been used extensively to favour industrial activities with high consumption of material inputs and non-renewable resources, or even natural resource extraction activities (hydrocarbons and mining). Also, many subsidies to the consumption of fuel, diesel, energy use and water in homes and industries have been applied in Mexico before and after the 2014 Reform [23].

Therefore, most of the subsidies and tax benefits presently in use do not qualify as environmental policy, but on the contrary benefit activities harmful to the environment. This increases the urgency of the challenge to redirect or drop these incentives, especially because of their harmful effects on the environment, as suggested by Stahel [82] and Martínez and Roca [3]. This argument should be distinguished from other questioning of subsidies and benefits. In fact, the OECD [48] recommends they be eliminated because international experience shows that such instruments diminish tax revenues and the effective reallocation of resources, while their effective impact on growth, productivity and investment remains unclear. Anyway, the redistributive effect of eliminating subsidies and implementing carbon taxes on household income in Mexico should be considered (some studies suggest a progressive impact, e.g., Rosas-Flores et al., [83]).

5. Results

The examination of the available information regarding environmental and environmental related fiscal instruments in Mexico since 1990, and particularly since the 2014 Public Financial Reform, allows us to highlight the following results:

Environmental taxes in Mexico are still underdeveloped when compared to European countries and other OECD countries, which also show a poor performance. The environment-related taxes are also quite lower than in European countries and, for years the OECD countries. It is worth mentioning that Mexico is reaching the mean OECD countries in the very last years, but its evolution is highly dependent on the variability of the main tax, IEPS-G&D, whose classification as an environmental-related tax is highly debatable.

Revenue volatility in IEPS-G&D is caused by its very design as an external-internal price adjustment mechanism for hydrocarbons. Such a sui generis design converts this tax into a pseudo-tax and, as a consequence of a political price strategy in the hydrocarbon sector, it includes implicit subsidies to reduce the effects of high prices for consumption. It is characterized as an environment-related tax because the product being taxed (fuel) has clear environmental impacts involving the use of natural resources and the polluting emissions resulting from its consumption. However, this tax does not incorporate explicit environmental criteria or objectives (concerning emissions or other impacts); neither its design nor its implementation respond to an environmental purpose.

Since the 2014 Reform, steps have been taken in the implementation of environmental fiscal instruments, both on the tax side and on the fiscal expenditure side, but they are modest in scope and reflect the approach adopted in the first generation of environmental taxation. Tax revenues from environmental taxes (CO_2 and pesticides) reveal a very limited tax base and extremely low prices that barely reached 0.036% of GDP on average between 2014 and 2019. In current currency, the carbon tax has evolved negatively (revenues in 2019 were only 53.3% of those obtained in its first year), while the pesticides tax has shown a positive but insignificant trend.

There is a positive correlation ($R^2 = 0.446$) between the CO_2 tax/GDP ratio and annual GDP growth, with a stronger fall in GDP growth than in the CO_2 tax, suggesting that economic growth is the main variable explaining the pace of emissions (instead of the tax).

Since the 2014 reform, some anti-environmental incentives and tax benefits have diminished, but they are still very broad and outweigh environmental incentives and benefits, which, while growing, remain at very low levels. Incentives and benefits which promote practices harmful to environment represented 1.66% of total revenue and 0.26% of GDP in 2018, of which almost half (0.12% of GDP) corresponded to activities with high CO_2 emissions. Meanwhile, environmental-oriented benefits scarcely accounted for 0.07% of revenues and 0.01% of GDP in 2018, which makes their positive evolution practically irrelevant.

The highest percentage of ecologically harmful tax waivers were granted under ISR-business (0.11% of GDP in 2018), mainly in the form of investment deductions (0.06%) or tax credits, or other credits and reductions (0.04%).

6. Conclusions and Future Research

Environmental fiscal policy is a potentially effective and fundamental instrument for achieving environmental objectives. Beyond public spending (green public investments, green public procurement, subsidies, etc) to finance actions to mitigate or repair environmental damage, we are focusing on the importance of other fiscal instruments such as taxation and tax expenditure (a large range of tax benefits as exemptions, reduced rates, depreciation, deferral, deductions, etc.). Taxes and tax benefits are flexible instruments that can alter the prices of goods or services, generating price signals and incentives to reorient the behaviour of economic agents (producer and consumers) towards sustainable patterns. They have the potential for transversal impact along the economy and they can generate resources for financing public environmental spending or compensating the negative redistributive effects of such taxes. Moreover, they are relatively easy to manage by the tax administration.

The analysis of the tax regulation shows that, since the 2014 Public Finance Reform, Mexico has taken steps in using fiscal instruments to respond to environmental policy objectives. Environmental taxes (CO_2 and pesticides) have been incorporated and tax benefits and incentives have been reduced for some specific polluting and harmful activities and consumption. However, the empirical evidence provided in this paper shows that the modest aims and commitments expressed in the 2014 Fiscal Reform are far from being substantiated. The detailed analysis of measures and results suggest some explanatory factors to the modest performance. On the one hand, environmental taxes have minimal incidence and revenue capacity due to the narrow range of activities subject to the tax and the extremely low prices of carbon emissions. To reach some effectiveness it should be necessary to expand substantially the tax base of the carbon tax, including all relevant emitters, and increase the tax rate at an accelerated pace (taking as reference some successful countries, e.g., Sweden (see Sterner [84]), or the expert recommendation to start at about EUR 50 per ton of CO_2 (see Edenhofer [85])). On the other hand, the anti-environmental incentives and benefits still outpace those designed to safeguard the environment. These are key results that allow us to understand the clearly negative environmental bias of the current tax system and, at the same time, to identify the type of problems and instruments on which to focus efforts to change tax policy in an environmentally friendly direction.

There are many reasons why Mexico and other developing countries have not given the same level of priority to the environmental agenda and to implementing the same advanced, systemic environmental fiscal policies as some developed countries (e.g., the urgent need to address the problems of poverty, welfare and job creation, or the awareness that the major global environmental problems are primarily the responsibility of the growth and consumption patterns of the more developed economies). However, it is increasingly evident that past and current fiscal architecture reinforces the unsustainability of the economic model, everywhere. Despite the progress of last three decades, the global balance sheet remains openly unsatisfactory, even for official international bodies [86]. A key conclusion of this work is that particular attention should be paid to the relevant environmental harmful benefits, a huge and hidden share of the current tax system. The problem is not exclusive to Mexico, but general in nature.

Significant progress in this direction requires taking firm steps towards systemic green tax reform that can align fiscal policy design with environmental and climate challenges. It will be necessary to focus more on taxation than on expenditure, because taxes (and tax expenditures) are among the most cost-effective environmental policy instruments [24,28,41,87]. The point is that taxes have the potential to change the relative prices of goods and services, but to do so changes must be significant in order to change the cost structure of the economy. Otherwise, the success in reshaping behaviours of economic agents towards sustainability will be very unlikely. Far from neutral, the tax shifts should have an uneven impact on different sectors, depending on the elasticity of supply and demand and the capacity of companies to transfer the tax impact to suppliers, customers or consumers.

The analysis of environmental taxes and tax expenditures in Mexico suggests the basis for an environmental tax reform in the short term: (a) significantly increase the tax base of the carbon tax, extending its application to all activities that are major emitters of CO_2, (b) significantly raise the tax levied on tonnes of CO_2 emitted, (c) eliminate in the short term all tax benefits for activities or products with clearly negative effects on the environment, and, (d) establish fair and generous tax benefits in VAT or corporate income tax for all labour-intensive circular economy activities such as repair, reuse, remanufacturing or remediation.

Of course, further research is needed to address the issue of the negative effects on income distribution and the design of the appropriate instruments to cushion it [88]. If the carbon tax were able to generate significant resources, these could be used to implement redistributive policies to benefit the poorest households or to balance significant VAT benefits for basic consumption. Going even further, the transition towards a sustainable circular economy requires a deeper study of the most profound changes needed in the architecture of the current tax system, which is directly or indirectly supported by the taxation of labour [89].

Author Contributions: Both authors contributed equally to the paper. All authors have read and agreed to the published version of the manuscript.

Funding: The authors belong to the ICEDE-Galician Competitive Research Group GRC ED431C2018/23 and to the "R2π: Transition from linear 2 circular: Policy and Innovation" H2020 Project. This research was supported by the Spanish Innovation Agency (AEI) through the project ECO-CIRCULAR: "La estrategia europea de transición a la economía circular: un análisis jurídico prospec-tivo y cambios en las cadenas globales de valor"—ECO2017-87142-C2-1-R. All these projects are co-funded by FEDER (UE). This research also received funding support from the National Council for Science and Research (CONACYT) México, PNPC 000625.

Institutional Review Board Statement: Not applicable.

Informed Consent Statement: Not applicable.

Data Availability Statement: Data Availability Statement:Publicly available datasets were analyzed in this study. This data canbe found here: CIATData Tax Expenditures "TEDLAC 2018 by Country". 2020. Available online: https://www.ciat.org/gastos-tributarios/ (accessed on 4 May 2020).

Acknowledgments: The authors are grateful to the comments and suggestions made by the three anonymous reviewers.

Conflicts of Interest: The authors declare no conflict of interest.

References

1. Stern, N. *El Informe de Stern: La Verdad Sobre El Cambio Climático*; Paidos Ibérica, S., Ed.; MONDE Diplomatique: Barcelona, Spain, 2007.
2. Rockström, J.; Steffen, W.; Noone, K. Planetary Boundaries: Exploring the Safe Operating Space for Humanity. *Ecol. Soc.* **2009**, *14*, 32. [CrossRef]
3. Martínez-Alier, J.; Jusmet, J.R. *Economía Ecológica y Política Ambiental*, 3rd ed.; Fondo de Cultura Económica: Mexico City, Mexico, 2013.
4. Secretaría de Mediio Ambiente y Recursos Naturales (SEMARNAT). Indicadores de Crecimiento Verde. Available online: https://apps1.semarnat.gob.mx:8443/dgeia/indicadores_verdes/indicadores/00_intros/intro.html (accessed on 10 October 2020).
5. IPCC. *Global Warming of 1.5 °C. An IPCC Special Report on the Impacts of Global Warming of 1.5 °C above Pre-Industrial Levels and Related Global Greenhouse Gas Emission Pathways, in the Context of Strengthening the Global Response to the Threat of Climate Change*; IPCC: Geneva, Switzerland, 2018.
6. International Monetary Fund (IMF). *Fiscal Policies for Paris Climate Strategies—From Principle to Practice*; International Monetary Fund: Washington, DC, USA, 2019; Volume 19, p. 1. [CrossRef]
7. Krogstrup, S.; Oman, W. *Macroeconomic and Financial Policies for Climate Change Mitigation*; International Monetary Fund: Washington, DC, USA, 2019; Volume 19. [CrossRef]
8. Schmidt, C.; Gebin, G.; Van Houten, F.; Van Close, C.; McGinty, D.B.; Arora, R.; Potocnik, J.; Ishii, N.; Bakker, P.; Kituyi, M.; et al. The Circularity Gap Report 2020. *Circ. Econ.* **2020**, *3*, 69.
9. Elkins, P.; Barker, P. Carbon Taxes and Emissions Trading. *J. Econ. Surv.* **2001**, *15*, 325–376. [CrossRef]
10. Georgescu-Roegen, N. Qué Puede Enserñar a Los Economistas La Termodinamica y La Biología? In *De la Economía Ambiental a la Economía Ecológica*; Aguilera, F., Alcántara, V., Eds.; Icaria: Barcelona, Spain, 1994; pp. 303–319.
11. Naredo, J.M. *Reíces Económicas Del Deterioro Ecológico y Social. Más Allá de Los Dogmas*; Siglo XXI de España Editores: Madrid, Spain, 2006.
12. Naredo, J.M. Fundamentos de La Economía Ecológica. In *De la Economía Ambiental a la Economía Ecológica*; Aguilera, F., Alcántara, V., Eds.; Icaria: Barcelona, Spain, 1994; pp. 235–252.
13. Smith, R. *Green Capitalism: The God That Failed*; World Economics Association, College Publications: Norcross, GA, USA, 2016.
14. Common, M.; Stagl, S. *Introducción a La Economía*, 1st ed.; Reverte: Barcelona, Spain, 2008.
15. International Energy Agency (IEA). Data and Statistics. Available online: https://www.iea.org/data-and-statistics (accessed on 4 May 2021).
16. Arlinghaus, J.; van Dender, K. *The Environmental Tax and Subsidy Reform in Mexico*; OECD: London, UK, 2017. [CrossRef]
17. Neri, A.F. Tributos Ambientales En México: Una Revisión de Su Evolución y Problemas. *Boletín Mex. Derecho Comp.* **2005**, *38*, 991–1020.
18. Azqueta, D.; Ramírez, A. *Introducción a la Economía Ambiental*; McGraw-Hill Interamericana: Madrid, Spain, 2007.
19. Secretaria de Hacienda y Crédito Público (SHCP). *Criterios Generales de Política Económica 2014*; Gobierno de México: Mexico City, Mexico, 2013; p. 194.
20. CEPAL. Panorama Fiscal de América Latina y el Caribe 2019: Políticas tributarias Para la Movilización de Recursos en el Marco de la Agenda 2030 Para el Desarrollo Sostenible. Available online: https://repositorio.cepal.org/handle/11362/44516 (accessed on 14 April 2021).
21. Lorenzo, F. *Inventario de Instrumentos Fiscales Verdes En América Latina*; LC/W.723; CEPAL: Santiago, Chile, 2016.
22. Beeks, J.C.; Lambert, T. Addressing Externalities: An Externality Factor Tax-Subsidy Proposal. *Eur. J. Sustain. Dev. Res.* **2018**, *2*, 1–19. [CrossRef]
23. ECLAC; OXFAM. *Tax Incentives in Latin America and the Caribbean*; LC/TS.2019/50; ECLAC: Santiago, Chile; OXFAM: Nairobi, Kenya, 2019.
24. Baumol, W.J. On Taxation and the Control of Externalities. *Am. Econ. Rev.* **1972**, *62*, 307–322.
25. Stiglitz, J. *La Economía del Sector Público*; Antoni Bosch: Barcelona, Spain, 2000.
26. Gago, A.; Labandeira, X.; López-Otero, X. *Las Nuevas Reformas Fiscales Verdes*; WP 05/2016; Economics for Energy: Vigo, Spain, 2016.
27. Freire-González, J. Environmental Taxation and the Double Dividend Hypothesis in CGE Modelling Literature: A Critical Review. *J. Policy Model.* **2018**, *40*, 194–223. [CrossRef]
28. Metcalf, G.E. On the Economics of a Carbon Tax for the United States. *Brook. Pap. Econ. Act.* **2019**, *1*, 405–484. [CrossRef]
29. Nordhaus, W.D. Carbon Taxes to Move toward Fiscal Sustainability. In *The Economists' Voice 2.0*; Columbia University Press: New York, NY, USA, 2010; Volume 7. [CrossRef]
30. Milne, J.E. Environmental Taxes. In *Elgar Encyclopedia of Environmental Law*; Faure, M., Ed.; Edward Elgar Publishing: Cheltenham, UK, 2020; pp. 170–182.

31. Aguilera, F.; Alcántara, V. *De la Economía Ambiental a la Economía Ecológica*; Icaria: Barcelona, Spain, 1994.
32. Fanelli, J.M.; Jiménez, J.P.; López, I. *La Reforma Fiscal Ambiental En América Latina*; LC/W.683; CEPAL: Santiago, Chile, 2015.
33. Barde, J.P. Reformas Tributarias Ambientales En Países de La Organización de Cooperación y Desarrollo Económicos (OCDE). In *Política Fiscal y Medio Ambiente. Bases Para una Agenda Común*; Acquatella, J., Bárcena, A., Eds.; Naciones Unidas: Santiago, Chile 2005; pp. 65–88.
34. Bosquet, B. Environmental Tax Reform: Does It Work? *Ecol. Econ.* **2000**, *34*, 19–32. [CrossRef]
35. Ekins, P.; Speck, S. *Environmental Tax Reform (ETR): A Policy for Green Growth*; Illustrate; Ekins, P., Speck, S., Eds.; Oxford University Press: Oxford, UK, 2011.
36. Gago, A.; Labandeira, X. *Impuestos Ambientales y Reformas Fiscales Verdes En Perspectiva*; 09/2010; Economics for Energy: Vigo Spain, 2010.
37. Banco Interamericano de Desarrollo (IBD); Cooperación Alemana (GIZ); Centro Interamericano de Administraciones Tributaria (CIAT). *Modelo de Código Tributario del CIAT: Un Enfoque Basado En la Experiencia Iberoamericana*; CIAT: Panama, Panama, 2015 p. 210.
38. Doshi, T.K. *Costs and Benefits of Market-Based Instruments in Accelerating Low-Carbon Energy Transition*; Anbumozhi, V., Kalirajan K., Kimura, F., Eds.; Springer: Singapore, 2018. [CrossRef]
39. Andrew, J.; Kaidonis, M.A.; Andrew, B. Carbon Tax: Challenging Neoliberal Solutions to Climate Change. *Crit. Perspect. Account.* **2010**, *21*, 611–618. [CrossRef]
40. Nordhaus, W.D. The Many Advantages of Carbon Taxes. *Tax. Debate Clim. Policy Copenhagen. Growth* **2009**, *61*, 64–70.
41. Nordhaus, W.D. To Tax or Not to Tax: Alternative Approaches to Slowing Global WarmingNo Title. *Rev. Environ. Econ. Policy* **2007**, *1*, 26–44. [CrossRef]
42. Pearse, R.; Böhm, S. Ten Reasons Why Carbon Markets Will Not Bring about Radical Emissions Reduction. *Carbon Manag.* **2014**, *5* 325–337. [CrossRef]
43. Coelho, R.S. *The High Cost of Cost Efficiency: A Critique of Carbon Trading*; Universidade de Coimbra: Coimbra, Portugal, 2015.
44. Böhm, S.; Misoczky, M.C.; Moog, S. Greening Capitalism? A Marxist Critique of Carbon Markets. *Organ. Stud.* **2012**, *33*, 1617–1638 [CrossRef]
45. Groothuis, F. *Tax as a Force for Good Rebalancing Our Tax Systems to Support a Global Economy Fit for the Future*; The Association of Chartered Certified Accountants: London, UK, 2018.
46. Smith, R. An Ecosocialist Path to Limiting Global Temperature Rise to 1.5 °C. *Real-World Econ. Rev.* **2019**, 149.
47. Nordhaus, W.D. *El Casino Del Clima. Por Qué No Tomar Medidas Contra El Cambio Climático Conlleva Riesgo y Genera Incertidumbre* 1st ed.; Deusto: Barcelona, Spain, 2019.
48. OECD. *OECD Economic Surveys: Mexico 2019*; OECD Publishing: Paris, France, 2019. [CrossRef]
49. Best, R.; Burke, P.J.; Jotzo, F. Carbon Pricing Efficacy: Cross-Country Evidence. *Environ. Resour. Econ.* **2020**, *77*, 69–94. [CrossRef]
50. Jiménez, J.P.; Podestá, A. *Inversión, Incentivos Fiscales y Gastos Tributarios En América Latina*; CEPAL: Santiago, Chile, 2009.
51. CEPAL; Oxfam Internacional. Los Incentivos Fiscales a Las Empresas En América Latina y El Caribe. In *Comisión Económica Para América Latina y el Caribe/Oxfam Internacional*; Naciones Unidadas y Oxfam: Santiago, Chile, 2019; pp. 1–81.
52. Ashiabor, H. *Tax Expenditures and Environmental Policy*; Edward Elgar Publishing: Cheltenham, UK, 2020. [CrossRef]
53. Peláez Longinotti, F. *Los Gastos Tributarios En Los Países Miembros del CIAT*; 2219-780X; DT-06-2019; CIAT: London, UK, 2019.
54. Agostini, C.; Jorratt, M. Política Tributaria Para Mejorar La Inversión y El Crecimiento En América Latina. In *Consensos y Confictos en la Política Tributaria de América Latina*; Gómez Sabaini, J.C., Jiménez, P., Martner, R., Eds.; Naciones Unidas: Santiago, Chile 2017; pp. 229–251.
55. Surrey, S.S.; Mcdaniel, P.R. Tax Expenditures and Tax Reform. *Vand. L. Rev.* **1985**, *38*, 1397–1414.
56. OECD. *Tax Expenditures in OECD Countries*; OECD Publishing: Paris, France, 2010. [CrossRef]
57. López Pérez, S.J.; Vence, X. Estructura y Evolución de Ingresos Tributarios y Beneficios Fiscales En México. Análisis Del Periodo 1990–2019 y Evaluación de La Reforma Fiscal de 2014. *Trimest. Econ.* **2021**, *88*, 373–417. [CrossRef]
58. CIAT-Interamerican Center of Tax Administration. CIAT Data Tax Revenues. Available online: https://www.ciat.org/base-de-datos-de-recaudacion-bid-ciat/ (accessed on 4 May 2020).
59. CIAT-Interamerican Center of Tax Administration. CIAT Data Tax Expeditures. Available online: https://www.ciat.org/gastos-tributarios/ (accessed on 4 May 2020).
60. Secretaria de Hacienda y Crédito Público (SHCP). *Nota Metodológica. Recaudación Del IEPS a Combustibles Fósiles*; Estadísticas Oportunas de Finanzas Pública: Mexico City, Mexico, 2018; p. 3.
61. Secretaria de Hacienda y Crédito Público (SHCP); Servicios de Administración Tributaria (SAT). Datos abiertos del SAT. Available online: http://omawww.sat.gob.mx/cifras_sat/Paginas/datos/vinculo.html?page=IngresosTributarios.html (accessed on 20 April 2020).
62. Zubillaga, J.M.A. *La Utilización Extrafiscal de Los Tributos y Los Principios de Justicia Tributaria*; Servicio Editorial de la Universidad del País Vasco: Bilbao, Spain, 2001.
63. Villela, L.; Lemgrumber, A.; Jorrat, M. *Gastos Tributarios*; CEPAL: Santiago, Chile, 2012; Volume 20.
64. Secretaria de Gobernación (SEGOB). *DECRETO Que Establece Los Estímulos Fiscales Para El Fomento de La Actividad Preventiva de La Contaminación Ambiental*; Diario Oficial de la Federación (DOF): Mexico City, Mexico, 1981; p. 5.

65. López Pérez, S.J.; Vence, X. Structure and Evolution of Tax Revenues and Tax Benefits in Mexico. Analysis of the 1990–2019 Period and Evaluation of the 2014 Fiscal Reform. *Trimest. Econ.* **2021**, *88*, 373–417. [CrossRef]

66. OECD. *The Political Economy of Environmentally Related Taxes*; OECD Publishing: Paris, France, 2006. [CrossRef]

67. Hernández, F.; Antón, A. *El Impuesto Sobre Las Gasolinas. Una Aplicación Para El Ecuador, El Salvador y México*; Estudios de Cambio Climático en América Latina; LC/W5978; CEPAL: Santiago, Chile, 2014.

68. Elizondo, A.; Pérez-Cirera, V.; Strapasson, A.; Fernández, J.C.; Cruz-Cano, D. Mexico's Low Carbon Futures: An Integrated Assessment for Energy Planning and Climate Change Mitigation by 2050. *Futures* **2017**, *93*, 14–26. [CrossRef]

69. OECD. *Effective Carbon Rates 2018: Pricing Carbon Emissions through Taxes and Emissions Trading*; OECD Publishing: Paris, France, 2018. [CrossRef]

70. Gómez Sabaíni, J.C.; Jiménez, J.P.; Morán, D. El Impacto Fiscal de La Explotación de Los Recursos Naturales No Renovables. In *Consensos y Conflictos en la Política Tributaria de América Latina*; Gómez Sabaíni, J.C., Jiménez, J.P., Morán, D., Eds.; Naciones Unidas: Santiago, Chile, 2017; pp. 393–413.

71. Altomonte, H.; Sánchez, R.J. *Hacia Una Nueva Gobernanza de Los Recursos Naturales En América Latina y El Caribe*; Libros de; Naciones Unidas: Santiago, Chile, 2016.

72. OECD; CIAT; IDB; ECLAC. *Revenue Statistics in Latin America and the Caribbean*; OECD Publishing: Paris, France, 2020. [CrossRef]

73. Cámara de Diputados del Congreso de la Union. *Ley Del Impuesto Especial Sobre Producción y Servicios (LIEPS)*; DOF 09-12-2019; Cámara de Diputados del Congreso de la Union: Mexico City, Mexico, 2019; p. 148.

74. Secretaria de Gobernación (SEGOB). *DECRETO por el que se Reforman, Adicionan y Derogan Diversas Disposiciones de la Ley del Impuesto Sobre la Renta, de la Ley del Impuesto Especial Sobre Producción y Servicios, del Código Fiscal de la Federación y de la Ley Federal de Presupuesto y Responsabilidad Hacendaria*; Diario Oficial de la Federación (DOF): Mexico City, Mexico, 2015; p. 48, Reforma 42: Ley del Impuesto Especial sobre Producción y Servicios. DOF 18-11-2015 (diputados.gob.mx).

75. Banco Mundial. DataBank. *Crecimiento del PIB (% Annual)—México*. Available online: https://datos.bancomundial.org/indicator/NY.GDP.MKTP.KD.ZG?locations=MX (accessed on 11 October 2021).

76. Galindo, L.M.; Beltrán, A.; Ferrer Carbonell, J.; Alatorre, J.E. *Potential Effects of a Carbon Tax on Gross Domestic Product in Latin American Countries: Preliminary and Hypothetical Estimates from a Meta-Analysis and Benefit Transfer Function*; S.17-00590; LC/TS.2017/58; CEPAL: Santiago, Chile, 2017.

77. Alatorre, J.E.; Beltrán, A.; Ferrer, J.; Galindo, L.M. *Reformas Fiscales Ambientales e Innovación y Difusión Tecnológicas En El Contexto de Las Contribuciones Determinadas (CDN): Una Visión Desde América Latina Gracias Por Su Interés En Esta Publicación de La CEPAL*; S.18-00469; LC/TS.2018/78; CEPAL: Santiago, Chile, 2018.

78. Stavins, R.N. The Future of U S Carbon-Pricing Policy. *Environ. Energy Policy Econ.* **2019**, *1*, 25912.

79. Cámara de Diputados del Congreso de la Union. *Ley General de Cambio Climático*; Cámara de Diputados del Congreso de la Union: Mexico City, Mexico, 2014; pp. 1–45.

80. Cámara de Diputados del Congreso de la Union. *Ley General Del Equilibrio Ecológico y La Protección Al Ambiente*; Cámara de Diputados del Congreso de la Union: Mexico City, Mexico, 2021; pp. 1–138.

81. Nachmany, M.; Fankhauser, S.; Townshend, T.; Collins, M.; Matthews, A.; Pavese, C.; Rietig, K. *The Globe Climate Legislation Study: A Review of Climate Change Legislation in 66 Countries*; GLOBE International and Grantham Research Institute: London, UK, 2014.

82. Stahel, W.R. Policy for Material Efficiency—Sustainable Taxation as a Departure from the Throwaway Society. *Philos. Trans. R. Soc. A Math. Phys. Eng. Sci.* **2013**, *371*, 20110567. [CrossRef] [PubMed]

83. Rosas-Flores, J.A.; Bakhat, M.; Rosas Flores, D.; Zayasm, J.L. Distributional Effects of Subsidy Removal and Implementation of Carbon Taxes in Mexican Households. *Energy Econ.* **2017**, *61*, 21–28. [CrossRef]

84. OECD; World Bank; United Nations Environment Programme. *Financing Climate Futures Rethinking Infrastructure*; OECD Publishing: Paris, France, 2018.

85. Sterner, T. The Carbon Tax in Sweden. In *Standing up for a Sustainable World*; ElgarOnline: Cheltenham, UK, 2020; pp. 59–67. [CrossRef]

86. Edenhofer, O.; Flachsland, C.; Kalkuhl, M.; Knopf, B.; Pahle, M. *Options for a Carbon Pricing Reform*; Mercator Research Institute on Global Commons and Climate Change (MCC): Berlin, Germany, 2019; pp. 1–9.

87. Bogacheva, O.V.; Fokina, T.V. *Tax Expenditures Management in OECD Countries*; OECD Publishing: Paris, France, 2017; Volume 61. [CrossRef]

88. Klenert, D.; Schwerhoff, G.; Edenhoofer, O.; Mattauch, L. Environmental Taxation, Inequality and Engel's Law: The Double Dividend of Redistribution. *Environ. Resour. Econ.* **2018**, *71*, 605–624. [CrossRef]

89. Vence, X.; López Pérez, S.D.J. Taxation for a Circular Economy: New Instruments, Reforms, and Architectural Changes in the Fiscal System. *Sustainability* **2021**, *13*, 4581. [CrossRef]

Article

A Future Study of an Environment Driving Force (EDR): The Impacts of Urmia Lake Water-Level Fluctuations on Human Settlements

Somayeh Mohammadi Hamidi [1,2,*], Christine Fürst [1], Hossein Nazmfar [2], Ahad Rezayan [3] and Mohammad Hassan Yazdani [2]

1 Department of Sustainable Landscape Development, Institute for Geosciences and Geography, Martin Luther University Halle-Wittenberg, 06120 Halle (Saale), Germany; christine.fuerst@geo.uni-halle.de
2 Department of Geography and Urban Planning, University of Mohaghegh Ardabili, Ardabil 009845, Iran; nazmfar@uma.ac.ir (H.N.); yazdani@uma.ac.ir (M.H.Y.)
3 Futures Studies Department, National Research Institute for Science Policy, Tehran 009821, Iran; rezayan@nrisp.ac.ir
* Correspondence: S_mohammadi@uma.ac.ir; Tel.: +49-(0)-16-3489-0422

Abstract: Lake Urmia, one of the world's largest salt lakes, is rapidly losing water and drying up. This environmental hazard has raised concerns about the consequences and impact on the surrounding communities. In this paper, we use a futuristic view (horizon of 10 years based on medium-term planning) to identify the main environmental drivers in the surrounding settlements of the Urmia Lake basin. A qualitative method, based on cross-impact analysis, was used as a means of future research. We also used a Delphi-based expert panel method to collect data and extract the environmental impacts of Urmia Lake. After the three rounds of the Delphi process, the expert panel reached a high level of agreement (100%) on the top 17 environmental consequences. Then, these consequences were classified by driving force and dependency using the MICMAC method. The results show that reducing pasture area, soil and water salinity, groundwater decline and depletion, and destruction of surrounding agricultural lands play a significant role in environmental change in Urmia Lake. Overall, any small change in these variables may lead to fundamental changes in the entire system.

Keywords: future study; driving force; MICMAC; Delphi; Urmia Lake

Citation: Hamidi, S.M.; Fürst, C.; Nazmfar, H.; Rezayan, A.; Yazdani, M.H. A Future Study of an Environment Driving Force (EDR): The Impacts of Urmia Lake Water-Level Fluctuations on Human Settlements. *Sustainability* **2021**, *13*, 1495. https://doi.org/10.3390/su132011495

Academic Editors: Jun Yang, Ayyoob Sharifi, Baojie He and Chi Feng

Received: 22 August 2021
Accepted: 9 October 2021
Published: 18 October 2021

Publisher's Note: MDPI stays neutral with regard to jurisdictional claims in published maps and institutional affiliations.

1. Introduction

Salt lakes are a significant part of the earth's inland aquatic ecosystems and are found worldwide. These lakes play a pivotal role in determining regional climate patterns, maintaining biotic productivity and diversity, sustaining ecological and human health, and providing recreational services, minerals, and other resources [1]. Therefore, the dynamics of salt lakes are of great importance to a wide range of stakeholders [2].

The Caspian Sea is the world's largest salt lake, accounting for 41% of all salt lake volume, and is home to a robust fishing, shipping, and mining sector. Other major hypersaline systems, such as the Great Salt Lake, offer various waterfowl habitats that can be exploited for minerals. Flamingos and other birds live in small Andean salars and lakes in the Middle East and Africa [3]. Around the world, we witness that salt lakes are shrinking. On closer inspection, environmental factors, namely almost all human activities, such as diversion of surface inflows, salinisation, mining activities, pollution, and climate change, threaten salt lakes [4]. Humans' increased water use, particularly for agricultural irrigation, is a major contributor to lakes drying up. Agricultural water consumption in the Aral Sea 2 watershed, for example, has reduced the lake's surface area by 74% and volume by 90% [5]. There is little doubt that, by 2025, the natural character of most of the world's salt lakes will have changed. Several recent 'vision' statements clearly point in this direction.

In central Asia, for example, the 'vision' proposed by UNESCO (2000) for the Aral Sea basin involves almost complete desiccation of the lake itself, and greatly increased 'development' of its catchment to support the growing populations of Kazahkstan, Uzbekistan, Turkmenistan, and other smaller states in the region [1]. Figure 1 shows the largest salt lakes in the world which are losing their water and drying up.

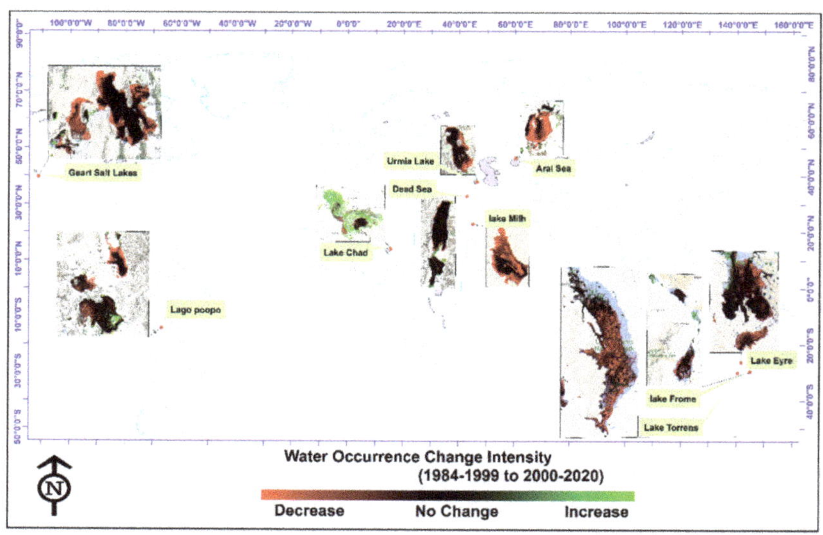

Figure 1. Examples of the observed decrease in surface water area in major salt lakes worldwide over the period 1984–2015 [6].

The drying up of salt lakes is not a new occurrence, and scientists have observed an alarming and accelerated loss in many of these essential ecosystems [7]. The Evans River diversion, for example, became the major source of pollution and particle matter in that section of the lake in 1926, creating asthma and other health concerns for nearly 40,000 people living there [8]. In a similar vein, the Utah Lake in the United States and the Aral Sea in Kazakhstan and Uzbekistan [9] have caused many disasters for the residents of these areas. The Tarim Basin, which resulted in the collapse of the Loulan Kingdom in 645 environmental consequences, is likely the oldest known direct human action that caused salt lakes to dry up [10]. Other consequences of increasing water consumption are more recent. Since 2000, the Salton Sea in California has dropped by over 7 m due to management measures that restricted water input to the lake [11].

Lake Urmia in Iran, like many other salt lakes on the continent except Antarctica, has experienced a similar fate. Nowadays, Lake Urmia exposes a salt desert that produces dust harmful to health and threatens crops and people [12]. According to similar occurrences involving Lake Aral, Lake Ebinur, and Lake Chad, the continued drying process of Urmia Lake would bring multiple social, environmental, and ecological disastrous consequences to the region [4] and pose a threat to the regional economy and human wellbeing.

The major goal of this paper is to (1) determine the environmental impacts of water level fluctuations of Urmia Lake; (2) to identify the environmental drivers (10-year horizon) in the study area; (3) analyse the impacts of these drivers on human settlements around the lake that are directly and indirectly affected by these impacts.

2. Study Design and Methods

2.1. Model Region: Lake Urmia Basin, Iran

The largest hyper-saline water body in Iran, Lake Urmia, is located in northwestern Iran and encompasses around 52,000 square kilometres (Figure 2).

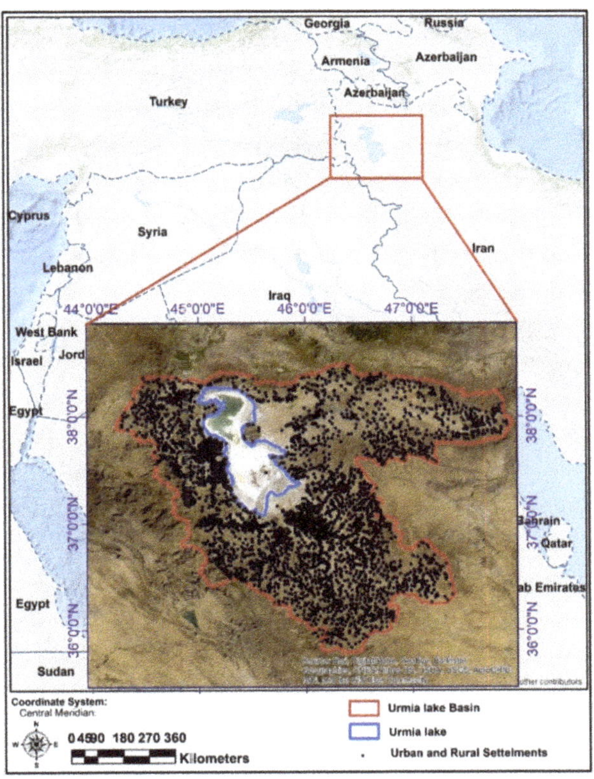

Figure 2. Location of the study area (Urmia Lake basin) and settlements.

The lake is located in a mountainous region between three provinces: East Azerbaijan (39 percent), West Azerbaijan (51 percent), and the Southern Province of Kurdistan (10 percent). The lake's mean annual precipitation ranges between 240 and 272 mm [13], with approximately 77 percent of the precipitation falling between December and May each year. Precisely speaking, it is the largest lake in the Middle East and the third largest salt lake globally [14], at 140 km long, 55 km wide, and 16 m deep [15].

After two decades (2000–2020) of intensive agricultural development, construction of reservoirs, and reduced river flows, the water level of Lake Urmia in northwestern Iran decreased by 6.5 m from a historical maximum of 1278 m in 2000 to 1271.5 m in July 2020, during which time the lake lost almost 45% of its area and 85% of its volume [16]. Figure 3 shows the trend of lake water depletion from 1975 to 2018.

The water area of Urmia Lake was estimated to be 5412 km^2 in 1975. UL's surface area decreased by more than 5000 km^2 between 1995 and 2014, despite the lake's water surface area growing after that. According to the Urmia Lake Restoration Program report (ULRP), in 2016, the increase in rainfall in the whole basin, especially in some stations, was about 43% compared to the long-term average) [17]. The water surface area increased by 1490 km^2 in 2016, whereas the salt body shrank by 348 km^2. In August 2018, the lake's water surface area was estimated to be 1624 km^2 (1852 km^2) [18].

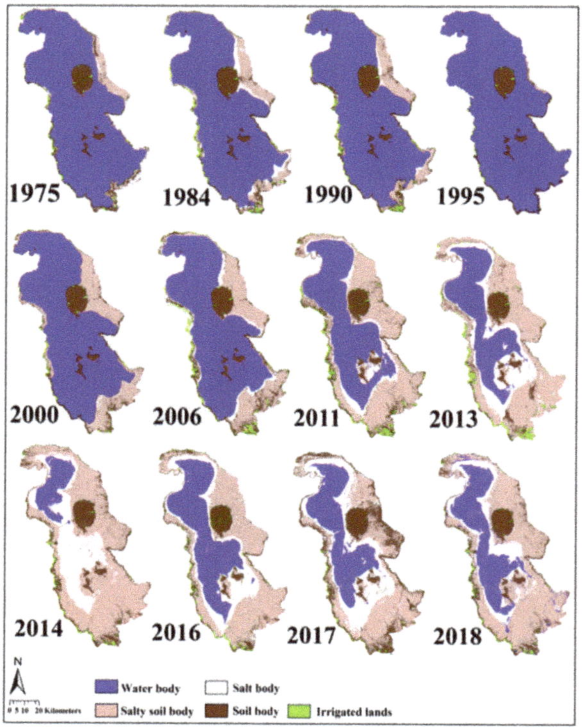

Figure 3. From 1975 to 2018, the salinisation and desertification of Urmia Lake progressed.

In the Urmia lake basin (Refare to Figure 2), more than 36 cities and 3150 villages have more than 5 million inhabitants (in this study, we have assumed these settlements to be an integrated system, whether or not environmental impacts have disrupted the system) Important demographic characteristics of the population within the ecological zone (a zone or area characterized by extensive but relatively homogeneous natural vegetation formations) [1], which is subject to an increasing old-age dependency ratio, a decreasing gender dependency ratio as a result of male labor migration, rising unemployment, and rapidly rising literacy [19]. Table 1 shows Urmia Lake's key ecological, hydrological, climate, social, and economic data characteristics.

One of the most important government programs to address the declining lake levels was the Urmia Lake Restoration Program. ULRP was launched in 2013 under the responsibility of the Ministry of Energy and in cooperation with the Ministry of Agriculture and the Iranian Ministry of Planning and Budget Organization to apply integrated solutions in the field of lake basin management and present several solutions at Sharif University of Technology. As Figure 4 shows, the vision of the ULRP is to reach ecological level (1274MCM).

Table 1. Economic-social and ecological characteristic of Urmia Lake.

	Attributes		Description	
Social	**Settlement**		**Population**	
	3186		5 million	
	Village	3150	City	69%
	City	36	Village	31%
Economic	**GDP percent share**		**Employment sectors**	
	Service	58	Service sector	39%
	Agriculture	15	Agriculture sector	36%
	Industry	27	Industry sector	25%
	Cultivation Status			
	Aquaculture	39.6%		
	Rain-fed Agriculture	60.4%		
	Conservation status	**Products**	**Services**	**Functions**
Ecological	Lake is a National Park, Protected since 1957 (No hunting area), Ramsar site, UNESCO Biosphere Reserve	Salt harvest, grazing for domestic animals, Artemia harvest, fishery) some satellite wetlands only), controlling saline underground waters, water-birds for hunting, reeds, fodder, medicinal herbs	Tourism/eco-tourism, recreation, education, training, research, therapeutic muds, cultural heritage	Lake: Biodiversity support, landscape, climatic moderation, pollution and sediment retention. Satellite Wetlands: Biodiversity support, landscape, ground water recharge, climatic moderation, pollutant and sediment retention

Climate	**Elevation (m)**	**Average annual daily temperature (°C)**			**Frost (Days)**	**Average Wind speed (Km/h)**	**Evapotranspiration (mm)**
	1361	Min	Ave.	Max	101.0	2.9	1626
		16.7	12.1	17.6			

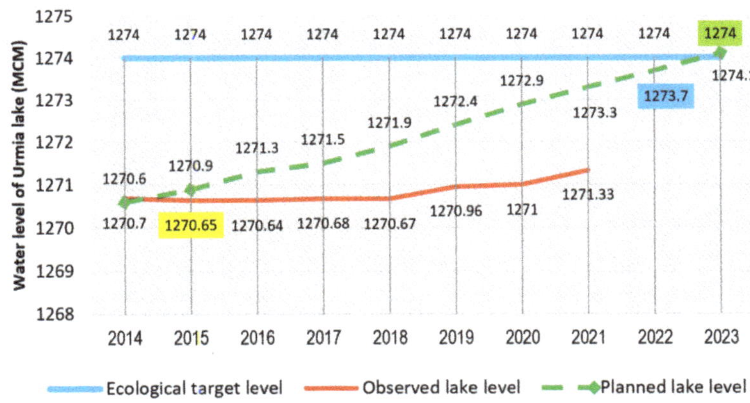

Figure 4. Lake Urmia restoration road map.

According to Figure 4, Lake Urmia restoration program has been planned in three steps, as follows: (1) the stabilisation period (2014–2016)—the main goal in this period is to stabilise Lake Urmia's level and implement projects to reduce the potential impacts of the persistence of Lake Urmia; (2) the restoration period (2016–2022)—the main objective of this period is to implement all solutions for the lake's water supply and gradually increase its level (1273MCM); and (3) final restoration (2023)—the expected purpose of this period is to stabilise the lake's restoration conditions and create the necessary conditions for the final restoration and sustainable maintenance of lake conditions (1274.1 MCM).

2.2. Methodology

This study is based on cross-impact analysis, which falls under the category of cross-impact analysis approaches. The most popular methods are Gordon's cross-impact method, SMIC, BASICS, MICMAC, KSIM, and the cross-impact balance approach [20]. Accordingly, this study used the Delphi Survey for data collection and MICMAC for analysis due to the qualitative database.

2.2.1. Delphi Survey: Foresight Desirability

Foresight departs from probabilistic prediction by fostering a notion of the future as variously imagined, shaped, and influenced through strategic planning and intervention [21]. Opening up foresight processes implies the advantage of integrating valuable sources of knowledge more effectively and systematically to make future decisions based on comprehensible facts [22]. Foresight methods allow individuals and organisations to imagine different future scenarios and plan for greater future resilience [23]. Consequently, foresight can identify benefits and constraints in terms of innovative solutions [24]. There are many methods of foresight or predicting the future. One of the leading tools in foresight is the Delphi method. Delphi is defined as a means of organizing a group communication process to allow a group of people to address a difficult problem as a whole. Through the four features of Delphi: (1) anonymity; (2) feedback; (3) statistical aggregation of group responses; and (4) iteration, experts can achieve more objective results, such as more objective probability estimates [23]. Delphi and other expert panel methods for planning the future are also widely used in environmental studies [25]. We performed a three-round Delphi study between November 2019 and June 2020, comprising three overarching phases as follows:

Round 1: In the first Delphi round, key experts were identified. There are no clear guidelines on the number of participants included in studies that use Delphi questioning as the sample is purposively selected, depending on the problem being studied [26]. Some studies have used 15 participants, while others have used 60 [27]. In this paper, the target panel of experts included 35 people (Table 2). They were selected based on published papers (also five people who have worked on the Aral Sea); recommendations by other scientists; and consultations with directors of scientific institutions, such as the Lake Urmia restoration program (ULRP), the Urmia Lake research Institute, and Urmia and Tabriz Universities. Except for five people (X11, 12, 13, 15, and 17), all experts had local knowledge (village level) and regional knowledge (research across districts).

Table 2. Expert panel characteristics.

Group	ID	Affiliation	Implementation
Urmia Lake Restoration centre Committee	X1	Chairman House Agriculture Committee	XP
	X2	Head of Urmia Lake Restoration centre, East Azerbaijan province Branch	XP
	X3	Member of the Social & Cultural Council of Urmia Lake Restoration centre	XP
	X4	head of the National Wheat Farmers Foundation	Q
	X5	Chairman of the Alternative Employment and Livelihood Committee of Urmia Lake Restoration centre	XP
	X6	Head of the Lake Urmia Monitoring Department of Tabriz University	XP
	X7	Chairman of the NGO Committee	XP
	X8	Member of the Social & Cultural Council of Urmia Lake Restoration centre	XP
	X9	Member of the Social & Cultural Council of Urmia Lake Restoration centre	XP
Researchers who have worked on Aral sea and Lake Urmia.	X10	Teaching Assistant at The University of British Columbia, As part of the NSERC ResNet Strategic Network	Q
	X11	institute of geology and geophysics, Uzbekistan Academy of Sciences	Q
	X12	Faculty of Geography, Lomonosov Moscow State University(PhD)	Q
	X13	Chief of Executing Office (Head), Agency for project implementation of the International Fund for the Aral Sea Saving	Q

Table 2. *Cont.*

Group	ID	Affiliation	Implementation
	X14	Head of Department, Urmia Lake Restoration Program-Sharif University of Technology	Q
	X15	Professor (Assistant), Faculty of Natural Resources, Urmia University, Urmia, Iran.	Q
	X15	Professor (Associate), KIMEP University, Kazakhstan	Q
	X16	Head of Department, Urmia University, Urmia Lake Research Institute	Q
	X17	Researcher, Department of Environmental Systems Science, ETH Zurich	Q
	X18	Department of Climatology, Faculty of Geography and Planning, University of Tabriz	Q
	X19	Department of Climatology, Faculty of Geography and Planning, University of Tabriz	Q
	X20	Head of Department, Associate Professor in Climatology at University of Tehran	Q
	X21	Professor of Climatology, University of Zanjan, Zanjan,	Q
	X22	Professor of Natural Geography Department, University Of Mohaghegh Ardabili	Q
	X23	Researcher, Department Of geography and urban planning, University of Zanjan	Q
	X24	Researcher, Department of Political Geography, Kharazmi University	Q
	X25	Researcher, Department of geography and urban-rural planning, University Of Mohaghegh Ardabili	Q
	X26	Active member of Lake Urmia Environmental Saviors Institute	Q
	X27	Active member of Yam Yashil Qushachai Institute	I
	X28	Active member of Tabriz Green Thinkers Association	I
	X29	Active member of Iranian Environmental fans Association	Q
NGOs members	X30	Active member of Yam Yashil Qushachai Institute	Q
	X31	Active member of Yam Yashil Qushachai Institute	Q
	P31	Active member of Green Bio Collaborators Association	Q
	P32	Active member of Green Hearts Association	Q
	P33	Active member of ' I'm Lake Urmia' campaign	Q
	P34	Active member of ' I'm Lake Urmia' campaign	Q
	P35	Active member of ' I'm Lake Urmia' campaign	Q

XP: An expert panel (official meeting) was held in the office of the Lake Urmia Restoration Center in East Azerbaijan Province. Led by a futurist researcher (Fellow at the National Research Institute for Science Policy. Futures Studies Postdoctoral Researcher, University of Tehran, Iran). I: Interview. Q: Questionnaire.

Due to the dispersion of experts, at first, a meeting was held in cooperation with the Lake Urmia Restoration Centre with committee members led by a futurist. In this meeting, members discussed the environmental consequences of Lake Urmia in the future. Then, a questionnaire was sent to other experts. The questionnaire included an open-ended question that was in the spirit of brainstorming. In this question, the experts were asked, "What will be the environmental impact of drying up Lake Urmia? (with a 10-year horizon)". A total of 35 (100% response rate) completed the first round. Subsequently, all responses were coded using the qualitative technique of grounded theory (open and axial coding). Coding of the open-ended questions of the questionnaire resulted in the identification of 80 general concepts, and axial coding took the form of 62 categories and 35 main outcome indicators.

Round 2: The second round was conducted with experts at Urmia Lake Restoration Centre. Nine experts (25.71% response rate) participated in the second round. In this round, the results of the first round were discussed. Finally, 17 environmental consequences were selected, which were agreed upon by the panel members.

Round 3: The third round was also via email. We emailed 17 environmental consequences (which had been agreed upon in Round 2) to all 35 participants, and 27 experts participated in Round 3 (77.14% response rate). After applying some changes by the respondents, the results of the expert responses were summarized in Table 3.

Table 3. Results of Round 3 of the Delphi survey (final environmental consequences).

N°	Long Label	Short Label
1	Increase Temperature	X1
2	Extinction of Wildlife	X2
3	Flood	X3
4	Changing the Cultivation Pattern	X4
5	Reducing Pasture area	X5
6	Dust Storm	X6
7	Soil and Water Salinity	X7
8	Groundwater Decline and Depletion	X8
9	Air pollution	X9
10	Plant Extinction Species	X10
11	Drought	X11
12	Destruction of Surrounding Agricultural Lands	X12
13	Incidence of Different types of Diseases (respiratory, skin diseases, and various cancers)	X13
14	Threat of the Food Chain	X14
15	Disruption in Ecosystem Structure of lake	X15
16	Desertification	X16
17	Decreased Livability in Surrounding Cities	X17

2.2.2. Cross-Impact Analysis: MICMAC

From a collection of variables first determined by a committee of experts, the MICMAC technique of structural analysis tries to find the most essential variables inside a system and their role [28]. The three steps of MICMAC are as follows:

1. Definition of system variables—the system variables are determined using expert opinions, brainstorming, and a literature review. At this point, the result is a jumbled list of consequenses.

2. Identification of the relationships among the variables—the second stage is to specify the link between the variables after they have been identified. Table 4 shows the matrix of direct influence (MDI) is the name given to this matrix. Each MDIi j cell illustrates how much variable I influences variable j. For this purpose, the initial matrix was forwarded to experts for classification. Five members (14.29%) of the expert panel participated in this step. The grading was carried out as follows.

   Influences range from 0 to 3, with the possibility to identify potential influences:
   0: No influence;
   1: Weak;
   2: Moderate influence;
   3: Strong influence;
   P: Potential influences.

3. Identification of key variables—in the final step, the results from the use of the MICMAC software establish influence–dependence (Figure 5) and indirect (Figure 3) relations between variables into grids [29].

Table 4. Direct influences matrix sample.

	X1	X2	X3	X4	X5	X6	X7	X8	X9	X10	X11	X12	X13	X14	X15	X16	X17
X1	0	2	1	2	0	1	0	2	0	0	0	0	1	1	0	0	2
X2	0	0	0	0	3	0	2	1	0	0	0	3	2	0	2	2	0
X3	0	1	0	0	3	2	3	1	2	3	3	2	2	3	3	3	1
X4	0	1	3	0	3	3	1	3	0	1	1	2	3	2	3	2	0
X5	2	0	2	3	0	3	1	3	0	2	2	0	3	3	0	0	2
X6	0	1	1	1	0	0	0	0	0	0	0	0	0	0	1	0	3
X7	0	1	0	3	1	2	0	2	0	0	0	0	1	1	0	2	0
X8	0	1	0	2	2	1	1	0	0	0	1	0	3	0	0	0	3
X9	0	0	1	0	0	0	0	0	0	0	2	0	0	1	0	1	p
X10	1	2	3	3	3	3	3	2	3	0	3	3	3	3	3	3	2
X11	2	1	2	0	1	0	0	0	0	0	0	0	0	2	2	3	2
X12	1	3	2	1	0	0	0	0	0	3	3	0	1	2	1	2	3
X13	0	2	2	3	2	2	0	2	0	1	1	2	0	1	2	1	3
X14	0	2	3	3	1	3	0	2	0	3	1	1	0	0	1	1	0
X15	2	1	2	1	0	0	0	0	1	1	1	2	0	1	0	0	2
X16	2	2	0	1	1	1	0	0	0	0	0	0	0	0	0	0	3
X17	0	0	0	0	0	0	0	0	0	0	2	0	0	0	0	0	0

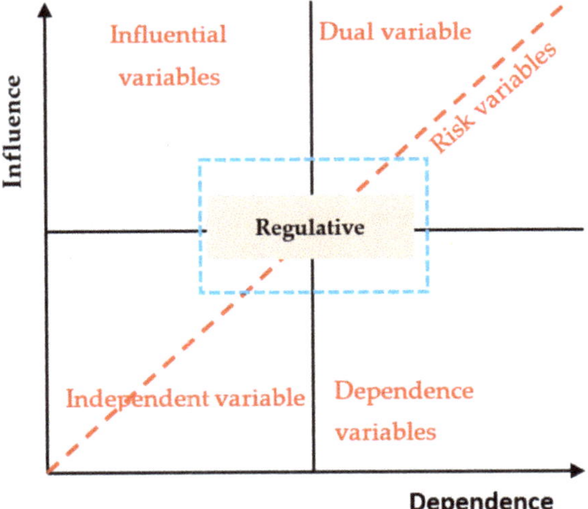

Figure 5. MICMAC influence–dependence map.

The MICMAC analysis is used to cluster consequences based on their type. Clustering is performed based on the driving and dependency power of each barrier [30]. Figure 5 shows the output of MICMAC and how to distribute the variables.

The results are discussed in the following section.

3. Results

Figure 6 shows a MICMAC diagram that was created by changing the reachability matrix. A MICMAC diagram is useful for determining driving and dependency power. If the value of dependence power is higher, other barriers should be addressed before the highest dependence power barrier is removed. With more driving power, other hurdles can be readily overcome after this one is removed.

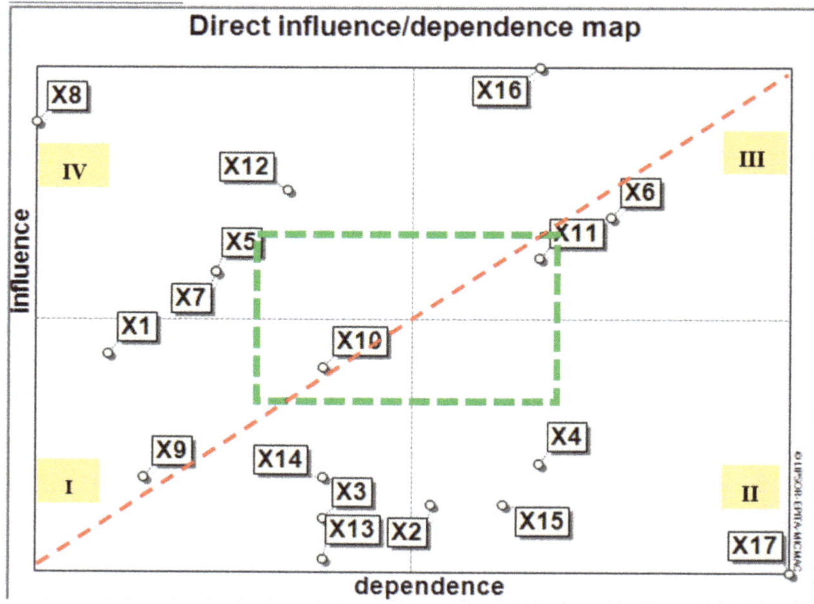

Figure 6. Direct influence–dependence map.

The grid is divided into four quadrants, each representing one of four different categories of variables. The differences between the variables are in the values for influence and dependence. Decreased livability in surrounding cities (X17) will have the largest dependence, and groundwater decline and depletion (X8) will have the most significant influence on the system (see page 4, row 103). Moreover, the other variables are as follows:

Autonomous barriers (I): Quadrant I represents autonomous environmental consequences and consists of increased temperature (X1), air pollution (X9), plant extinction species (X10), threats to the food chain (X14), flooding (X3), and incidence of different types of diseases (respiratory, skin diseases, and various cancers) (X13). Dependence and conductivity are low, and a change in these variables does not lead to a serious change in the system.

Dependent barriers (II): Quadrant II shows the dependent barriers, which are disturbance in the ecosystem structure of the lake (X15), extinction of wildlife (X2), decreased viability in surrounding cities (X17), and changes in cultivation patterns (X4). They have strong dependence and poor conductivity. These variables basically have high influence but less impact on the system.

Linkage barriers (III): Quadrant III shows the linkage barriers, including dust storm (X6), drought (X11), and desertification (X16), which are all possible.

Independent barriers (IV): Quadrant IV consists of driving factors with a strong driving force but a weak dependent force. In these barriers, four environmental consequences—reducing pasture area (X5), soil and water salinity (X7), and groundwater decline and depletion (X8), and destruction of surrounding agricultural lands (X12)—were found to play a controlling role in environmental changes in the case study area. Overall, any small change in these variables leads to fundamental changes in the entire system as Figure 7 shows. We will continue research into these variables.

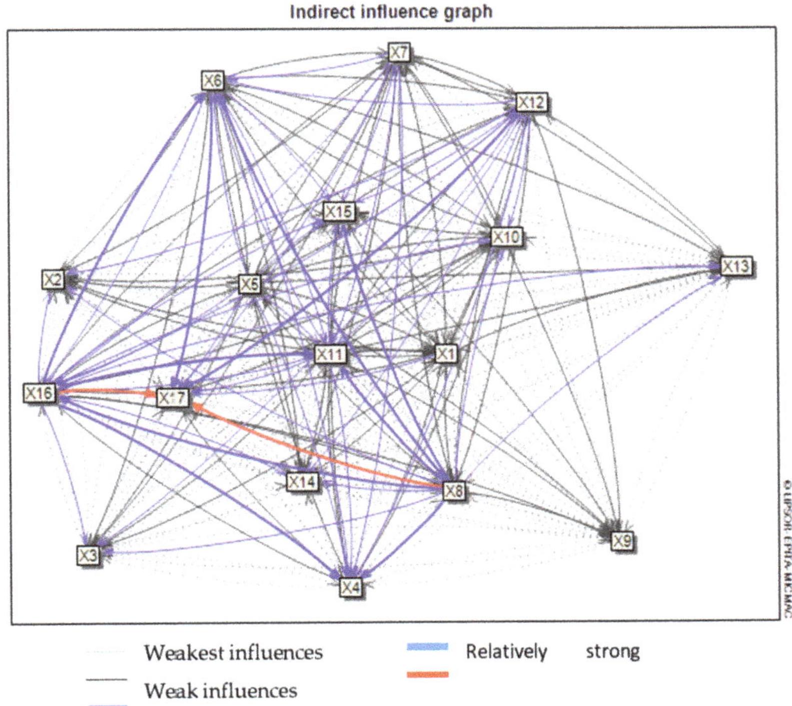

Figure 7. Direct influence graph.

Groundwater decline (X8) and destruction of surrounding agricultural lands (X12) are the most important and effective consequences in this system. Desertification (X16) and groundwater decline (X8) also have the strongest influence on the quality of life in cities (X17).

4. Discussion

Planning for the future predicted by our current data, accepting uncertainty, and preparing for the improbable gives us a chance to select a better plan. According to the results of this study, four driving forces were identified, which means that by controlling and monitoring these driving forces, other consequences of the drying up of Lake Urmia can be counteracted in the future. These driving forces are the driving factors that will disrupt the ecosystem of Lake Urmia.

The environmental driving force of Urmia lake dessication:

Seventeen significant factors were identified in this paper. Among the 17 factors, four factors were identified as driving forces, as seen in Figure 6.

Groundwater decline and depletion: The construction of numerous dams on rivers flowing into Lake Urmia has led to a shortage of water for agriculture, and farmers have extracted groundwater by digging deep and semi-deep wells (Figure 8) to provide water for agriculture.

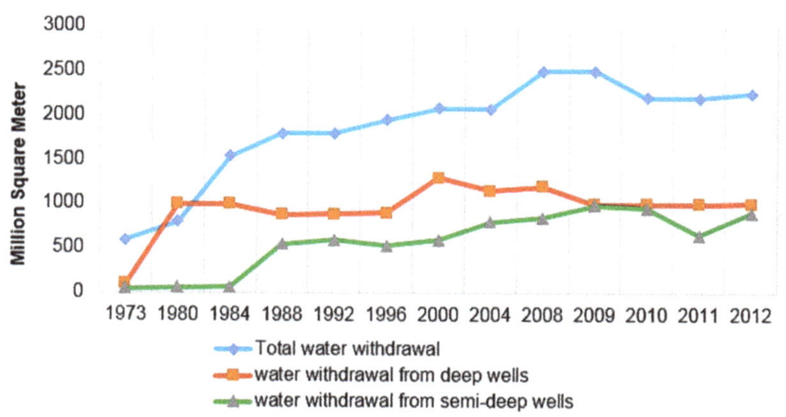

Figure 8. The amount of groundwater extracted in the Urmia Lake basin has changed dramatically over the last four decades.

According to ULRP statistics (2019), the number of legal wells in the Lake Basin is more than 40,000. Figure 9 shows the distribution of wells in Lake Urmia's basin. The number of semi-deep wells in the basin more than quadrupled in 2012 compared to 1982, and their respective discharges increased about sixfold. It should be mentioned that the majority of these wells are illegally drilled and used by garden and farm owners [17]. Furthermore, because there are many illegal wells, the government has limited control over groundwater abstraction, indicating a major data gap. Field-level water use is not included in current data collected near ULRP, and the existing data are either obsolete or held in inaccessible archives.

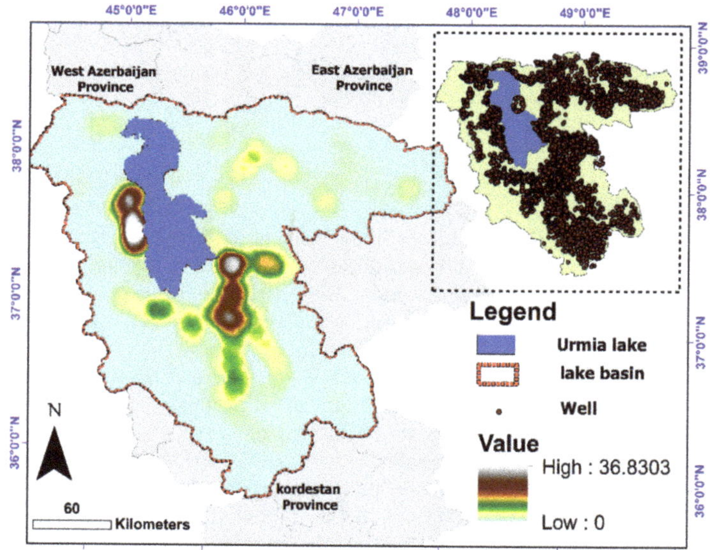

Figure 9. The distribution and locations of water extraction wells in Lake Urmia's basin. The majority of the lake basin wells are located towards the southeast and west of the lake.

As a result, we are skeptical of the ULRP's target of reducing agricultural water use by 40%, as we believe it will be challenging to accomplish and will have severe consequences for farmers and people in the basin. The idea could even exacerbate societal tensions; for example, a proposal to limit or restrict local farmers' illicit use of groundwater would be met with criticism.

Reducing Pasture area: The second and strongest driving force is the loss of a large area of pasture and vegetation. Furthermore, because of the diversity of soils and topography, the Urmia Lake islands possess a diverse flora. The first report on the flora of Lake Urmia was carried out in 1990–1994 (1998) (when the lake level was at the highest water level), and the second report was in 2016–2017 [31], but when the lake level was at its lowest, this rate was reached). During this period, various events caused many changes in flora, plant communities and vegetation levels. Comparing the results of these two reports shows that the number of plant species in these salt marshes decreased from 201 species to 93 species during these two decades. Out of 28 plant genera, only 24 members remain [30]. The majority of the plants in this genus are salinity-sensitive and have been wiped out due to habitat shifts away from the lake's edge. They play an essential role in the ecosystem's food chain. Therefore, this consequence does not directly affect human settlements, but is the root of other natural consequences in the field.

Soil and water salinity: One of the most major concerns promoting desertification in Iran's northwest is soil salinisation caused by the drying of Urmia Lake and dust produced from the lake's dried bottom (Figure 10). It results in a decrease in biomass output and a decrease in soil productivity.

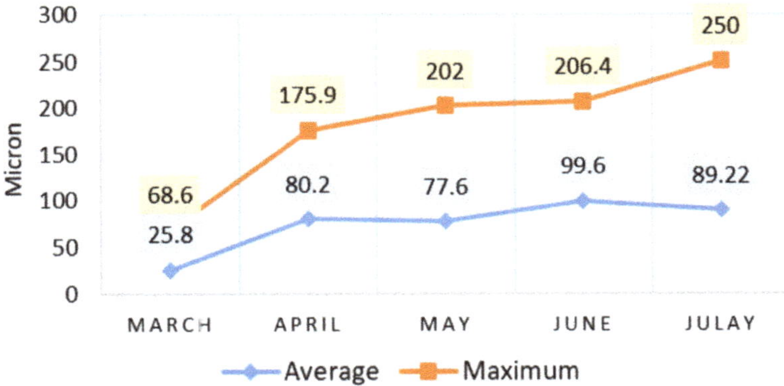

Figure 10. Monthly variations in the average and maximum suspended particle concentrations less than 10 microns at the Islamic Island (the largest island in Urmia Lake) station.

As a result, soil salinity looks to be the most serious hazard to this region's agricultural areas. As a result, the majority of the lake has been transformed into useless land in recent years. As the drought continues and the lake's salt concentration increases, the extinction of the lake's wildlife has increased (http://www.iew.ir/ accessed on 22 August 2021). Nikjoo et al.'s (2017) findings show that the environmental situation was better before 2009 when approaching the shore of Lake Urmia; however, after 2009, it has improved by moving away from the shore, and the situation at the lakeshore has become much worse [32].

Destruction of surrounding agricultural lands: Agriculture is a major source of employment in the region. However, it has been particularly hard hit by the drying up of the lake, a major water deficit, and climate change. Salt deposits, dry wells, extreme spring rains, and long droughts have negatively impacted agricultural production.

Several dams and dikes have been built in the Lake Urmia Basin, primarily to divert and storage for agricultural purposes. Although the first and second downstream canals are modern, old irrigation systems are still used in the region and water is distributed through open canals constructed by farmers. The number of farms around Lake Urmia is shown in Figure 11.

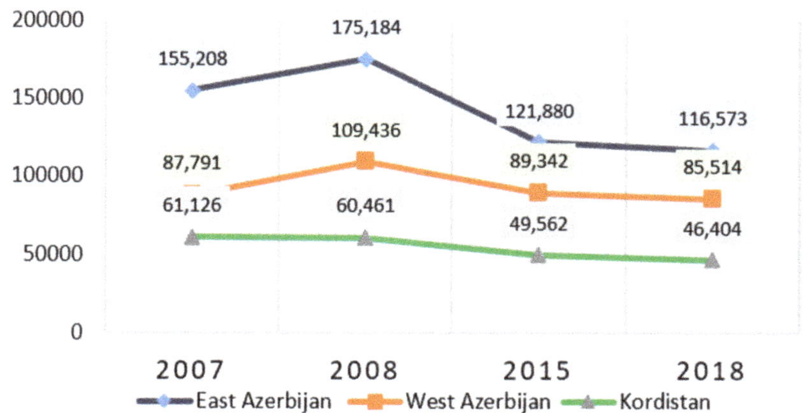

Figure 11. The number of farms with horticultural activity in three provinces around Lake Urmia.

From 2007 to 2018, 38,635 farmers lost their jobs (majority in the agriculture sector in the province of East Azerbaijan, 2277 in West Azerbaijan, and 14,722 in Kurdistan Furthermore, according to Soleimani Zeyveh's (2010) results, the drying up of Urmia Lake is damaging thousands of hectares of agricultural land in the lake area, resulting in a rise in unemployment and migration of roughly 3 million people to the neighbouring areas [33]

According to Mohammadi Hamidi et al. (2019), declining production and changing the type of some agricultural products are the main negative consequences that will lead to the migration of people from this region. Additionally, Ebrahimzadeh et al. (2014 argued that the least important consequence of the drying up of Lake Urmia would be the unemployment of more than three million people in the surrounding provinces. The research findings are summarised as follows [34]. Figure 12. shows an summarizes of research findings and the cause-and-effect relationships between outcomes and ultimately their impact on human habitats, which were the focus point of this paper.

Increasing unemployment due to loss of agricultural land and the related social and economic problems [35], the creation of salt winds as a threat to agricultural lands on the shores of the lake, and the inhalation of salt dust [36], which could cause cancer and lung problems [37], are some of the greatest environmental disasters, resulting in the emigration of tens of thousands of people from the region. For example, in the past 50 years, East Azerbaijan province (located on the eastern side of Lake Urmia and directly influenced by the water fluctuations of Lake Urmia due to western winds) has experienced a surge in emigration and is ranked top among Iran's other immigrant sending provinces, as seen in Figure 13.

According to the census results, during the 2011 to 2016 census, a total of 171,892 people left the province. Currently, East Azerbaijan province has 3083 villages, 138 of which have been depleted and are void of inhabitants compared to the 1996 census. In 2006, the number of depleted villages reached 378 (Figure 14). Finally, according to the last census in 2016 494 villages in this province were depopulated and multiplied since 1996.

Considering the prevailing wind direction in the region, coming from the west, the probability of damage to the settlements of eastern and northeastern Urmia due to salt storms is very high. Figure 15 shows that most of the settlements in this area are depopulated.

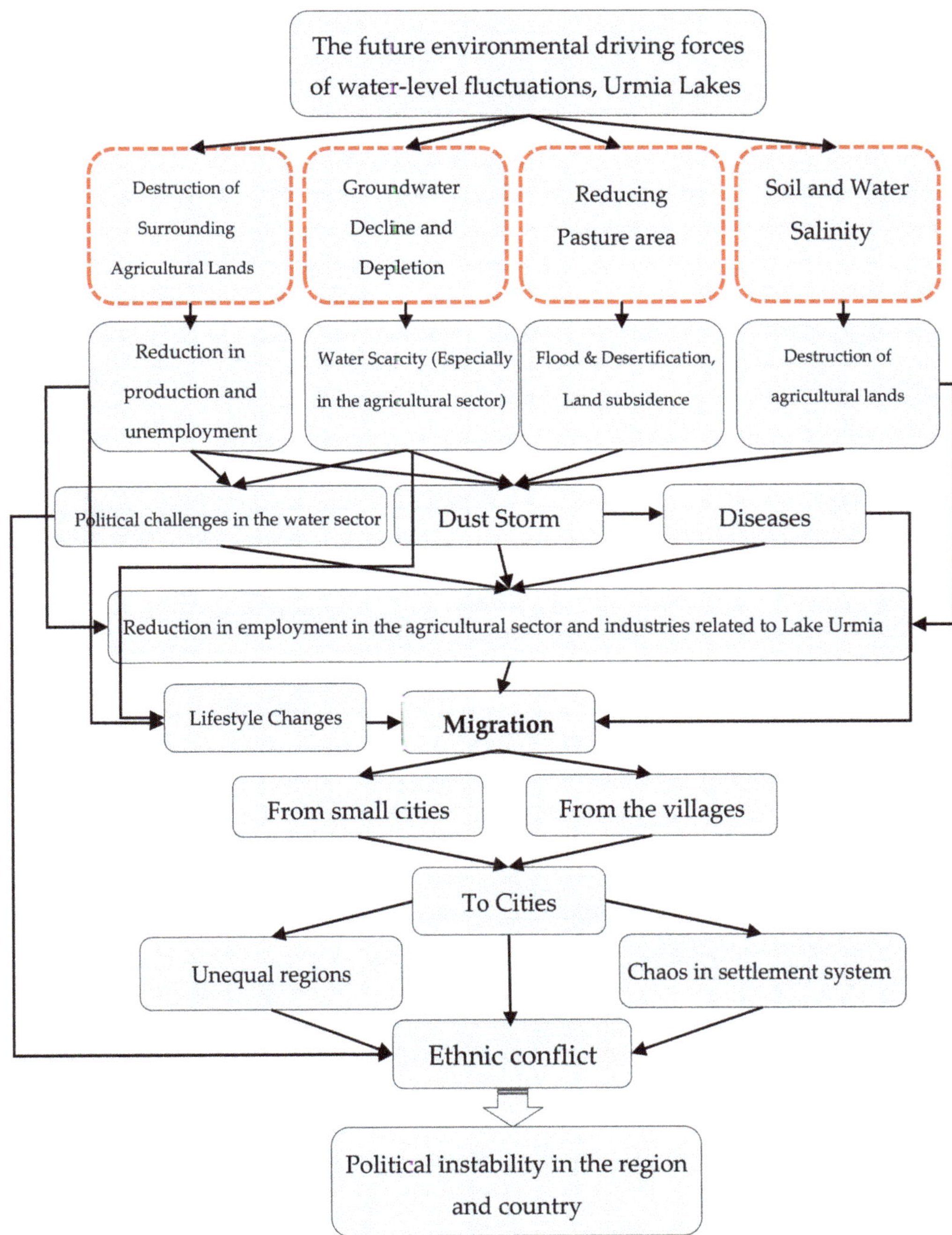

Figure 12. The decreased water level of Lake Urmia and its consequences in the surrounding settlements.

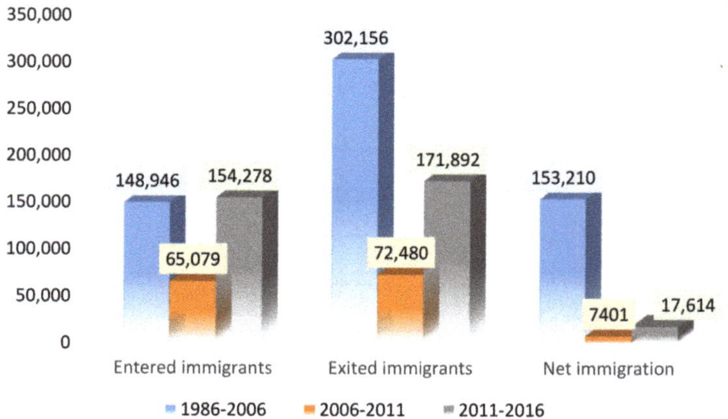

Figure 13. Migrants entered and exited during consecutive census (1986–2006, the Iran–Iraq war caused people to migrate from this border province) (Iran Statistical Yearbook).

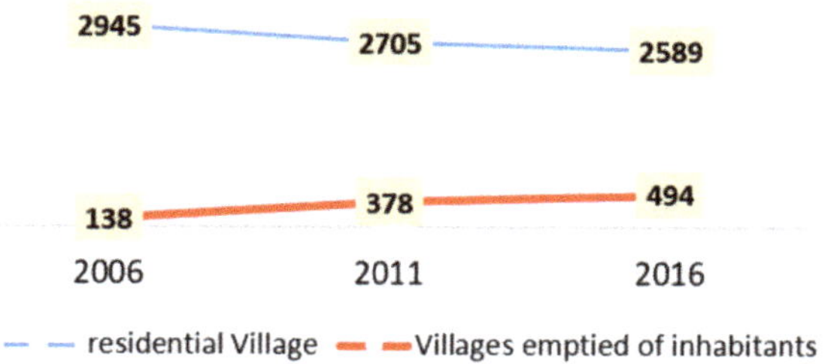

Figure 14. The populated and depopulated villages in the East Azerbaijan province according to the consecutive census.

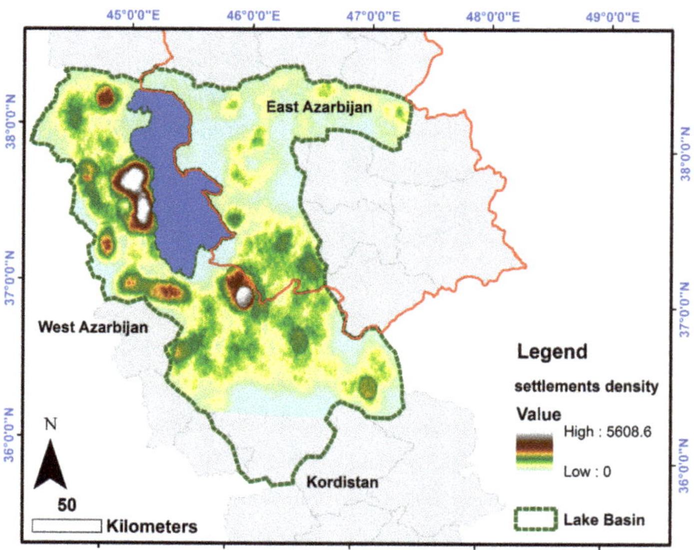

Figure 15. Distribution of settlements in the Lake Urmia basin.

According to Figure 15, most of the settlements are on the western and southern shores of the lake. Most of the settlements on the lake's eastern shore (East Azerbaijan province) were abandoned by the inhabitants in recent decades.

In response to those threats, the government's action was the formation of the Lake Urmia restoration centre, which, despite the passage of more than 80% of the program, has not yet achieved its goals, and this lake is still considered a dangerous threat to this region. Thus, the wave of migration that began in the region in recent decades will continue gradually, causing problems for cities and political conflicts in the region.

5. Conclusions

In this paper, the consequences of the drying of Lake Urmia on human settlements were investigated. The results of this study are as follows.

(1) There were 17 significant environmental consequences of lake water fluctuation identified. (2) Among the 17 consequences, decreased livability in surrounding cities was identified as the most dependent, and groundwater decline and depletion were identified as the most significant and effective environmental consequences in the system. (3) Finally, four environmental consequences— reducing pasture area, soil and water salinity, and groundwater decline and depletion, and destruction of surrounding agricultural lands— were identified.

Furthermore, this effort emphasized the importance of finding long-term solutions to mitigate environmental disasters and provide opportunities for local residents as soon as possible. However, many of the present efforts to restore the lake's natural equilibrium have proven to be ineffective.

The northwestern regions of Iran (around Lake Urmia) have strategic conditions due to the border. In addition, due to the impact of water scarcity in Lake Urmia Basin, the migration process from these areas has accelerated in recent years. Given the region's geopolitical status, this population loss will have long-term security ramifications. As a result, it appears that planners and policymakers must address the climatic, sociological, demographic, and political consequences of this change. Although programs and actions have been undertaken in recent years to revitalise the lake and take steps toward environmental sustainability, these trends are so slow that they are far off from achieving the desired sustainability conditions.

According to the Department of Energy, the basin is home to five million people, with average daily water use of 220 litres per person per day. With appropriate water management, the average water use in many urban areas (particularly in Europe) is around 150 litres per person per day. The urban population must play a role in the solution. Therefore, a public education program to reduce urban water use is needed. Some initiatives, such as reducing water pressure within the system, introducing a rebate program for low flush toilets, and installing water-efficient appliances in all new homes, could help to maintain the lake's water level. Agricultural growth has had a significant influence on the marshes surrounding the lake (cultivation and grazing). Wetlands are excellent water storage and filtration systems that can help the lake recover. They need to be better preserved and maintained. While technological capability is strong, there appears to be a lack of coordination among government agencies. ULRP staff work only in the headwaters but should be able to work in the agricultural area. To launch a collaborative plan, all concerned agencies must work together. The ULRP appears to have been tasked with taking the lead, promoting cooperation and integration, and putting together an action plan.

Conducting a groundwater resource analysis includes mapping aquifers and well inventories, groundwater use, extraction vs recharge rates, and groundwater and surface water interactions, which are all crucial for drought management.

There appears to be much disagreement among specialists about how much groundwater contributes to the lake, and there is a lot of information on how it is used in agriculture. Identification and mapping of key aquifers, an inventory of groundwater wells, estimates of groundwater use, determining if extraction rates are in balance with recharge rates, and research to determine where groundwater–surface water interactions are occurring within the watershed should all be undertaken. Considering all of these, it should be emphasized that these solutions represent crucial information as some of the few solutions that are accessible and executable during droughts.

Author Contributions: S.M.H., C.F., H.N., A.R. and M.H.Y., participated in the writing and editing of the manuscript and the preparation of the figures. All authors have read and agreed to the published version of the manuscript.

Funding: This research received no external funding.

Institutional Review Board Statement: Not applicable.

Informed Consent Statement: Not applicable.

Data Availability Statement: All data generated or analysed during this study are included in this published article.

Conflicts of Interest: The authors declare no conflict of interest.

References

1. Williams, W.D. Environmental threats to salt lakes and the likely status of inland saline ecosystems in 2025. *Environ. Conserv* **2002**, *29*, 154–167. [CrossRef]
2. Hassani, A.; Azapagic, A.; D'Odorico, P.; Keshmiri, A.; Shokri, N. Desiccation crisis of saline lakes: A new decision-support framework for building resilience to climate change. *Sci. Total. Environ.* **2020**, *703*, 134718. [CrossRef] [PubMed]
3. Messager, M.L.; Lehner, B.; Grill, G.; Nedeva, I.; Schmitt, O. Estimating the volume and age of water stored in global lakes using a geo-statistical approach. *Nat. Commun.* **2016**, *7*, 1–11. [CrossRef] [PubMed]
4. Heydari, N.; Jabbari, H. Worldwide Environmental Threats to Salt Lakes. *Int. J. Des. Nat. Ecodynamics* **2012**, *7*, 292–299. [CrossRef]
5. Wurtsbaugh, W.A.; Miller, C.; Null, S.E.; DeRose, R.J.; Wilcock, P.; Hahnenberger, M.; Howe, F.; Moore, J. Decline of the world's saline lakes. *Nat. Geosci.* **2017**, *10*, 816–821. [CrossRef]
6. Pekel, J.F.; Cottam, A.; Gorelick, N.; Belward, A.S. High-resolution mapping of global surface water and its long-term changes. *Nature.* **2016**, *540*, 418–422. [CrossRef]
7. Gross, M. The world's vanishing lakes. *Curr. Biol.* **2017**, *27*, 43–46. [CrossRef]
8. Nicoll, K.; Hahnenberger, M.; Goldstein, H.L. 'Dust in the wind 'from source-to-sink: Analysis of the 14–15 April 2015 storm in Utah. *Aeolian Res.* **2020**, *46*, 100532. [CrossRef]
9. Micklin, P. The future Aral Sea: Hope and despair. *Environ. Earth Sci.* **2016**, *75*, 844. [CrossRef]
10. Mischke, S.; Liu, C.; Zhang, J.; Zhang, C.; Zhang, H.; Jiao, P.; Plessen, B. The world's earliest Aral-Sea type disaster: The decline of the Loulan Kingdom in the Tarim Basin. *Sci. Rep.* **2017**, *7*, 1–8. [CrossRef]
11. Case, H.L.I.; Boles, J.; Delgado, A.; Nguyen, T.; Osugi, D.; Barnum, D.A.; Decker, D.; Steinberg, S.; Steinberg, S.; Keene, C.; et al *Salton Sea Ecosystem Monitoring and Assessment Plan*; Open File Report 2013,1133; U.S. Geological Survey: Reston, VA, USA, 2013
12. Stone, R. Saving Iran's great Salt Lake. *Science* **2015**, *349*, 1044–1047. [CrossRef]
13. Ministry of Energy (Iran). Daily Rainfall Report for First and 2nd Level Catchments. 2021. Available online: http://wrs.wrm.ir/m3/gozaresh_print.asp (accessed on 22 August 2021).
14. Jahanbakhsh, S.; Adalatdost, M.; Tadayoni, M. Urmia Lake: The Relationship between Sunspots and Climate on the Northwestern of Iran. *Geogr. Res. Q.* **2011**, *25*, 16656–16684.
15. Zoljoodi, M.; Didevarasl, A. Water-Level Fluctuations of Urmia Lake: Relationship with the Long-Term Changes of Meteorological Variables (Solutions for Water-Crisis Management in Urmia Lake Basin). *Atmos. Clim. Sci.* **2014**, *4*, 358–368. [CrossRef]
16. Sima, S.; Rosenberg, D.E.; Wurtsbaugh, W.A.; Null, S.E.; Kettenring, K.M. Managing Lake Urmia, Iran for diverse restoration objectives: Moving beyond a uniform target lake level. *J. Hydrol. Reg. Stud.* **2021**, *35*, 100812. [CrossRef]
17. Urmia Lake Restoration National Committee. *Necessity of Lake Urmia Resuscitation, Causes of Drought and Threats*; Report No: ULRP-6-4-3-Rep 1; Urmia Lake Restoration National Committee: Tehran, Iran, 2015.
18. Ghale, Y.A.G.; Baykara, M.; Unal, A. Investigating the interaction between agricultural lands and Urmia Lake ecosystem using remote sensing techniques and hydro-climatic data analysis. *Agric. Water Manag.* **2019**, *221*, 566–579. [CrossRef]
19. Mohammadi Hamidi, S.; Nazmfar, H.; Ahad, R.; Yazdani, M.H. Futurology of the Economic Drivers of Urmia Lake Water Level Fluctuations on the Spatial Unbalanced. *J. Spat. Plan.* **2020**, *24*, 69–97.
20. Panula-Ontto, J.; Luukkanen, J.; Kaivo-Oja, J.; O'Mahony, T.; Vehmas, J.; Valkealahti, S.; Björkqvist, T.; Korpela, T.; Järventausta, P.; Majanne, Y.; et al. Cross-impact analysis of Finnish electricity system with increased renewables: Long-run energy policy challenges in balancing supply and consumption. *Energy Policy* **2018**, *118*, 504–513. [CrossRef]
21. Buehring, J.; Bishop, P.C. Foresight and Design: New Support for Strategic Decision Making. *She Ji Des. Econ. Innov.* **2020**, *6*, 408–432. [CrossRef]

22. Jouan, J.; Ridier, A.; Carof, M. Legume production and use in feed: Analysis of levers to improve protein self-sufficiency from foresight scenarios. *J. Clean. Prod.* **2020**, *274*, 123085. [CrossRef]
23. Gariboldi, M.I.; Lin, V.; Bland, J.; Auplish, M.; Cawthorne, A. Foresight in the time of COVID-19. *Lancet Reg. Health-West. Pac.* **2021**, *6*, 100049. [CrossRef]
24. Ecken, P.; Gnatzy, T.; Heiko, A. Desirability bias in foresight: Consequences for decision quality based on Delphi results. *Technol. Forecast. Soc. Chang.* **2011**, *78*, 1654–1670. [CrossRef]
25. Aschemann-Witzel, J.; Perez-Cueto, F.J.; Niedzwiedzka, B.; Verbeke, W.; Bech-Larsen, T. Transferability of private food marketing success factors to public food and health policy: An expert Delphi survey. *Food Policy* **2012**, *37*, 650–660. [CrossRef]
26. Shariff, N. Utilising the Delphi survey approach: A review. *J. Nurs. Care* **2015**, *4*, 246. [CrossRef]
27. Powell, C. The Delphi technique: Myths and realities. *J. Adv. Nurs.* **2003**, *41*, 376–382. [CrossRef] [PubMed]
28. Swarnakar, V.; Tiwari, A.K.; Singh, A.R. Evaluating critical failure factors for implementing sustainable lean six sigma framework in manufacturing organisation. *Int. J. Lean Six Sigma* **2020**. [CrossRef]
29. Villacorta, P.J.; Masegosa, A.D.; Castellanos, D.; Lamata, M.T. A new fuzzy linguistic approach to qualitative Cross Impact Analysis. *Appl. Soft Comput.* **2014**, *24*, 19–30. [CrossRef]
30. Kinker, P.; Swarnakar, V.; Singh, A.R.; Jain, R. Identifying and evaluating service quality barriers for polytechnic education: An ISM-MICMAC approach. *Mater. Today Proc.* **2020**, *46*, 9752–9757. [CrossRef]
31. Asem, A.; Eimanifar, A.; Djamali, M.; De los Rios, P.; Wink, M. Biodiversity of the hypersaline Urmia Lake national park (NW Iran). *Diversity* **2014**, *6*, 102–132. [CrossRef]
32. Nikjoo, B.; Abdeshahi, A.; Yazdanpanah, M. Prioritising the economic, social and environmental consequences on rural areas of Malekan township of the drying of lake Urmia. *Environ. Sci.* **2017**, *15*, 27–44. (In Persian)
33. Soleimani Ziveh, M. Health and Environmental Consequences of Lake Urmia Water Loss, Comparative Comparison with Similar cases. In Proceedings of the Thirteenth National Conference on Environmental Health, Kerman, Iran, 2 November 2010. (In Persian).
34. Ebrahimzadeh, A.; Hssani, A.R.; Farnoodfar, R. Investigating the drying of Lake Urmia and environmental pollutions and its impact on the economy. In Proceedings of the First Conference of Environmental Pollution, Ardabil, Iran, 13 May 2014. (In Persian).
35. Mohammadi Hamidi, S.; Nazmfar, H.; Yazdani, M.H.; Rezayan Ghyeh Bashi, A. An Investigation and Analysis of the Effect of Urmia Lake Water Level Reduction on the Development Levels of Surrounding Counties. *Town Ctry. Plan.* **2019**, *11*, 285–309. (In Persian)
36. Asghari-Kaljahi, E.; Hoseinpour, S.; Nadiri, A.O. Evaluation of salt dust occurrence potential in the Northeast Zone of Urmia Lake. *Environ. Eros. Res.* **2018**, *8*, 42–61. (In Persian)
37. Musapour, J.; Dastgiri, S.; Asghari Jafarabadi, M.; Ziasarabi, P.; Khamnian, J. The environmental health catastrophe in Lake Urmia and asthma disease: A cohort study. *Electron. J. Gen. Med.* **2019**, *16*, em147. (In Persian) [CrossRef]

 *sustainability*

Article

The Effects of the Spatial Extent on Modelling Giant Panda Distributions Using Ecological Niche Models

Ziye Huang [1], Anmin Huang [1,*], Terence P. Dawson [2] and Li Cong [3]

1   Department of Human Geography and Urban and Rural Planning, College of Tourism, Huaqiao University, Quanzhou 362021, China; ziye.huang@kcl.ac.uk
2   Department of Geography, King's College, London WC2R 2LS, UK; terry.dawson@kcl.ac.uk
3   Department of Tourism, College of Architecture Landscape, Beijing Forestry University, Beijing 100083, China; lisacong@bjfu.edu.cn
*   Correspondence: amhuang@hqu.edu.cn

Abstract: Climate change and biodiversity loss have become increasingly prominent in recent years. To evaluate these two issues, prediction models have been developed on the basis of ecological-niche (or climate-envelope) models. However, the spatial scale and extent of the underlying environmental data are known to affect results. To verify whether the difference in the modelled spatial extent will affect model results, this study uses the MaxEnt model to predict the suitability range of giant pandas in the Min Mountain System (MMS) area through modelling performed (1) at a nationwide scale and (2) at a restricted MMS extent. The results show that, firstly, both models performed well in terms of accuracy. Secondly, extending the modelling extent does help improve the modelling results when the distribution data is incomplete. Thirdly, when environmental information is insufficient, the qualitative analysis should be combined with quantitative analysis to ensure the accuracy and practicality of the research. Finally, when predicting a suitability distribution of giant pandas, the modelling results under different spatial extents can provide management agencies at the various administrative levels with more targeted giant panda protective measures.

Keywords: MaxEnt model; spatial extent; ecological niche; *Ailuropoda melanoleuca*; Min Mountain System

Citation: Huang, Z.; Huang, A.; Dawson, T.P.; Cong, L. The Effects of the Spatial Extent on Modelling Giant Panda Distributions Using Ecological Niche Models. *Sustainability* **2021**, *13*, 11707. https://doi.org/10.3390/su132111707

Academic Editors: Baojie He, Ayyoob Sharifi, Chi Feng and Jun Yang

Received: 16 September 2021
Accepted: 20 October 2021
Published: 22 October 2021

Publisher's Note: MDPI stays neutral with regard to jurisdictional claims in published maps and institutional affiliations.

1. Introduction

Climate change and human activities have caused massive loss and fragmentation of animal and plant habitats and a sharp decline in the numbers of many rare species [1]. The wild giant panda (*Ailuropoda melanoleuca*) is one of the most globally endangered species, and its survival is uncertain. In 1869, the French priest Pere Armand David first discovered the giant panda in Sichuan province, China, and it attracted scholars' attention worldwide [2]. According to fossil evidence, documentary records, and quantitative surveys, the distribution of wild giant pandas has gradually shrunk; they have disappeared from most parts of China and from Vietnam, Laos, Myanmar, and other regions, and now remain only in parts of southwestern China and northwestern China [3,4]. Today, they are an endangered species unique to China and only distributed in six major Chinese mountain systems: the Qinling Mountain System, the Min Mountain System (MMS), the Qionglai Mountain System, the Daxiangling Mountain System, the Xiaoxiangling Mountain System, and the Liang Mountain System. These mountains are located in southwestern China, crossing Shaanxi, Gansu, and Sichuan provinces [5].

The Fourth National Giant Panda Survey Report (2015) reports the status of wild giant pandas in recent years [6]. The number of wild giant pandas increased by 268 individuals compared with the previous survey, reaching 1864 [7]. Habitats and potential habitats rose by 11.8% and 6.3%, respectively [6]. Even though the current survival situation of giant pandas has not continued to deteriorate based on data, the fragmentation of the habitat is a hidden danger that needs continuous attention. The existing giant pandas are divided into 33 local populations, and there has been a decrease in connectivity between these populations and some of the habitats are of poor quality. Therefore, although the number of wild giant pandas has increased, the ecological isolation between local populations and poor quality of some habitats still poses a significant threat to the giant pandas' future survival [3,4]. According to Linderman et al. (2004), human influence in the next 30 years will lead to a 16% loss of giant panda habitat [8,9].

Therefore, it is essential to figure out what has caused this species to become endangered and to understand the factors that interfere with their natural environment to ensure the sustainable development of this endangered species; moreover, such issues need to be resolved urgently [10]. To ensure the suitability of the wild giant pandas' habitat and promote the formulation of protection measures, many scholars are devoted to studying the factors that give rise to the fragmentation of habitats and cause other threats to the survival of the population. The conclusion of these research results identify four categories of factors that primarily affect the giant pandas' survival and habitat fragmentation: (1) human activities, logging, reclamation, construction of cities and roads, grazing, and poaching; (2) ecological limitations of giant pandas, such as the special requirement for food and low reproduction rate [11]; (3) natural disturbances, such as earthquakes, and bamboo blossoming [12,13]; (4) climate change, the changes of plants' growth structure, temperature, rainfall, and other factors affected by climate warming.

However, due to the imbalance of scientific research structure, most scientific research work only focus on the aspect of phenomenon analysis, and relatively little work has been carried out from the perspective of sustainable development [10]. Presently, to increase the population and prevent the reduction of giant pandas' habitat, there is an urgent need for quantifying how different populations of giant pandas are predicted to respond to different climate change scenarios. At the same time, corresponding response measures should be made based on the model results to ensure the sustainable survival of giant pandas. Ecological niche models (ENMs) are currently used in geo-biological research, and are also the main modelling tool for predicting and solving sustainable ecological development [11]. Due to the practicality and accuracy of such models in solving ecological problems, ENMs have been widely used in environmental research, including issues such as climate change, the impact of species invasion on another species, and for predicting the potential distribution areas of different species [14,15]. The popular ENMs in recent years include: Genetic Algorithm for Rule-Set Prediction (GARP), Ecological Niche Factor Analysis (ENFA), Bioclimate Analysis and Prediction System (BIOCLIM), and MaxEnt [16]. Of these, the MaxEnt model is the most widely used in studying the geographic-biological connection.

Few studies have examined the influence of spatial scale on the predictions generated by MaxEnt modelling. Of these, some studies have shown that different scales have minimal influence on model predictions [17], whereas other studies have concluded that variation in scale-dependent effects is influenced by environmental variable selection [5,18]. However, there remain few concrete recommendations on how to deal with these issues in order to improve MaxEnt modelling.

Based on the concern about the development of the giant pandas' living conditions and interest in further research on modelling extents, this study aims to examine how the spatial extent of the modelling affects the niche model result. Taking spatial extent as a breakthrough point to technically reduce the errors in the study of endangered species models and will make it possible to provide suggestions for future monitoring and protection methods for giant pandas and thereby ensure the survival of giant pandas and alleviate the conditions that threaten their survival. To accomplish this aim, the research has the following objectives: (1) to quantify some significant human disturbances and incorporate them into the environmental variables in the MaxEnt model; (2) to predict the accuracy of the model to ensure the reliability of follow-up research; (3) to model suitable habitats for giant pandas in different spatial ranges and identify some differences between them; (4) to follow the results with further discussion and suggest how this research provides practical help for the future.

2. Materials and Methods

2.1. Introduction to Model

The MaxEnt model (Philips et al., 2006) takes the Maximum Entropy Theory proposed by Jaynes in 1957 as theoretical guidance [19]. It is a widely used method with high performance and is commonly used in climate prediction, biological protection, species invasion, and other biological research [5,20,21]. The principle of MaxEnt is that the 'dissipation' of the system increases the entropy. Until the species and the environment reach the maximum entropy, the environment will be in equilibrium [22]. The model predicts species' distribution by calculating the state parameters when the entropy is maximum [22]. The MaxEnt models exhibit little sensitivity or change to model accuracy with significant changes in data quantity, although processing times tend to be accelerated and model accuracy is higher when models are based on presence-only data [23]. Thus, follow-up research is based on the MaxEnt model with high accuracy, easy operation, and low data quantity requirements.

2.2. Case Study Area and Research Extent

Wild Chinese giant pandas are distributed in six major mountain systems of the Sichuan, Shaanxi, and Gansu provinces. The area and habitat ratio of each mountain system are shown in Table 1 [13]. The MMS is located in southwestern China, running through northern Sichuan province and southern Gansu province. The geographical position is 102° 70′–105.60° E and 31°40′–33.70′ N, transitioning from the Yangtze River's upper reaches to the Tibetan Plateau [24]. The giant pandas in the MMS are mainly distributed in the mountainous dark coniferous forest belt at 2100 to 3400 m. The subalpine dark coniferous forest belt is at a height of 3000 to 3900 m in the alpine valley [13]. There are 27 established nature reserves in the MMS, where 44% of China's pandas are distributed [21].

Table 1. Statistics of the habitat area of giant pandas in each mountain system.

Mountain Systems	Area/hm^2	Percentage of Total Habitat/%	Administrative Regions/Province
The Qinling Mountain	352,914	15.31	Shaanxi, Gansu
The Min Mountain	960,313	41.66	Sichuan, Gansu
The Qionglai Mountain	610,122	26.47	Sichuan
The Daxiangling Mountain	81,026	3.52	Sichuan
The Xiaoxiangling Mountain	80,204	3.48	Sichuan
The Liang Mountain	220,412	9.56	Sichuan
Total	2,304,991	100	

Source from: Zhang, Q., 2009. Study on Habitat Selection of Giant Panda in the Min Mountain, Gansu. Master. Northwest Normal University.

Since the MMS region is the most critical habitat for giant pandas among the six mountain systems, this area was selected as the target area for modelling. A national extent is selected as a control group, which contains all the known global distributional data for the giant panda. Comparing the model result under the MMS region and the nationwide extent (subsequently clipped to the MMS range) enabled us to evaluate variation in predicted suitable habitat areas. The specific modelling area comparison shows in Figure 1.

2.3. Data Collection

Species distribution data were derived from two sources; firstly the Global Biodiversity Information Faculty [25] and National Specimen Information Infrastructure [26]; and secondly, from all relevant peer-review literature records of giant pandas' occurrence. A total of 260 occurrence sites were obtained from these two sources. Google Earth was used to filter the distribution information to gain the relevant latitude and longitude coordinates, as well as helping to remove 'double records' and locations with unknown geographic coordinates. This resulted in 89 occurrence records, of which 65 were from the MMS.

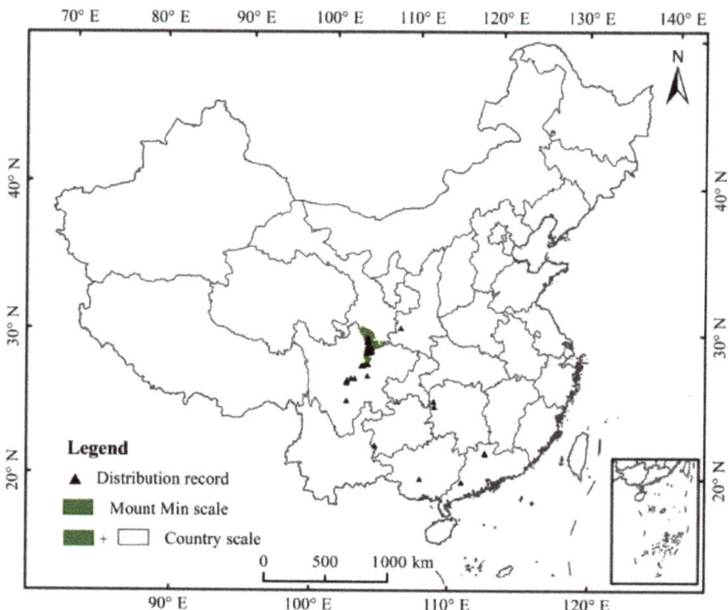

Figure 1. Modelling areas of spatial extent and distribution sites of the wild giant panda.

All climate and environmental data were obtained from WORLDCLIM [27]–a global database based on meteorological information recorded by weather stations worldwide from 1970 to 2000 (Table 2). It uses interpolation to generate global climate raster data with a spatial resolution of 30 arc-seconds (1 km^2) [17]. The roads, rivers, and residential areas data were derived from OpenStreetMap then calculated by the Euclidean distance tool in ArcGIS to acquire the distance to the residential area, the distance to the road, and the distance to the river, with resampling of 30 arc-seconds (1 km^2). The data of land use types in 2018 came from the Resource and Environment Data Center of the Chinese Academy of Sciences [28], with a resolution of 30 arc-seconds (1 km^2).

However, the redundancy and overfitting of variables can bias the data results [29]. To avoid overfitting, only when the environmental variables are reduced to a reasonable number can the accuracy and predictive ability of the model be improved [30,31]. Therefore, the environmental variable data values of 89 sample points were extracted by DIVA-GIS software. The Pearson correlation coefficient was used to test the multicollinearity between climate variables in a set of climate variables with high correlation (r > 0.8). According to the contribution rate, only one variable closely related to the species distribution or convenient for model interpretation was selected for model prediction. In this case, 11 environmental variables were ultimately screened at two extents, respectively, for model construction.

2.4. Model Construction

Firstly, giant pandas' distribution and environmental data were imported into the MaxEnt version 3.4.1 to model potential distributions of giant pandas at both spatial extents [32]. The maximum number of iterations (500 times) and the maximum number of background points (10,000) were kept recommended default settings. The cross-validation method was used to test the robustness of each model. The study used 25% distribution points as the test set and 75% distribution points as the training set and ran these 10 times repeatedly to make the data utilization rate higher [33]. Jackknife and percentage rate were selected to evaluate the relative importance of each environmental factor on the giant pandas' distribution. The response curve was chosen to obtain specific values of environmental variables most suitable for the survival of giant pandas.

Table 2. 19 bioclimatic variables.

Index	Description
Bio1	Mean annual temperature
Bio2	Mean diurnal air temperature range
Bio3	Isothermality (Bio2/Bio7 × 100)
Bio4	The standard deviation of temperature seasonality
Bio5	Max temperature of the warmest month
Bio6	Min. temperature of the coldest month
Bio7	Temperature annual range (Bio5-Bio6)
Bio8	Mean temperature of the wettest quarter
Bio9	Mean temperature of the driest quarter
Bio10	Mean temperature of the warmest quarter
Bio11	Mean temperature of the coldest quarter
Bio12	Annual precipitation
Bio13	Precipitation of the wettest month
Bio14	Precipitation of the driest month
Bio15	Coefficient of variation of precipitation seasonality
Bio16	Precipitation of the wettest quarter
Bio17	Precipitation of the driest quarter
Bio18	Precipitation of the warmest quarter
Bio19	Precipitation of the coldest quarter

The area under the curve (AUC) of receiver operating characteristics (ROC) was used to predict the model performance. When the ROC curve cannot indicate which classifier performs better, AUC provides a more intuitive way to predict the accuracy of models, which is a single measure of overall accuracy that does not depend on a particular threshold [34]. The value of AUC is between 0 and 1. The larger the AUC value, the better the prediction effect, and the higher the prediction accuracy. The general evaluation standard is less than 0.5 is model meaningless, 0.5–0.6 is poor, 0.6–0.7 is fair, 0.7–0.8 is more accurate, 0.8–0.9 is very accurate, and 0.9–1.0 is extremely accurate [35]. When the AUC value is more than 0.7, the prediction result of the MaxEnt model can be credible.

3. Results

3.1. Model Prediction Accuracy Test

The ROC curve was obtained after running the MaxEnt model 20 times repeatedly to ensure the model's prediction results' stability. AUC values of the MMS region (Figure 2) and the national extent (Figure 3) were 0.879 and 0.987, respectively, suggesting that the MaxEnt models for both scales performed well, and the national extent scaled model is more accurate than the MMS regional model.

3.2. Distribution of Predicted Suitable Areas for the Giant Panda

The suitable habitat distribution map under the MMS (Figure 4) and the country extent (Figure 5) was directly obtained through modelling. To make the information in the map of the national range clearer, it was clipped to the range of the MMS (Figure 6). There is a big difference in the suitability distribution map under the two areas modelling. Geographically, modelling at the region of the MMS, the highly suitable areas are only scattered in the core area of the MMS, such as the Wanglang Reserve, Hulu ditch, Wenxian ditch, and Xi ditch. However, when modelling on a national size, most of the MMS areas are suitable for wild giant pandas to survive, and at least half of the regions are highly suitable. In the whole country, only the MMS area has a highly suitable area. Except for the Gansu and Sichuan provinces covered by the MMS, only a few areas at the junction of Tibet and Yunnan provinces have some suitable regions.

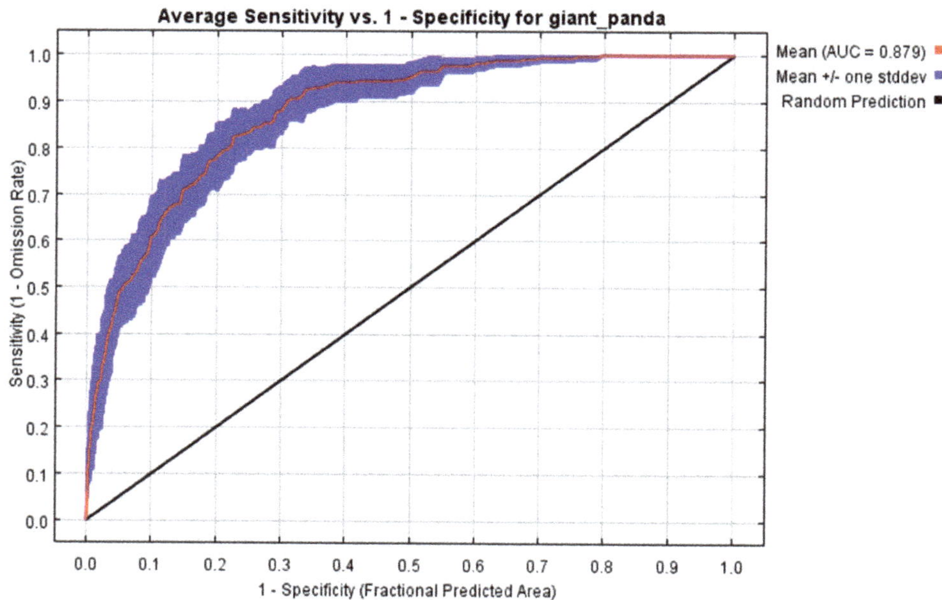

Figure 2. Receiver operating characteristic curve of Min Mountain System and area under the curve (AUC).

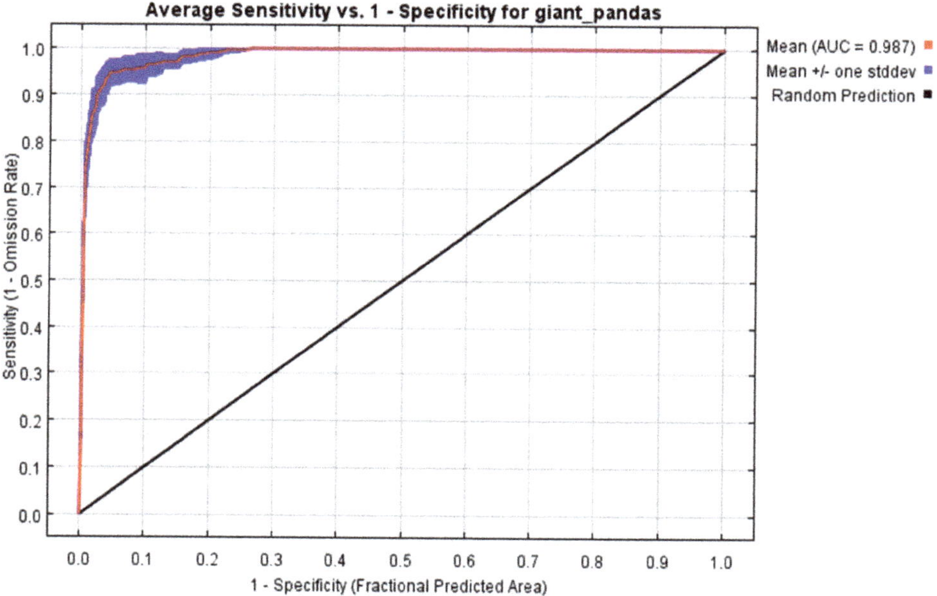

Figure 3. Receiver operating characteristic curve of country extent and area under the curve (AUC).

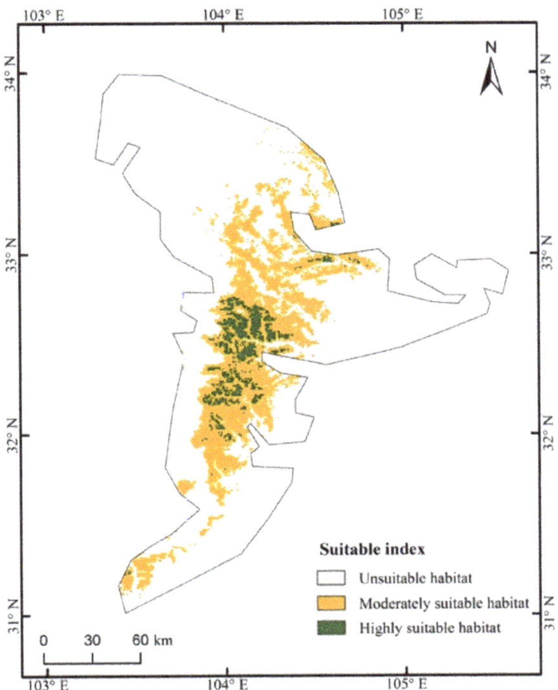

Figure 4. The distribution of suitable habitat of wild giant panda in the Min Mountain System obtained by modelling on the Min Mountain extent.

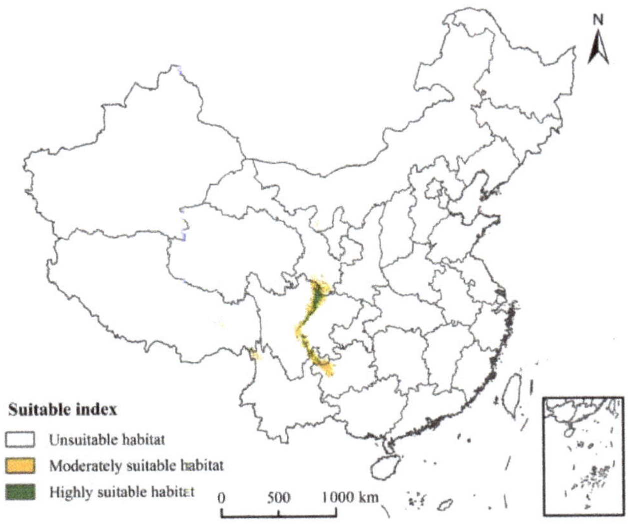

Figure 5. Prediction of the distribution probability of wild giant pandas in the Min Mountain System obtained by modelling on the country extent (not clipped).

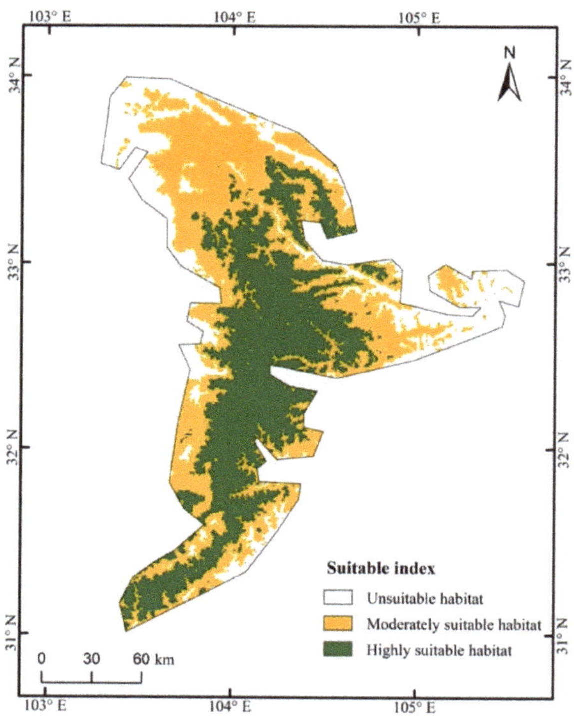

Figure 6. Prediction of the distribution probability of wild giant pandas in the Min Mountain System obtained by modelling on the country extent (clipped).

3.3. Contribution and Importance of Environmental Predictors

The result of jackknife under the MMS area (Figure 7) and the national range (Figure 8), as well as the percentage contribution rate under two areas (Table 3), are used to test the contribution rate of environmental variables. According to the results of the two methods, under the MMS area, bio15 (coefficient of variation of precipitation seasonality), bio2 (mean diurnal air temperature range), bio14 (precipitation of the driest period) and distance to a stream are the dominant factors affecting the spatial distribution of giant pandas. Under country area, bio12 (annual precipitation), bio3 (isothermality Bio2/Bio7 × 100), bio4 (standard deviation of temperature seasonality) and slope are the dominant factors affecting the geographic distribution of giant pandas.

Table 3. Contribution of environmental variables for Wild Giant Panda insignis under the Min Mountain System and Country extent.

Region	Variables	Contribution (%)	Region	Variables	Contribution (%)
Min Mountain extent	Bio2	25	Country extent	Bio3	23.4
	Bio5	0.8		Bio4	3.6
	Bio14	5.4		Bio6	2.4
	Bio15	30.1		Bio10	1.1
	Bio18	0.8		Bio12	34.4
	Slo	6		Slo	19.6
	Alt	9.1		Alt	14.3
	DTC	2		DTC	0.2
	DTS	12.9		DTS	0.3
	DTR	5.3		DTR	0.1

Slo = slope; Alt = altitude; DTC = distance to community; DTS= distance to a stream; DTR= distance to road.

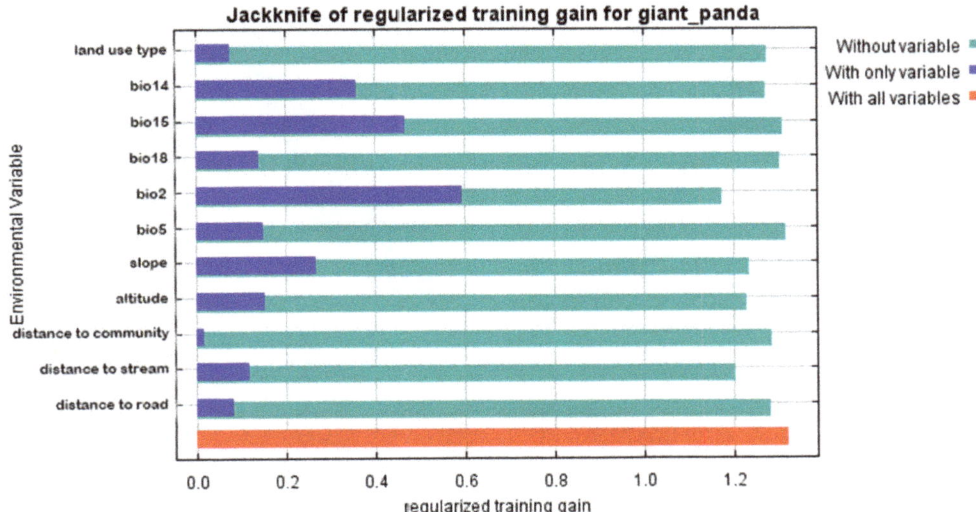

Figure 7. Results of jackknife evaluation of the environmental variables concerning regularized training gain for the Min Mountain System.

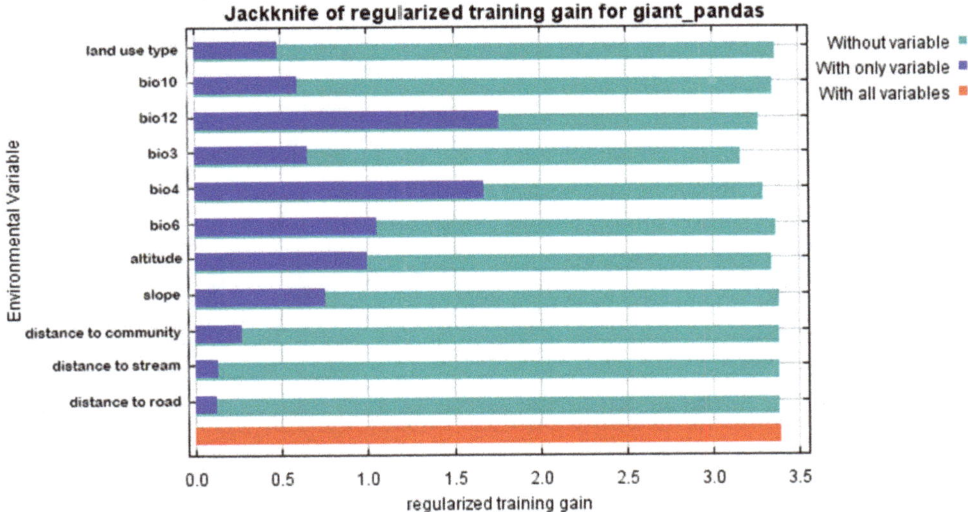

Figure 8. Results of jackknife evaluation of the environmental variables concerning regularized training gain for the country extent.

3.4. Response Curve

Four factors were selected from the percentage rate and jackknife on the MMS region (Figure 9) and the national region (Figure 10) to conduct the response curve further. The highest point in the figure represents the variable value that is most suitable for the survival of giant pandas. On the one hand, under the MMS area, the results show that the most suitable Mean diurnal air temperature range is about 9.5 °C, and 6–8 mm is the best range of precipitation of the driest month. In addition, 0.15–0.16 (unit: 10 km) is the most suitable distance from the giant pandas' habitat to a stream, and 65–72 mm is the best range of the value of the precipitation seasonality.

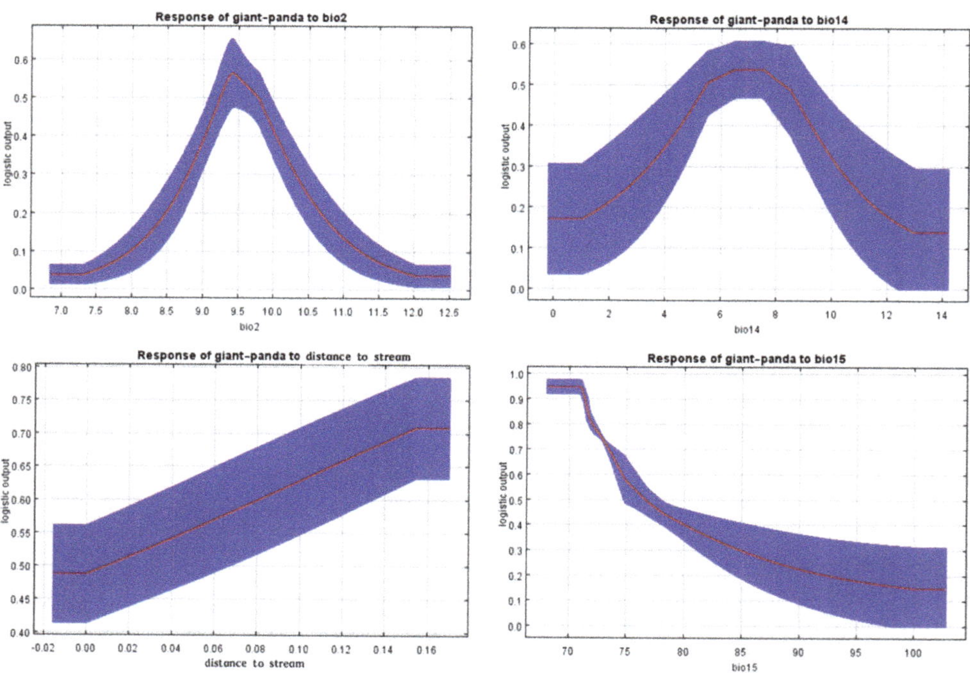

Figure 9. The response curve of the Min Mountain extent with the variables bio15, bio2, bio14, and DTS.

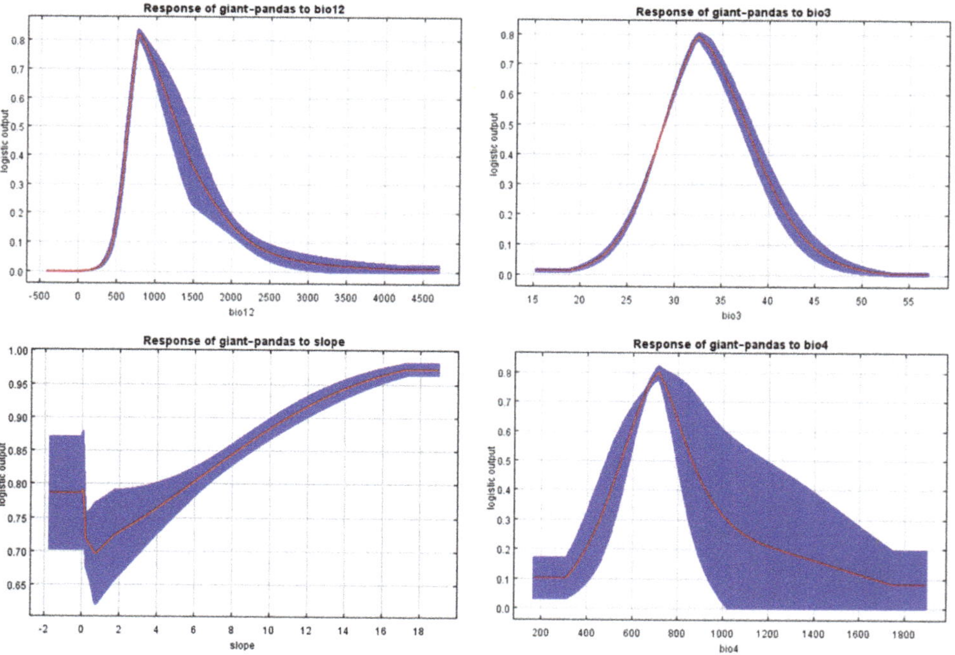

Figure 10. The response curve of the country extent with the variable bio12, bio3, bio4, and slope.

On the other hand, under the country extent, an ideal annual precipitation value for giant pandas' survival is about 800 mm, while isothermality (Bio2/Bio7 × 100) is about 32. The slope's value tends to be ideal at 17 ° and above, and the optimal value of the standard deviation of temperature seasonality is about 700.

4. Discussion

4.1. Model Results Will Be Affected by Errors in the Data Collection

From the process of the data collection and modelling, long data collection time and a lack of environmental and species location data will impact the model's accuracy. Many sites in the MMS areas are underdeveloped, the actual measurement points of the data are sparse, and the observation years of various types of data are not the same, which significantly increases the possibility of data errors [5]. For example, the climate data is from 1970 to 2000, the land type is based on the classification in 2018, and the species distribution data spans several decades. From the process of data collection and acquisition, time inequality is inevitable. Thus, timely updates and data supplements will be necessary to minimize the errors as much as possible and provide better data support for niche modelling.

However, this error has little effect on this study. In terms of the predicted suitable area, the research focus was not to obtain a specific geographical distribution area of giant pandas but to compare the differences in the distribution of the two regions. This error is simultaneously functioning on the two ranges. Therefore, the impact of the time inequality of the data on the research is reduced to some extent.

4.2. The Predicted Suitable Areas Are Different at Different Modelling Extents

The suitable habitat maps are quite different at the two spatial extents. However, the highly suitable area in a more extensive range of modelling results almost overlaps with the suitable area (highly suitable area plus moderately suitable area) in the target area modelling results. Thus, the suitable range under larger-extent modelling is more extensive than that under smaller-extent modelling for some reason.

Combined with the accuracy-test results, the model result on a national scale in this study shows a higher prediction accuracy. However, there is an assumption in extreme cases: assuming that the occurrence site data and environmental data are complete and accurate, there is no doubt that the most accurate results will be obtained when the target area is the modelling area. In other words, if the site data and environmental data are very accurate, the model established under the MMS area should be the most accurate.

When the actual results of the model do not match the assumptions, reviewing some phenomena that have not been taken into account in the modelling process can help explain these unexpected results. When the research object is a kind of animal, for example, the giant panda, their mobility and rarity often lead to a lack of data and uncertainty in obtaining distribution sites [5]. Furthermore, giant pandas are sensitive to humans, and many distribution records do not entirely accurately represent the areas where pandas have lived for a long time [5]. Therefore, the lack of distribution sites under the MMS modelling area will underestimate the range of potentially suitable distribution areas [17]. At the same time, modelling at a range more extensive than the target area will overestimate the suitable spatial range. It is easier to obtain more distribution sites and environmental data under a larger spatial scale, which will provide systematic and comprehensive information for establishing a species–environment relationship [17]. Thus, if there is a lack of species distribution sites, appropriately expanding the modelling scope will help offset the possible deviation in modelling at the target area scale.

However, when the research object is a plant, its distribution site will not move, so it is relatively easy to collect. The suitable distribution map from different spatial scales should be identical [17]. Therefore, the degree of movement of the species and the difficulty of data collection is different, and the range that may need to be expanded during modelling is also different.

Based on the above analysis, the MaxEnt model can provide adequate technical support for endangered or difficult-to-track species. For one, the MaxEnt model has low requirements for the number of data. For the other, this study concluded that appropriately expanding the modelling range and clipping out the appropriate scope of the target area may offset the error caused by the lack of distribution sites or environmental variables. The MaxEnt model makes it possible to predict the living conditions of endangered species, and the study of the spatial extent of modelling reduces the possible errors of the model. This provides technical support and guarantees that relevant personnel and agencies can monitor the living conditions of endangered species and intervene if necessary. However, how to expand the scope of the modelling to compensate for the errors caused by the missing sites and variables and to achieve the most accurate results further studies are needed.

4.3. Human Interference Factors Need to Be Considered

None of the human interference factors quantified in this study is significant under contribution rate analysis. For one, it may be because human interference shows a gradual weakening trend as the extent increases. Connor et al. (2019) mentioned that human interference, such as the negative impact of roads, is evident at more minor scales but becomes smoother as the total range increases to more than 500 km^2 [18]. The target area in this study is much larger than 500 km^2, so the influence of human interference is weakened. For the other, the MMS area has more than 30 protected reserves, and there are fewer permanent residents than in other regions [13]. In addition, because fewer roads are connected to the outside in each nature reserve, human disturbance caused by residents' activities and road traffic is not substantial [36].

However, on a more microscopic scale, there are still many human variables that should be considered. For example, the MMS area is the settlement of the Baima Tibetan ethnic minority in China. Grazing horses and yaks have been the traditional production method of this ethnic group for a long time. In recent years, the grazing range has expanded, causing domestic animals and giant pandas to compete for bamboo as a food source, further compressing wild pandas' food and living space [36]. Therefore, grazing activities have become the most influential human disturbance in the MMS [5].

Many empirical studies have mentioned the impact of human interference [8,9]. In order to more accurately predict the habitat area of the giant panda, a standardized quantification system of human interference should be established in the future, or at least combined qualitative analysis to ensure the comprehensiveness of research.

4.4. The Contribution Rate of Environmental Variables and Response Curves under Different Modelling Spatial Extents Can Provide New Ideas for Protecting Giant Pandas

Together with geographic extents, the response curve can provide data support for the cross-scale protection of wild giant pandas. It provides references for protection departments of different regions and levels when implementing the protection of giant pandas:

(1) The local managers of each protected area or county in the MMS should pay attention to the management and restriction of human activities on a small scale. Even though the setting scale of this study is too big to identify the influence of human interference, it can be seen from other literature combined with the model result that human interference has actual impacts on the habitat of giant pandas. For example, the effect of increased grazing needs to be taken seriously by local managers to alleviate the compression of the giant pandas' natural habitat.

(2) Managers of relevant departments of the entire MMS area can refer to the environmental contribution rate, and response curve at the MMS extent to take protective measures. For instance, they can monitor the dynamic changes from the factors of 'changes in daily average temperature' and 'precipitation in the driest month' to ensure that the giant pandas' habitat has the correct living conditions. The monitoring and control of these environmental variables can more effectively improve the protection capability within the scope of the Min Mountain System. In addition, monitoring

points can be set up in areas that are within the pandas' suitable distance to a water source to improve wild panda investigation and monitoring efficiency.

(3)   From a national macro-scale, national environmental protection organisations and national core ecological protection agencies should adopt macro-monitoring of factors with high contribution rates at large scales to control the fragmentation and degradation of the giant pandas' habitat, for example, monitoring annual precipitation and annual temperature differences.

In future relevant research, if protection or management suggestions need to be made for a certain level of administrative agencies, some data can be collected within their corresponding geographic level. This will help clarify the administrative divisions of giant panda protection agencies and make the division of labor between various administrative groups clearer [10].

5. Conclusions

Due to climate change and human disturbance activities, the habitats of endangered wild animals continue to be degraded and fragmented, so their survival is threatened. The research of endangered species can be effectively supported by ENMs, especially the MaxEnt model, which can maintain the stability of model results even when sample data is extremely scarce. At present, the application of the MaxEnt model is mature, but there is still a lack of research on its modelling extent. This study attempts to analyze how changes in the modelling spatial range affect the results of the MaxEnt model, with a view to reducing errors in model research and providing technical support and management suggestions for the sustainable development of giant pandas and their habitats. Taking the MMS region and the MMS region modelled on a national scale as a mutual control group, the main conclusions are as follow: (1) When an endangered species that difficult in data collection of sites is the research object, the modelling range should be appropriately expanded and clipped into the target area to offset errors caused by missing sites; (2) Human factors are generally more significant at a small scale, but human interference at any scale is a factor that cannot be ignored; (3) The contribution rates of environmental variables at different scales are different. Ecological protection departments at different levels should pay attention to changes in the most significant environmental factors at corresponding scales that help control and intervene in potential negative impacts to achieve efficient protection.

The study only suggests that appropriately expanding the modelling scope can help offset the errors caused by the lack of data, but how to accurately offset the errors caused by missing data by changing the extent of modelling requires further research. In addition, the supervision and intervention of species protection agencies are vital to ensure the sustainable development of endangered species, such as giant pandas, and maintain the stability of their habitat. Thus, the idea that administrative agencies at all levels distinguish the management priorities of endangered species through the spatial scope can be further applied to practice to make the effects of supervision and intervention more efficient.

Author Contributions: Z.H., conceptualization, data Curation, writing—original draft preparation, validation, formal analysis, visualization; T.P.D., methodology, data curation, supervision, writing—review and editing; L.C., formal analysis, writing—review and editing; A.H., writing—review and editing. All authors have read and agreed to the published version of the manuscript.

Funding: This research received no external funding.

Institutional Review Board Statement: Not applicable.

Informed Consent Statement: Not applicable.

Data Availability Statement: Not applicable.

Acknowledgments: Thanks to Harriet Dawson (University of Edinburgh) for her review and suggestions for improvement on earlier drafts of this manuscript.

Conflicts of Interest: The authors declare no conflict of interest.

References

1. Zang, C.; Cai, L.; Li, J.; Wu, X.; Li, X.; Li, J. Preparation of the China Biodiversity Red List and Its Significance for Biodiversity Conservation within China. *Biodivers. Sci.* **2016**, *24*, 610–614. [CrossRef]
2. Wei, F.; Zhang, Z.; Hu, J. Research Advances and Perspectives on the Ecology of Wild Giant Pandas. *ACTA Ecol. Sin.* **2011**, *31*, 412–421.
3. Hu, J.; Schaller, G.; Pan, W.; Zhu, J. *Wolong Giant Panda*; Sichuan Science and Technology Press: Chengdu, China, 1985; pp. 25–40.
4. Zhou, S.; Qu, Y.; Huang, J.; Huang, Y.; Li, D.; Zhang, H. A Summary of Researches on the Wild Giant Panda Population Dynamics. *J. Sichuan For. Sci. Technol.* **2017**, *38*, 17–30. [CrossRef]
5. Zhen, J. Fine-Scale Evaluation of Giant Panda Habitats and Countermeasures against the Impacts of Future Climate Change. Ph.D. Thesis, Chinese Academy of Sciences University, Beijing, China, 2018.
6. Forestry and Grassland Administration Results of The Fourth National Giant Panda Survey Announced_Current Affairs Information_State Forestry and Grassland Administration Government Website. Available online: http://www.forestry.gov.cn/main/304/20150302/758246.html (accessed on 4 August 2020).
7. WWF Topics in The Fourth National Giant Panda Survey. Available online: http://www.wwfchina.org/specialdetail.php?pid=205&page=13 (accessed on 4 August 2020).
8. Hull, V.; Zhang, J.; Zhou, S.; Huang, J.; Viña, A.; Liu, W.; Tuanmu, M.-N.; Li, R.; Liu, D.; Xu, W.; et al. Impact of Livestock on Giant Pandas and Their Habitat. *J. Nat. Conserv.* **2014**, *22*, 256–264. [CrossRef]
9. Linderman, M.A.; An, L.; Bearer, S.; He, G.; Ouyang, Z.; Liu, J. Modeling the Spatio-Temporal Dynamics and Interactions of Households, Landscapes, and Giant Panda Habitat. *Ecol. Model.* **2005**, *183*, 47–65. [CrossRef]
10. Zhang, Z.; Li, W.; Zhang, M.; Liu, D. Study on Sustainable Development Strategies of the Giant Panda Nature Reserve in New Period. *For. Resour. Manag.* **2018**, *5*, 1–7. [CrossRef]
11. Peng, W.; Wang, X. Concept and connotation development of niche and its ecological orientation. *Ying Yong Sheng Tai Xue Bao J. Appl. Ecol.* **2016**, *27*, 327–334.
12. Hu, J. The "Most" of Giant Panda. *Wild Anim.* **1984**, 1–5. [CrossRef]
13. Zhang, Q. Study on Habitat Selection of Giant Panda in Minshan, Gansu. Master's Thesis, Northwest Normal University, Lanzhou, China, 2009.
14. Li, R.; Xu, M.; Wong, M.H.G.; Qiu, S.; Li, X.; Ehrenfeld, D.; Li, D. Climate Change Threatens Giant Panda Protection in the 21st Century. *Biol. Conserv.* **2015**, *182*, 93–101. [CrossRef]
15. Kumar, S.; Graham, J.; West, A.M.; Evangelista, P.H. Using District-Level Occurrences in MaxEnt for Predicting the Invasion Potential of an Exotic Insect Pest in India. *Comput. Electron. Agric.* **2014**, *103*, 55–62. [CrossRef]
16. Qiao, H.; Huang, J.; Hu, J. Theoretical Basis, Future Directions, and Challenges for Ecological Niche Models. *Sci. Sin. Vitae* **2013**, *43*, 915–927. [CrossRef]
17. Zhuang, H.; Zhang, Y.; Wang, W.; Ren, Y.; Liu, F.; Du, J.; Zhou, Y. Optimized Hot Spot Analysis for Probability of Species Distribution under Different Spatial Scales Based on MaxEnt Model: Manglietia Insignis Case. *Biodivers. Sci.* **2018**, *26*, 913–920. [CrossRef]
18. Connor, T.; Viña, A.; Winkler, J.A.; Hull, V.; Tang, Y.; Shortridge, A.; Yang, H.; Zhao, Z.; Wang, F.; Zhang, J.; et al. Interactive Spatial Scale Effects on Species Distribution Modeling: The Case of the Giant Panda. *Sci. Rep.* **2019**, *9*, 14563. [CrossRef]
19. Liao, Y.; Wang, X.; Zhou, J. Suitability Assessment and Validation of Giant Panda Habitat Based on Geographical Detector. *J. Geo-Inf. Sci.* **2016**, *18*, 767–778.
20. Luo, M.; Wang, H.; Lyu, Z. Evaluating the performance of species distribution models Biomod2 and MaxEnt using the giant panda distribution data. *Ying Yong Sheng Tai Xue Bao J. Appl. Ecol.* **2017**, *28*, 4001–4006. [CrossRef]
21. Tang, C. Giant Panda Habitat, Where Do Giant Pandas Live in China, Panda Habitat Protection. Available online: https://www.chinahighlights.com/giant-panda/habitat.htm (accessed on 16 July 2020).
22. Xu, Z.; Peng, H.; Peng, S. The Development and Evaluation of Species Distribution Models. *Acta Ecol. Sin.* **2015**, *35*. [CrossRef]
23. Phillips, S.J.; Anderson, R.P.; Schapire, R.E. Maximum Entropy Modeling of Species Geographic Distributions. *Ecol. Model.* **2006**, *190*, 231–259. [CrossRef]
24. Liu, Y. Effect of Climate Change on Giant Pandas and Habitats in the Minshan Mountains. Master's Thesis, Beijing Forestry University, Beijing, China, 2012.
25. Global Biodiversity Information Facility. Available online: https://www.gbif.org/ (accessed on 20 June 2020).
26. National Specimen Information Infrastructure. Available online: http://www.nsii.org.cn/2017/home.php (accessed on 20 June 2020).
27. WORLDCLIM. Available online: https://www.worldclim.org/ (accessed on 20 June 2020).
28. Resource and Environment Data Center of the Chinese Academy of Sciences. Available online: http://www.resdc.cn/ (accessed on 20 June 2020).
29. Synes, N.W.; Osborne, P.E. Choice of Predictor Variables as a Source of Uncertainty in Continental-Scale Species Distribution Modelling under Climate Change: Predictor Uncertainty in Species Distribution Models. *Glob. Ecol. Biogeogr.* **2011**, *20*, 904–914. [CrossRef]
30. Guisan, A.; Zimmermann, N.E. Predictive Habitat Distribution Models in Ecology. *Ecol. Model.* **2000**, *135*, 147–186. [CrossRef]

31. Guisan, A.; Thuiller, W. Predicting Species Distribution: Offering More than Simple Habitat Models. *Ecol. Lett.* **2005**, *8*, 993–1009. [CrossRef]
32. Phillips, S.J.; Dudík, M.; Schapire, R.E. Maxent Software for Modeling Species Niches and Distributions. Available online: https://biodiversityinformatics.amnh.org/open_source/maxent/ (accessed on 19 May 2020).
33. Bradie, J.; Leung, B. A Quantitative Synthesis of the Importance of Variables Used in MaxEnt Species Distribution Models. *J. Biogeogr.* **2017**, *44*, 1344–1361. [CrossRef]
34. Fielding, A.H.; Bell, J.F. A Review of Methods for the Assessment of Prediction Errors in Conservation Presence/Absence Models. *Environ. Conserv.* **1997**, *24*, 38–49. [CrossRef]
35. Liu, C.; White, M.; Newell, G. Measuring and Comparing the Accuracy of Species Distribution Models with Presence-Absence Data. *Ecography* **2011**, *34*, 232–243. [CrossRef]
36. Yang, N.; Ma, D.; Zhong, X.; Yang, K.; Zhou, Z.; Zhou, H.; Zhou, C.; Wang, B. Habitat Suitability Assessment of Blue Eared-Pheasant Based on MaxEnt Modeling in Wanglang National Nature Reserve, Sichuan Province. *Acta Ecol. Sin.* **2020**, *40*, 1–9.

sustainability

MDPI

Article

Public's Intention and Influencing Factors of Dockless Bike-Sharing in Central Urban Areas: A Case Study of Lanzhou City, China

Wei Ji [1,2], Chengpeng Lu [1,2,*], Jinhuang Mao [1,2], Yiping Liu [1], Muchen Hou [1,2] and Xiaoli Pan [1,2]

1   Institute of County Economic Development & Rural Revitalization Strategy, Lanzhou University, Lanzhou 730000, China; jiw19@lzu.edu.cn (W.J.); maojh@lzu.edu.cn (J.M.); liuyiping@lzu.edu.cn (Y.L.); houmch20@lzu.edu.cn (M.H.); panxl20@lzu.edu.cn (X.P.)
2   School of Economics, Lanzhou University, Lanzhou 730000, China
*   Correspondence: lcp@lzu.edu.cn

Abstract: Taking the main district in Lanzhou city of China as an example, the questionnaires were designed and distributed, and then the effects of five factors, i.e., behavioral attitude, subjective norm, perceived behavioral control, perceived ease of use and perceived usefulness, on the behavioral intention of dockless bike-sharing (DBS) use were empirically analyzed based on the integrated model of technology acceptance model (TAM) and the theory of planned behavior (TPB) as well as the structural equation model. Results show that the five factors all impose significantly positive effects on the public's behavioral intention of DBS use but differ in influencing degrees. Behavioral attitude, subjective norm and perceived behavioral control can all directly affect the public's behavioral intention of DBS use, with direct influence coefficients of 0.691, 0.257 and 0.198, while perceived ease of use and perceived usefulness impose indirectly effects on behavioral intention, with indirect influence coefficients of 0.372 and 0.396. Overall, behavioral attitude imposes the most significant effect, followed by perceived ease of use, perceived usefulness and subjective norm, and finally perceived behavioral control. This indicates that the public's behavioral intention of DBS use depends heavily on their behavioral attitude towards the shared bikes. In view of the limited open space of the main district in Lanzhou, the explosive growth of shared bikes, oversaturated arrangements, disordered competition, unclear and unscientific divisions of parking regions, and hindrance of traffic, this study proposes a lot of policy suggestions from the research results. A series of supporting service systems related to DBS should be formulated. The shared bikes with different characteristics should be launched for different age groups, gender groups and work groups. The corresponding feedback platform for realtime acquisition, organization, analysis and solution of data information, as well as the adequate platform feedback mechanism, should be established.

Keywords: dockless bike-sharing (DBS); behavioral intention; influencing factors; theory of planned behavior (TPB); Lanzhou City

Citation: Ji, W.; Lu, C.; Mao, J.; Liu, Y.; Hou, M.; Pan, X. Public's Intention and Influencing Factors of Dockless Bike-Sharing in Central Urban Areas: A Case Study of Lanzhou City, China. *Sustainability* **2021**, *13*, 9265. https://doi.org/10.3390/su13169265

Academic Editors: Baojie He, Ayyoob Sharifi, Chi Feng and Jun Yang

Received: 26 July 2021
Accepted: 17 August 2021
Published: 18 August 2021

1. Introduction

Urban traffic problem now has gradually become one of the main factors that affect the improvement of urban built environment in China [1]. Green travel is exactly the reflection and reformation of current urban traffic development [2]. Bike sharing, as a novel travel mode featured by the Internet and sharing, offers a new transport mode for short trips and enhances the connections with other modes such as bus and subway [3–6]. Bike sharing can not only satisfy the heavy demand for short trips but also contribute to solving a lot of urban stubborn diseases such as traffic jams and environmental pollution [7], thereby injecting new vitality into the urban traffic system. Bike sharing can make urban traffic services more diversified and intelligent [8] and play quite an active role in establishing a green travel system. However, with the rapid development of mobile information

269

communication technology and full penetration into social life, time, space and distance have been highly compressed, leading to the transformation of daily lives of residents [9]. Meanwhile, temporal-spatial elasticity and flexibility of some activities such as shopping, leisure and commuting have been enhanced [10], and the public's intention and choice to use shared bike services also show some new features [11].

Over the last several years, China has experienced a fast development in DBS, which provides great convenience for travelers [12]. It is a crucial supplement for other traffic modes in cities. Understanding the travel mobility and traffic demand of DBS is conducive to many urban problems such as urban development and traffic management [13–15]. Currently, scholars have mainly investigated cycling trips on the planning of public bike docks and the arrangement model [16,17] and explored the influencing factors of public bike trips from the perspectives of the built environment, land utilization and user characteristics [18–24]. The DBS has been poorly investigated to date. Most existing studies have laid the research emphasis on temporal-spatial distribution characteristics [25–27], use travel features [28] and the influencing factors based on the spatial scale [29]. Accompanied with increasing urban traffic pressure and environmental pressure, bike sharing has become one of the most basic daily living consumptions for the public. Investigating the use of shared bikes by residents should include both in-depth studies at a macro spatial scale and the discussions of transportation behavioral intentions of urban residents and the related influencing factors from the micro individual perspective.

Currently, some studies have revealed that individual psychological factors imposed significant effects on the residents' intention to use public bikes [30,31]. Accordingly, whether individual psychological factors also significantly affect the use of DBS. In view of this, this study selected the main urban zone in Lanzhou, China, i.e., Chengguan District, Qilihe District, Xigu District and Anning District as research areas. By taking DBS as an example, the integrated model of the public's intentions of DBS use was established based on the technology acceptance model and the theory of planned behavior. In combination with the structural equation model, the subjective psychological factors that affect the public's intentions of DBS use were explored so as to make up for the shortcomings of investigating travel behaviors in geographic space. Meanwhile, this study can provide decision-making basis for the governmental administration departments to adequately guide green travel.

2. Study Area and Methods

2.1. Study Area

Lanzhou city is located at the geometric center of China's land territory, which is also an important center city, industrial base and comprehensive transportation junction in the northwest of China. As also shown in Figure 1, Lanzhou is a key node city in Silk Road Economic Belt. Five districts (Chengguan, Qilihe, Xigu, Anning and Honggu), five counties (Yongdeng, Yuzhong and Gaolan), 1 national-level new district (Lanzhou New District), and two national-level development zones (Lanzhou High-Tech Development Zone and Lanzhou Economic and Technological Development Zone) are under the administration. Lanzhou covers a total area of 13,100 km^2 and an urban area of 1631.6 km^2. According to the 7th National Census data, the permanent resident pollution in Lanzhou reaches up to 435.94 million. DBS has successfully appeared in Lanzhou since March 2017. Currently, over 300,000 shared bikes and electrical bikes owned by different companies such as Hellobike, Qingju and Mobike were put on the market in Lanzhou. Additionally, 1 subway line, 27 intercity bus lines, 52 urban-rural public bus lines and 11 micro-bus lines operate normally in Lanzhou. According to The Investigation Report of Xinhua Green Travel Index (2017), Lanzhou ranks first in terms of green travel satisfaction degree [32]. The respondents are satisfied with the infrastructure construction, service and policies of urban green travel. The shared bikes can provide the residents with great convenience owing to the design concept of go and stop at any time. However, due to limited urban spatial resources, unclear and unscientific division of bike parking zones can not only cause the wasting of resources and affect urban appearance but also result in the congestion on sidewalk road and hinder the traffic [33].

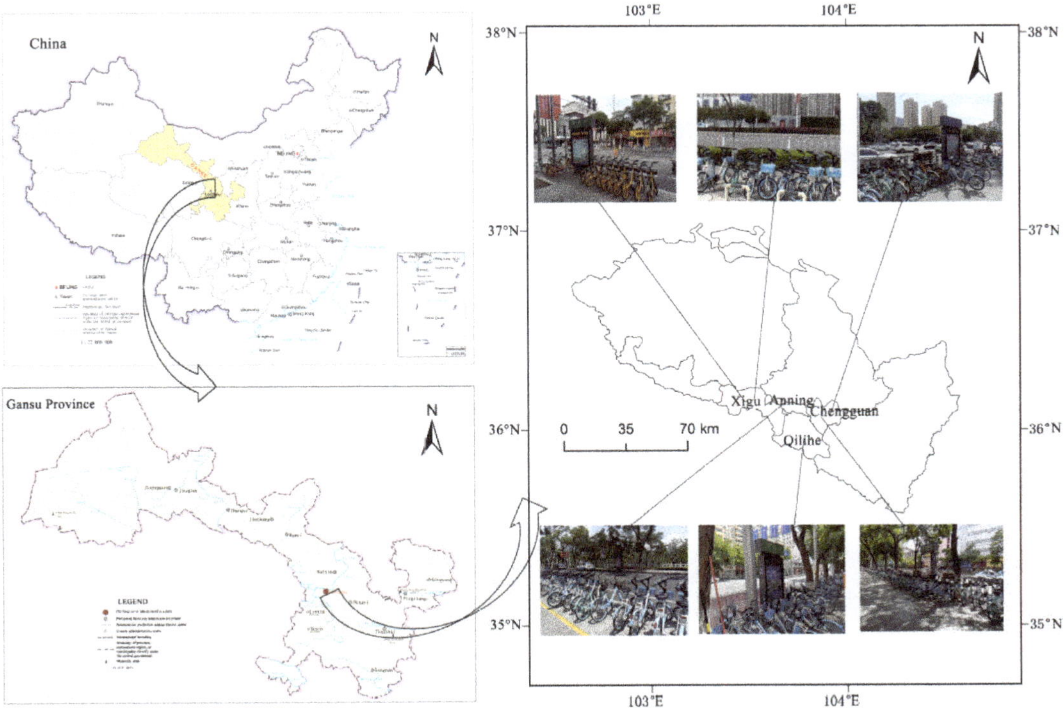

Figure 1. Location of the Lanzhou City.

2.2. Integrated Model

Theory of reasoned action (TRA) is one of the most basic theories to investigate cognitive behaviors, which can be used for predicting behaviors and behavioral intention including both general social activities and consumption activities [34]. TRA advocates that individual behaviors are subjected to behavioral intention while the behavior intention is codetermined by attitude and subjective norms. However, TRA is only applicable to predicting the behaviors fully controlled by the mind. Therefore, on the basis of TRA, Ajzen added a new predictive variable, i.e., perceived behavioral control, and established the Theory of Planned Behavior (TPB) [35]. Nevertheless, TPB still exists limitations in the adoption of new technologies. For this reason, Davis added two factors, i.e., perceived ease of use and perceived usefulness, to investigate users' using behaviors of information system, and proposed the technology acceptance model (TAM), which holds the opinion that both perceived case of use and perceived usefulness can affect the behavioral attitude and thereby affect the behavioral intention [36]. On the basis of TRA, the integrated model organically combines TPB and TAM, in which behavioral intention is regarded as the outcome variable, and behavioral attitude, subjective norm, perceived behavioral control, perceived case of use and perceived usefulness are five causal variables affecting the behavioral intention [37].

2.3. Structural Equation Model (SEM)

The structural equation model, integrating factor analysis and path analysis, can be used for analyzing both direct and indirect relations among variables. A complete structural equation includes two equations, i.e., the measurement equation that describes the relation between the latent variable and measured variable, and the structural equation that describes the relationship between latent variables [38,39]. According to the structural relationship among variables and the definitions of variables as shown in Table 1, the model including 6 latent variables and 23 measured variables was established, as shown in Figure 2.

Table 1. Variable definition.

Latent Variable	Measured Variable	Number of Questions	Label
Behavioral attitude(X_1)	It is convenient to use the shared bike.	4	X_{11}
	It is comfortable to use the shared bike.		X_{12}
	It is interesting to use the shared bike.		X_{13}
	It is valuable to use the shared bike.		X_{14}
Subjective norm (X_2)	Family members think we should use the shared bike.	3	X_{21}
	Friends think we should use the shared bike.		X_{22}
	Schoolmates or workmates think we should use the shared bike.		X_{23}
Perceived behavioral control(X_3)	Possess the mobile phone skills of using the shared bike.	4	X_{31}
	Possess the riding skills of using the shared bike.		X_{32}
	Possess the physical fitness of using the shared bike.		X_{33}
	Possess the psychological quality of using the shared bike.		X_{34}
Perceived usefulness (X_4)	Using the shared bike can protect the environment.	5	X_{41}
	Using the shared bike can avoid the traffic jam.		X_{42}
	Using the shared bike can enhance travel efficiency.		X_{43}
	Using the shared bike can contribute to taking exercise.		X_{44}
	Using the shared bike can save resources.		X_{45}
Perceived ease of use (X_5)	The registration procedure for the use of shared bike is easy and convenient.	4	X_{51}
	The shared bike can park conveniently.		X_{52}
	Payment for the use of shared bike is easy and economical.		X_{53}
	The shared bike possesses excellent performance.		X_{54}
Behavioral intention (X_6)	With the intention to use the shared bike under the current condition.	3	X_{61}
	With the intention to use the shared bike in the future.		X_{62}
	With the intention to recommend the shared bike to other people.		X_{63}

Conclusively, based on TAM and TPB, this study employed SEM for empirically analyzing the effects of five factors, i.e., behavioral attitude, subjective norm, perceived behavioral control, perceived ease of use and perceived usefulness, on the public's behavioral intention of DBS use. The whole research process is shown in Figure 3.

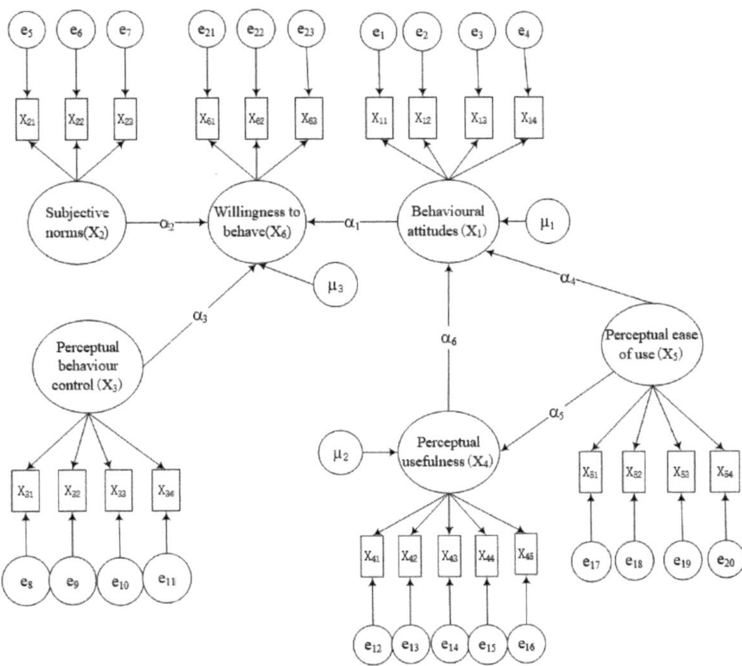

Figure 2. Path analysis results of structure equation model.

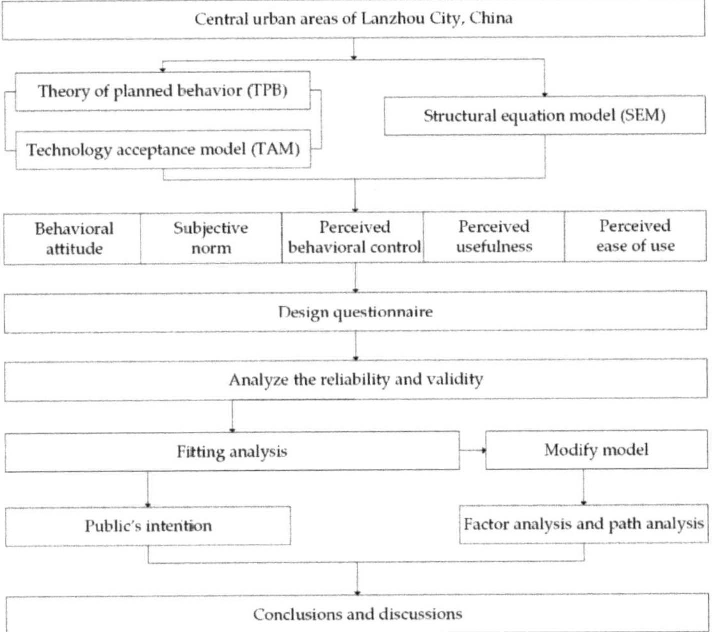

Figure 3. Technology roadmap.

2.4. Data Sources

Based on TPB and the scale of TAM [40,41], the questionnaire was designed by consulting the related references [42–45]. The present questionnaire was implemented by an authoritative survey company in China, WJX.cn. In order to ensure the respondents have ever used the shared bikes, the jumping question was set in the questionnaire. For example, the respondent can end the questionnaire if he or she gives the answer 'Never use the shared bike'. In other words, the users that never use the shared bike can jump to the end of the questionnaire after filling in the basic information. The detailed contents of the questionnaire are shown in Table 2. The distribution and collection of the questionnaires were performed from May to July 2021. To be specific, 700 questionnaires were distributed and 680 questionnaires were collected, of which 52 questionnaires were judged as invalid. Therefore, 628 valid questionnaires were collected, while 52 questionnaires from the respondents who gave the answer 'Never use the shared bike' or answered the question within a short time were considered invalid. Therefore, 628 valid questionnaires were collected. Table 3 shows the basic characteristics of the sample data. Among all respondents, 54% were female, that is, the female respondents slightly exceeded male respondents. In terms of age, 14.04% of the respondents were within the age range of 12~18. According to the Regulations for the Implementation of Road Traffic Safety Law, the cyclists should be aged 18 or above. The respondents under the age of 12 were rejected with the answer 'Never use the shared bike'. An overwhelming majority of respondents (with a proportion of 81.21%) were 19~50 years old, which may be due to the fact that the residents in Lanzhou with an age of over 50 mainly use non-smart phones. Since the survey was performed in central downtown, the respondents showed favorable education degrees. A total of 66.08% of the respondents had a Bachelor degree or above, while 55.09% had a monthly income of below CNY 4000. This suggests that DBS is more preferred by low-income groups. In terms of the frequency of utilization, approximately 57.97% of users used the shared bikes 1~4 times in one week, while 7.96% of users used the shared bikes over 16 times in one week; approximately 38.06% of users selected to use the shared bikes to reach railway stations, motor stations or subway stations. Obviously, DBS played an indispensable role in public transport connection. Overall, the structural characteristics of the respondents in this study fit well with the sociological population characteristics of the residents who use the shared bikes in previous studies [46,47], which confirms the favorable representativeness of the samples in this study.

Table 2. Questionnaire design.

Question	Options
1. Have you ever used the shared bikes?	A. Yes B. No (If you choose B, skip to the end)
2. Your gender	A. Male B. Female
3. Your age	A.$\leq$17 B.12–18 C.19–30 D.31–50 E.>50
4. Your education background	A. Primary school or below B. Middle school C. Undergraduate D. Graduate student or above
5. Your occupation	A. Student B. Civil servant C. Worker D. Liberal profession E. Others
6. Your income	A. Below 2000 yuan B. 2000–4000 yuan C. 4000–6000 yuan D. Above 6000 yuan
7. Your frequency of DBS use in one week	A. 1–4 times B. 5–8 times C. 9–12 times D. 13–16 times E. Over 16 times
8. Where is your cycling destination?	A. School or work unit B. Shopping mall or entertainment venue C. Railway station, subway station or bus station D. Others
9.Your attitude towards the use of shared bikes	A. It is convenient to use the shared bikes. B. It is comfortable to use the shared bikes. C. It is interesting to use the shared bikes. D. It is valuable to use the shared bikes.

Table 2. *Cont.*

Question	Options
10. Who imposes great influence on your selection of the shared bikes?	A. Family member B. Friend C. Schoolmate or colleague
11. Which kind of ability is needed for your use of shared bikes?	A. Mobile phone performance B. Cycling skill C. Physical fitness D. Psychological quality
12. What are the advantages of DBS use?	A. DBS use can protect the environment. B. DBS use can avoid traffic jams. C. DBS use can enhance traveling efficiency. D. DBS use can provide physical exercise. E. DBS use can save resources.
13. What is the greatest convenience in the use of shared bikes?	A. It is convenient to register the usage procedure. B. It is convenient to park the shared bikes. C. It is convenient to pay the cost of using shared bikes. D. The shared bikes have good performance.
14. Your future attitude towards the use of shared bikes	A. I'm willing to use the shared bikes under current conditions. B. I'm willing to use the shared bikes in future. C. I'm willing to recommend shared bikes to others.

Table 3. Basic characteristics of sample data.

Statistical Indicator	Classification Indicator	Number of People	Proportion in the Valid Samples (%)	Statistical Indicator	Classification Indicator	Number of People	Proportion in the Valid Samples (%)
Gender	Male	289	46.02	Profession	Student	208	33.12
	Female	339	53.98		Civil servant	214	34.08
Age	12–18	88	14.01		Worker	37	5.89
	19–30	378	60.19		Liberal profession	84	13.38
	31–50	132	21.02		Others	85	13.54
	>50	30	4.78	Income	Below CNY 2000	220	35.03
Degree of education	Primary or below	18	2.97		CNY 2000–4000	126	20.06
	Middle school	195	31.05		CNY 4000–6000	188	29.94
	Undergraduate	286	45.54		Above CNY 6000	94	14.97
	Postgraduate or above	129	20.54				

3. Results

3.1. Factor Analysis of Residents' Behavioral Intention of Using the Shared Bikes

The reliability and validity of the questionnaire were analyzed by SPSS23.0. The Cronbach's α coefficient was selected as the measuring index of reliability [47]. Through measurements, the Cronbach's α coefficients of behavioral attitude, subjective norm, perceived behavioral control, perceived usefulness, perceived ease of use, and behavioral intention were 0.742, 0.765, 0.801, 0.831, 0.859 and 0.728, respectively, while the overall Cronbach's α coefficient was 0.907, indicating the high reliability of both the whole questionnaire and various dimensions. The validity was measured via the KMO test and Bartlett sphericity test [48]. The overall KMO value of six latent variables was 0.907, which has also passed the Bartlett sphericity test. Accordingly, factory analysis can be performed on the questionnaire. The correlation coefficient matrix was then constructed by Amos23.0, and the path coefficient in the model was estimated via maximum likelihood estimation. Before parameter estimation, the model was first fitted. The goodness of fit of the model was evaluated by 11 specific indexes. In the initial model fitting, the Chi-square freedom degree ratio (c2/df) was 4.563, suggesting the model could not adequately reflect the questionnaire and needed to be improved [49]. Based on the principle of reasonableness and specification, the model was modified in accordance with the Amos correction index so that various fitting indexes can finally satisfy the test requirements (see Table 4). Overall, the fitting was favorable, suggesting high compatibility between the model and the present questionnaire. The integrated model based on TAM and TPB is applicable to the studies in the traffic domain, which can well explain the public's behavioral intentions of using the shared bikes.

Table 4. Model fitting indexes and the fitting results.

Fitting Index	Specific Index	Ideal Value	Model Estimated Value	Test Result
Measure of absolute fit	GFI	>0.90	0.917	Accepted
	AGFI	>0.90	0.927	Accepted
	SRMR	<0.05	0.043	Accepted
	RMSEA	<0.08	0.068	Accepted
Measure of incremental fit	NFI	>0.90	0.920	Accepted
	TLI	>0.90	0.930	Accepted
	CFI	>0.90	0.942	Accepted
	IFI	>0.90	0.915	Accepted
Measure of simple fit	PGFI	>0.50	0.641	Accepted
	PNFI	>0.50	0.683	Accepted
	NC (Chi-square freedom degree ratio)	1<NC<3	2.830	Accepted

The standard loading factors between 23 observable variables in parameter test results and the corresponding latent variables ranged from 0.545 to 0.891, which all exceeded 0.5, suggesting a favorable basic fit measure. The values of C.R all exceeded 1.96, that is, the parameter estimated values all reached the significance level of 1% (Table 5). The standard factor loads of four measured variables regarding behavioral attitude were 0.690, 0.776, 0.635 and 0.862, respectively, suggesting the convenience, comfort, interesting degree and value of using the shared bikes can significantly affect the public's opinions and behavioral intentions of DBS use. The standard factor loads of three measured variables regarding subjective norm were 0.651, 0.891 and 0.862, respectively, suggesting that residents are subjected to positive influences from family members, friends and schoolmates (work-mates) when using the shared bikes. The standard factor loads of four measured variables regarding perceived behavioral control were 0.545, 0.710, 0.765 and 0.621, respectively, sug-gesting positive effects of the mobile skill, riding skill, physical fitness and psychological quality, especially physical fitness, on the use of shared bikes. The standard factor loads of five measured variables regarding perceived usefulness were 0.767, 0.679, 0.701, 0.595 and 0.644, respectively, suggesting that DBS use can bring about useful experiences such as en-vironmental protection, enhancement of efficiency, bodybuilding and energy conservation, among which environmental protection was most remarkable. The standard factor loads of four measured variables regarding perceived ease of use were 0.690, 0.801, 0.778 and 0.791, respectively, suggesting that the registration program, the parking convenience, payment and program and the performance of the shared bikes affect the public's perceiving ease degree of DBS use; in particular, the parking convenience imposes most significant effect. The standard factor loads of three measured variables regarding behavioral intention were 0.596, 0.674 and 0.713, respectively, indicating that the residents not only are willing to use the shared bikes but also encourage and advocate the other people to use the shared bikes.

Table 5. Model fitting indexes and the fitting results.

Latent Variable	Measured Variable	P	C.R Value	Standard Factor Load
Behavioral attitude	X_{11}			0.690
	X_{12}	***	14.882	0.776
	X_{13}	***	12.314	0.635
	X_{14}			0.632
Subjective norm	X_{21}	***	12.453	0.651
	X_{22}	***	15.428	0.891
	X_{23}	***	15.270	0.862
Perceived behavioral control	X_{31}			0.545
	X_{32}	***	14.360	0.710
	X_{33}	***	14.782	0.765
	X_{34}	***	11.917	0.621
Perceived usefulness	X_{41}			0.767
	X_{42}	***	12.835	0.679
	X_{43}	***	13.378	0.701
	X_{44}	***	11.666	0.595
	X_{45}	***	12.350	0.644
Perceived ease of use	X_{51}			0.690
	X_{52}	***	14.986	0.801
	X_{53}	***	14.827	0.778
	X_{54}	***	14.892	0.791
Behavioral intention	X_{61}			0.596
	X_{62}	***	12.827	0.674
	X_{63}	***	14.253	0.713

Note: *** suggests the significance at the level of 1%, and C.R value equals the t value (the same below).

3.2. Analysis of the Influencing Factors of the Public's Behavioral Intention of DBS Use

Figure 4 shows the relations among six latent variables, i.e., behavioral attitude, subjective norm, perceived behavioral control, perceived usefulness, perceived ease of use and behavioral intention. Table 6 displays the direct, indirect and total effects of the former five factors on behavioral intention. The following conclusions can be drawn.

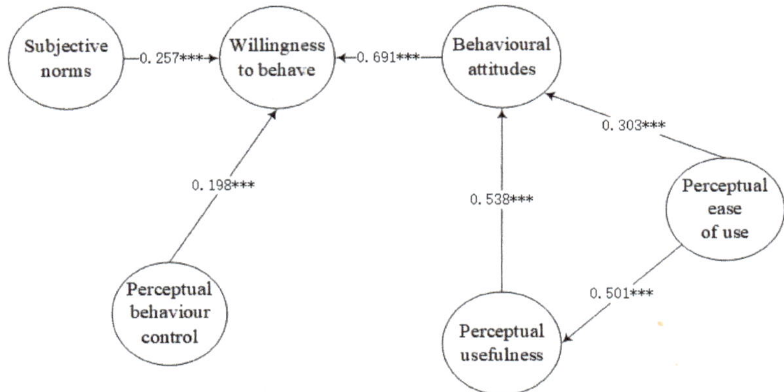

Figure 4. Path analysis results of structure equation model. Note: *** suggests the significance at the level of 1%.

Table 6. Effects of different variables on the willingness to use shared bikes.

Variables	Behavioral Intention		
	Direct Effect	Indirect Effect	Total Effect
Behavioral attitude	0.691	-	0.691
Subjective norm	0.257	-	0.257
Perceived behavioral control	0.198	-	0.198
Perceived usefulness	-	0.372	0.372
Perceived ease of use	-	0.396	0.396

- Behavioral attitude can directly affect behavioral intention, with an influence value of 0.691. Owing to the convenience, comfort, interestingness and value, the public shows positive attitudes towards DBS use, among which convenience and comfort impose most significant effects. Therefore, enhancing the appearance and performance of the shared bikes and placing the shared bikes at the appropriable positions can contribute to enhancing the public's intention of using shared bikes.

- Subjective norms impose a direct effect on behavioral intention, with an influence value of 0.257. The opinions of acquaintances show important references for the public's selection of shared bikes. The sense of trust in the surrounding acquaintances and both collective consciousness and sense of community in daily life can impose imperceptible effects on the public's psychological actives and use behaviors. The individual behavioral intentions are inclined to stay in step with the people around. Therefore, the business should consider the difference in different user groups and the propaganda of DBS when making operating strategies. Formulating different operating schemes based on population differences can enhance the irradiating effect on more users.

- Perceived behavioral control can directly affect behavioral intention, with an influence value of 0.198. Both physical and psychological quality and riding skills can impose positive effects on the public's choice of DBS. The riding skill and physical fitness can significantly affect the perceived behavioral control. According to the questionnaires, it can be found that some residents are anxious about collisions with motor vehicles or passers-by under bad weather or traffic conditions. In addition to skill and physical fitness, residents still doubted the use of shared bikes. It is therefore recommended to add a bike lane and advocate the comity to pedestrians to alleviate the risk to a certain degree, thereby enhancing the residents' behavioral intention of using shared bikes.

- Perceived usefulness can affect the behavioral intention indirectly but impose a direct effect on the behavioral attitude, with influence values of 0.372 and 0.538, respectively. Behavioral attitude plays a mediating role in directly affecting behavioral intention. This means that residents still enjoy both interestingness and convenience in DBS in addition to environmental protection, alleviation of traffic jams and the enhancement of traveling efficiency. Owing to the multiple advantages of riding the shared bikes, residents are more inclined to use the shared bikes.

- Perceived ease of use can indirectly affect behavioral intention while directly affect perceived usefulness and behavioral attitude, with influence values of 0.396, 0.501 and 0.303, respectively. Perceived ease of use imposes indirect effect on behavioral intention via the following two paths: (1) Using behavioral attitude as the mediating variable and imposing indirect effect on the behavioral intention with an influence value of 0.209, and (2) using perceived usefulness and behavioral attitude as two mediating variables and imposing direct effect on the behavioral intention with an influence value of 0.187. DBS, as a new form of Internet bike renting mode, has the greatest advantage in freeing users from the parking stations and paying fees via the app on the mobile phone. Great convenience and flexibility change the residents' opinion and attitude towards the use of shared bikes and enhance the use intention.

4. Conclusions and Discussion

By taking the use of shared bikes in the main distribution of Lanzhou, China as an example, this study established the integrated model based on TAM, distributed 628 valid questionnaires and systematically analyzed the effects and the related influencing mechanism of five factors (i.e., behavioral attitude, subjective norm, perceived behavioral control, perceived ease of use and perceived usefulness) on the residents' behavioral intention of using shared bikes. The main conclusions are given in detail below.

Firstly, these five factors, i.e., behavioral attitude, subjective norm, perceived behavioral control, perceived ease of use and perceived usefulness, all have significantly positive effects on the public's behavioral intention of DBS use. To be specific, behavioral attitude, subjective norm and perceived behavioral control can directly affect the residents' behavioral intention, with direct influence coefficients of 0.691, 0.257 and 0.198, respectively, while perceived usefulness and perceived ease of use impose indirect effects on behavioral intention, with indirect influence coefficients of 0.372 and 0.396, respectively. Secondly, different subjective psychological factors affect behavioral intention to varying degrees. Overall, behavioral attitude imposes most significant effect, followed by perceived ease of use, perceived usefulness and subjective norm, while perceived behavioral control imposes the least effect. The public's behavioral intention of DBS use depends greatly on their behavioral attitude towards the shared bikes. Finally, behavioral attitude is an important bridge that integrating both TAM and TPB since perceived usefulness and perceived ease of use can indirectly affect the public's behavioral intention of DBS use via the mediating effect of behavioral attitude. Meanwhile, perceived ease of use can also indirectly affect the public's behavioral intention of DBS use via two mediating variables, namely, perceived usefulness and behavioral attitude.

This study focused on the behavioral intention of DBS use and analyzed the main influencing factors, which is expected to provide theoretical guidance for DBS enterprises in product design and marketing plans. The combination of TPB and SEM in urban sharing economy study is a beneficial attempt, which shows stronger explanatory power than a single use of TPB and SEM. An in-depth investigation of the main distribution of Lanzhou can reflect the overall characteristics of urban residents in DBS use to a certain degree but still needs to be improved. In future studies, we can expand the research range, prolong the research period and perform multi-scale and dynamic analysis. Next, this study laid the emphasis on the effects of the public's psychological factors on behavioral intention. By taking into account the built environment [50,51], topographic features [52] and climate factors [53], future studies will overall evaluate the effects of both subjective and objective factors on the public's behavioral intention of DBS use so as to obtain more strict and abundant conclusions.

Considering the limited open space of the main district in Lanzhou, the explosive growth of shared bikes, oversaturated arrangements, disordered competition, unclear and unscientific divisions of parking regions and hindrance of traffic, the following policy suggestions were proposed in combination with the present conclusions. Firstly, enterprises should work on technological innovation, optimization and upgrade, enhance the fit measure between software and hardware, and formulate a series of supporting service systems related to shared bikes. To be specific, the shared bikes can be equipped with some devices such as speed variators, shock attenuation devices, shelves and back seats, and some protective devices such as helmets and gloves to provide better riding experiences under the premise of ensuring bike quality. Secondly, the shared bikes with different characteristics should be launched for different age groups, gender groups and work groups. For example, pink bikes can be designed for women, the bikes with animal themes can be designed for students, while the elderly electric shared bikes should be added for the middle and old people. Thirdly, enterprises should set the corresponding feedback platform for real-time acquisition, organization, analysis and solution of data information, establish adequate platform feedback mechanism, build their WeChat official accounts and official websites and achieve good management, so as to reinforce the communication with

users and the interaction between enterprises and users. Fourthly, the government sh
perfect the social credit system and incorporate DBS into the individual and enter
credit system. On one hand, the enterprise credit system should be strengthened, that is
credit system should be formulated to govern vicious competition among enterprises
to encourage fair competition among enterprises, ensure the consumers' usage experie
and protect the related rights and interests of consumers. On the other hand, indivi
credit system should be established. For example, some improper illegal behaviors sh
be integrated into individual credit system via Internet, and then the users with insuffi
credit scores will be restricted to a certain degree, which can better restrain user's DBS
behaviors and achieve civilized DBS use. Finally, as regard to the parking of shared b
the enterprise dominants should take reasonable distribution and grid administra
Various operators should reasonably allocate and distribute the shared bikes in accord
with usage density of shared bikes and population density, and increase the num
of shared bikes at some public regions such as railway stations, subway stations,
stations, schools, shopping malls and entertainment venues, which can facilitate connec
with other transportation tools. Some supervision departments, such as the Burea
Transport, the Bureau of Urban Administration and the Bureau of Market Supervi
and Administration, should act in close coordination and strengthen communication
coordination to make concerted efforts and further standardize the parking region
shared bikes. The parking system should set the function of temporary parking to pro
the users with great convenience. Moreover, the convenient query function whether
destination can be parked or not should be added to avoid the extra dispatch fee
exceeding the designated parking area.

Author Contributions: Conceptualization, C.L.; methodology, W.J. and J.M.; formal analysis,
Y.L.; investigation, W.J., J.M., Y.L., M.H., X.P. and C.L.; writing—original draft preparation, W.J.
C.L.; writing—review and editing, C.L. All authors have read and agreed to the published versic
the manuscript.

Funding: This research was funded by National Natural Science Foundation of China (41701
and Fundamental Research Funds for the Central Universities (2020jbkyjc002).

Institutional Review Board Statement: Not applicable.

Informed Consent Statement: Not applicable.

Data Availability Statement: The data used to support the findings of this study are available f
the corresponding author upon reasonable request.

Conflicts of Interest: The authors declare no conflict of interest.

References

1. Guo, X.; Lu, C.; Sun, D.; Gao, Y.; Xue, B. Comparison of Usage and Influencing Factors between Governmental Public Bicy
   and Dockless Bicycles in Linfen City, China. *Sustainability* **2021**, *13*, 6890. [CrossRef]
2. Bocker, L.; Anderson, E.; Uteng, T.P.; Throndsen, T. Bike sharing use in conjunction to public transport: Exploring spatiotempo
   age and gender dimensions in Oslo, Norway. *Transp. Res. Part A Policy Pract.* **2020**, *138*, 389–401. [CrossRef]
3. Molinillo, S.; Ruiz-Montanez, M.; Li'ebana-Cabanillas, F. User characteristics influencing use of a bicycle-sharing system integra
   into an intermodal transport network in Spain. *Int. J. Sustain. Transp.* **2020**, *14*, 513–524. [CrossRef]
4. Zuo, T.; Wei, H.; Chen, N.; Zhang, C. First-and-last mile solution via bicycling to improving transit accessibility and advanc
   transportation equity. *Cities* **2020**, *99*, 102614. [CrossRef]
5. Shi, J.G.; Si, H.; Wu, G. Critical factors to achieve dockless bike-sharing sustainability in China: A stakeholder-oriented netw
   perspective. *Sustainability* **2018**, *10*, 2090. [CrossRef]
6. Shi, Y. *Research on Demand Forecasting and Scheduling Methods for Shared Bicycles*; Beijing Jiaotong University: Beijing, China, 2
7. Bo, W.; Feng, Z.; Zongcai, W. The research on characteristics of urban activity space in Nanjing: An empirical analysis based
   big data. *Hum. Geogr.* **2014**, *29*, 14–21.
8. Zhen, F.; Wei, Z.; Yang, S. The impact of information technology on the characteristics of urban resident travel: Case of Nanj
   Geogr. Res. **2009**, *28*, 1307–1317.
9. Yang, Y.X.; Heppenstall, A.; Turner, A.; Comber, A. A spatial and graph-based analysis of dockless bike sharing pattern
   understand urban flows over the last mile. *Comput. Environ. Urban Syst.* **2019**, *77*, 101361. [CrossRef]

Han, S.S. The spatial spread of dockless bike-sharing programs among Chinese cities. *J. Transp. Geogr.* **2020**, *86*, 102782. [CrossRef]

Xing, Y.Y.; Wang, K.; Lu, J.J. Exploring travel patterns and trip purpose of dockless bike sharing by analyzing massive bike-sharing data in Shanghai, China. *J. Transp. Geogr.* **2020**, *87*, 102787. [CrossRef]

Chai, Y.; Shen, J.; Zhao, Y. Activity-based approach for urban travel behavior research. *Sci. Online* **2015**, *5*, 402–409.

Wan, F.; Yang, G.; Li, X. A study of Hangzhou urban residents green travel choice in metro era. *J. Green Sci. Technol.* **2012**, *8*, 203–206.

Bai, K.; Li, C.; Zhang, C. Reference group influence and self-perceived value judgment of Xi'an urban residents' green travel behavior. *Hum. Geogr.* **2017**, *32*, 37–46.

Newton, P.; Meyer, D. Exploring the attitudes-action gap in household resource consumption: Does "Environmental Lifestyle" segmentation align with consumer behavior. *Sustainability* **2013**, *5*, 1211–1233. [CrossRef]

Ran, L.; Li, F. An analysis on characteristics and behaviors of traveling by bike-sharing. *J. Transp. Inf. Saf.* **2017**, *35*, 93–100.

Huang, A. *Study on Structure and Dynamic Behaviors in Weighted Complex Public Transit Network Based on Passenger Flow*; Beijing Jiaotong University: Beijing, China, 2014.

Wang, B.; Zhou, T.; Zhou, C. Statistical physics research for human behaviors, complex networks, and information mining. *J. Univ. Shanghai Sci. Technol.* **2012**, *34*, 103–117.

Zheng, J.; Zhang, B.; Cheng, Y. Grop choice behavior in green travel based on scale-free network. *Chin. J. Manag. Sci.* **2019**, *27*, 198–208.

Valkila, N.; Saari, A. Attitude behavior gap in energy issues: Case study of three different finish residential areas. *Energy Sustain. Dev.* **2013**, *17*, 24–34. [CrossRef]

Chardon, C.M.D.; Caruso, G. Estimating bike-share trips using station level data. *Transp. Res. Part B* **2015**, *78*, 260–279. [CrossRef]

Lin, J.R.; Yang, T.H. Strategic design of public bicycle sharing systems with service level constraints. *Transp. Res. Part E* **2011**, *47*, 284–294. [CrossRef]

Caperello, N.D.; Kurani, K.S. Households' stories of their encounters with a plugin hybrid electric vehicle. *Environ. Behav.* **2012**, *44*, 493–508. [CrossRef]

Faghih-Imani, A.; Eluru, N.; El-Geneidy, A.M. How land-use and urban form impact bicycle flows: Evidence from the bicycle sharing system (BIXI) in Montreal. *J. Transp. Geogr.* **2014**, *41*, 306–314. [CrossRef]

Saneinejad, S.; Roorda, M.J.; Kennedy, C. Modelling the impact of weather conditions on active transportation travel behaviour. *Transp. Res. Part D Transp. Environ.* **2012**, *17*, 129–137. [CrossRef]

Gebhart, K.; Noland, R.B. The impact of weather conditions on bikeshare trips in Washington, DC. *Transportation* **2014**, *41*, 1205–1225. [CrossRef]

Lee, K.H.; Ko, E.J. Relationships between neighbourhood environments and residents' bicycle mode choice: A case study of Seoul. *Int. J. Urban Sci.* **2014**, *18*, 383–395. [CrossRef]

González, F.; Melo-Riquelme, C.; Grange, L.D. A combined destination and route choice model for a bicycle sharing system. *Transportation* **2016**, *43*, 407–423. [CrossRef]

Vogel, M.; Hamon, R.; Lozenguez, G. From bicycle sharing system movements to users: A typology of Vélo'v cyclists in Lyon based on large-scale behavioural dataset. *J. Transp. Geogr.* **2014**, *41*, 280–291. [CrossRef]

Gao, F.; Li, S.; Wu, Z. Spatial-temporal characteristics and the influencing factors of the ride destination of bike sharing in Guangzhou city. *Geogr. Res.* **2019**, *38*, 2859–2872.

Wei, Z.; Mo, H.; Liu, Y. Spatial-temporal characteristics of bike-sharing: An empirical study of Tianhe District, Guangzhou. *Sci. Technol. Rev.* **2018**, *36*, 71–80.

The Investigation Report of Xinhua Green Travel Index; China Economic Information Service: Beijing, China, 2017. Available online: http://www.xinhuanet.com//politics/2017-09/27/c_1121731198.htm (accessed on 1 August 2021).

Ma, J.H.; Li, J.B.; Liu, B. Solving the persistent problems of urban governance and helping to build a sophisticated Lanzhou City. *Shanghai Bus.* **2021**, *5*, 88–90.

Carstensen, T.A.; Olafsson, A.S.; Bech, N.M. The spatial-temporal development of Copenhagen's bicycle infrastructure 1912–2013. *Geogr. Tidsskr. Dan. J. Geogr.* **2015**, *115*, 142–156.

Bao, J.; Xu, C.C.; Liu, P. Exploring bike sharing travel patterns and trip purposes using smart card data and online point of interests. *Netw. Spat. Econ.* **2017**, *17*, 1231–1253. [CrossRef]

Mo, H.; Wei, Z.; Zhai, Q. Travel behaviors and influencing factors of bike sharing in old town: The case of Guangzhou. *South Archit.* **2019**, *1*, 7–12.

Zainuddin, N.B.; Min, L.H.; Teng, C.S. Sustainable transportation scheme in university: Students' intention on bike sharing system: An empirical approach. *J. Glob. Bus. Soc. Entrep.* **2016**, *2*, 144–163.

Kaplan, S.; Manca, F.; Nielsen, T.A.S. Intentions to use bike-sharing for holiday cycling: An application of the theory of planned behavior. *Tour. Manag.* **2015**, *47*, 34–46. [CrossRef]

Fishbein, M.; Ajzen, I. Belief, Attitude, Intention, and Behavior: An Introduction to Theory and Research. *Contemp. Sociol.* 1975. Available online: http://worldcat.org/isbn/0201020890 (accessed on 1 August 2021).

Ajzen, I. The theory of planned behavior. *Organ. Behav. Hum. Decis. Process.* **1991**, *50*, 179–211. [CrossRef]

Davis, F.D. Perceived usefulness, perceived ease of use, and user acceptance of information technology. *MIS Q.* **1989**, *13*, 319–340. [CrossRef]

42. Chen, Y.; Zha, Q.; Jing, P. Modeling and analysis of autonomous technology acceptance considering age heterogeneity. *J. Jia* *Univ. Nat. Sci. Ed.* **2021**, *42*, 131–138.
43. Ju, P.; Zhou, J.; Chen, X.G. A study of the intent of shared use of automobiles based on TAM and TPB integration models. *Manag.* **2016**, *36*, 82–85.
44. Wang, Y.; Wang, Q. Factors affecting Beijing residents' buying behavior of new energy vehicle: An integration of technol acceptance model and theory of planned behavior. *Chin. J. Manag. Sci.* **2013**, *21*, 691–698.
45. Kraft, P.; Rise, J.; Sutton, S. Perceived difficulty in the theory of planned behavior: Perceived behavioural control or affec attitude? *Br. J. Soc. Psychol.* **2005**, *44*, 479–496. [CrossRef]
46. Zhang, W.; Li, G. Ecological compensation, psychological factors, willingness and behavior of ecological protection in the Q ecological function area. *Resour. Sci.* **2017**, *39*, 881–892.
47. Lao, K.F.; Wu, J. Research on influencing mechanism of consumer green consumption behavior referring to TPB. *Financ. I* **2013**, *2*, 91–100.
48. Yang, H.L.; Cao, X.S.; Li, T. Analysis of willingness and influence factors of urban residents to use shared bikes: A case stud Xi'an. *J. Arid. Land Resour. Environ.* **2019**, *33*, 78–83.
49. Shao, P.; Wang, Q.; Zhao, C. Research on the factors influencing shared bicycle green use behavior and intention. *J. Arid. I Resour. Environ.* **2020**, *34*, 64–68.
50. Chen, Y.; Wang, H. Comparison and analysis of influential on the degree of satisfaction of city bike: Based on the rese involving six main districts in Beijing. *Econ. Probl.* **2018**, 105–112.
51. Qian, J.; Wang, D.; Niu, Y. Analysis of the influencing factors of urban public bikes: A case study of Suzhou. *Geogr. Res.* **2014** 358–371.
52. Wu, M.L. *Structural Equation Modeling: Operation and Application of AMOS*; Chongqing University Press: Chongqing, China, :
53. Ma, J.; Yang, Y.; Yang, D. Influence of urban morphological characteristics on thermal environment. *Sustain. Cities Soc.* **2** *72*, 103045.

MDPI

St. Alban-Anlage 66

4052 Basel

Switzerland

Tel. −41 61 683 77 34

Fax +41 61 302 89 18

www.mdpi.com

Sustainability Editorial Office

E-mail: sustainability@mdpi.com

www.mdpi.com/journal/sustainability

Climate Change and Environmental Sustainability-Volume 2